The Dynamic Nature
of Ecosystems

The Dynamic Nature of Ecosystems

Chaos and Order Entwined

Claudia Pahl-Wostl

Swiss Federal Institute of Technology (ETH)
Zürich, Switzerland

and

Swiss Federal Institute for Environmental Science and Technology (EAWAG)
Dübendorf, Switzerland

JOHN WILEY & SONS

Chichester · New York · Brisbane · Toronto · Singapore

Copyright © 1995 by John Wiley & Sons Ltd,
Baffins Lane, Chichester,
West Sussex PO19 1UD, England

Telephone National Chichester (01243) 779777
International (+44) 1243 779777

Other Wiley Editorial Offices

John Wiley & Sons Inc., 605 Third Avenue,
New York, NY 10158-0012, USA

Jacaranda Wiley Ltd, 33 Park Road, Milton,
Queensland 4064, Australia

John Wiley & Sons (Canada) Ltd, 22 Worcester Road,
Rexdale, Ontario M9W 1L1, Canada

John Wiley & Sons (SEA) Pte Ltd, 37 Jalan Pemimpin #05-04,
Block B, Union Industrial Building, Singapore 2057

British Library Cataloguing in Publication Data

A catalogue record for this book is available from the British Library

ISBN 0-471-95570-1

Typeset in 10/12pt Times by Vision Typesetting, Manchester
Printed and bound in Great Britain by Antony Rowe Ltd, Chippenham, Wiltshire

This book is printed on acid-free paper responsibly manufactured from
sustainable forestation, for which at least two trees are planted for
each one used for paper production.

To my bicycle "Fury"

Contents

Preface

Is it possible to understand life and evolution within the materialistic, mechanistic framework of the natural sciences? If not, what are the consequences?

Of course, it is possible to reply to this question by saying that this is what the science of biology is all about. However, I would be more hesitant to give such a straightforward answer. Tackling such a fundamental question might well involve more than a lifetime's work for any scientist, and in addition would require an interdisciplinary dialogue involving philosophers and social scientists. Definitions of understanding, evolution, and even life would have to be found. There is no need to worry, however, that I am going to indulge in theorizing about these issues throughout the remainder of this book. Nevertheless, I do wish to make clear here that this book originated in feelings of doubt, in an uneasiness with current thinking and methods in the natural sciences, the inadequacy of which becomes particularly obvious when confronted with our inability to cope with the pressing environmental problems facing us today. It seems to me that by accumulating more and more detailed knowledge about single mechanisms we are unable to see the wood for the trees. To counter this, we need a conceptual base that has to exceed the traditional realm of the natural sciences. This book may therefore be somewhat different in tone to more usual scientific texts. Even when I take recourse to traditional methods, these are interpreted from a different perspective. This book represents an ongoing process, and as such cannot be considered to yield a congealed, coherent whole. I invite you, the reader, to share with me my thoughts and doubts, my ideas and hopes.

Self-organization, chaos, order – are these only new buzzwords, vacuous, with no real content? The current popularity of these terms and their application to subjects as far apart as fluid convection and business management may nurture such a suspicion. Far too often ecology seems to have been dominated by one or other of various theoretical bandwagons – ecologists might rightly be very reluctant to accept any new panacea. However, if we strip off all the glitter currently surrounding these terms, we are left with important messages: systems are highly dynamic and variable, organization arises from context. The current fascination of many scientists with these new ideas may simply arise from a frustration with traditional concepts that prove to be inadequate when dealing with complex systems, which environmental systems typically are. New concepts are badly needed at a time when ecological knowledge is increasingly required to cope with the global environmental problems facing us. But what really is ecological knowledge? Depending on whether you ask an ecologist studying single organisms, single populations, communities or ecosystems, and further whether this ecologist studies lakes, marine systems, forests or grasslands, the answer may differ greatly. This is only natural, you may object. Even though all ecologists study organisms in their natural environment, a multitude of different approaches had to arise to do justice to the plethora of systems they had to deal with. I do not claim that we need a single unifying theory embracing all disciplines of ecology. On the contrary, I argue for pluralism and tolerance, which, to be honest, are not prominent attributes of most scientific disputes. However, we do need a coherent theoretical framework to counteract the overwhelming emphasis on predictability and quantification that has arisen within the context of the mechanistic world view. Instead of always aiming to reduce uncertainties, we have also to learn to deal with those uncertainties that cannot be reduced, that are inherent in natural systems. And these are many – as I am going to show throughout this book. We have to draw a clear distinction between areas within which statements with a specific degree of certainty can be made and areas within which certainty simply cannot be obtained. To communicate our results, to convert them into useful knowledge, we have to develop tools to deal with uncertainties, to convey our often qualitative knowledge in structured terms that transcend mere description. Ecological knowledge has not proven to be particularly "useful" because society does not esteem scientific knowledge that is not expressed in terms of clearly quantifiable cause-and-effect relationships.

What are the goals of this book? Rejecting a purely mechanistic perception, its main emphasis is on developing a new framework within which the dynamical nature and organization of ecosystems can be understood. Many traditional ecological concepts such as that of diversity and of the ecological niche gain new meaning in the light of a dynamic perspective on ecosystems. Seemingly contradictory terms such as population/ecosystem and part/whole can be united. However, this cannot be enough. In a historical perspective I attempt to show that many of the problems that have faced ecology since its beginnings are epistemological. With regard to this, I have to admit that I cannot yet proffer answers to many of the questions I am going to pose. Essential for a new understanding of both research and environmental management will be an evolutionary perspective: an evolutionary perspective emphasizing the indeterminate nature of the future and changes in relationships and interactions in what must finally be perceived to be a common human and ecological system.

Discussions with numerous people contributed to the completion of this book. I should like to thank the following people for their interest and active support: Martin Büssenschütt, Ursula Gaedke, Giulio Genoni, Sandra Harding, Dieter Imboden, Carlo Jaeger, Simon Levin, David Livingstone, Gregor Nickel, Colin Reynolds, Stanly Salthé, Paul Schmidt-Hempel, Dietmar Straile, Robert Ulanowicz, Matthias Wächter.

Names are given in alphabetic order since quantity and quality are not identical – one single remark can sometimes have unprecedented effects. It is impossible to judge in retrospect. However, one person deserves special thanks: my husband Wolfgang, for whom the writing of this book constituted a massive upheaval in his private life. Nevertheless, he was courageous enough to spend much time and effort fighting his way through piles of material he definitely was not very familiar with.

Last, but not least, I should presumably explain why I am dedicating this book to one of my bicycles. Imagine yourself cycling through a pleasant and peaceful landscape. Suddenly your bike decides to throw you off and you make the unpleasant discovery that the ground is rather hard, at least harder than yourself. That is what happened to me. To my dismay, I found myself immobilized at home at the height of the cycling season – what a disaster! Somehow I had to get rid of all my excess energy – and that is how the whole book project took off. In this light, even a cycling accident may have its benefits – or possibly not. Please judge for youself!

C. Pahl-Wostl
September 1994

Chapter One

Setting the stage

1.1 ECOLOGY AT THE CROSSROADS?

Nowadays ecology faces an unprecedented challenge. The needs to ameliorate the rapidly deteriorating state of the environment and to enhance its capacity to sustain an ever increasing human population have become paramount. Up to now the contribution of ecology as a scientific discipline to environmental discourse has not been as significant as we might expect it to have been – neither with regard to philosophical perspective nor with regard to problem solutions. On the one hand, this may derive from the fact that ecology does not offer unifying solutions to complex environmental problems or produce clear-cut predictions upon which management decisions can be based. On the other hand, ecology's internal state of fragmentation and its apparent inability to speak with one voice reflects the complexity and the diversity of the objects studied. A diversity of approaches should essentially be welcomed. However, it is not always advantageous to the pursuit of possible common goals, especially when communication among different groups is virtually absent.

The lack of a more unifying framework has been deplored by many researchers (e.g. Allen and Hoekstra, 1992; McIntosh, 1985; 1987; Peters, 1991). The community of ecologists is split into subdisciplines not only with regard to the objects of study – plants versus animals, aquatic versus terrestrial, populations versus communities, to name some of the broad categories – but also with regard to the theoretical background of the research endeavour. However, the object chosen and the theoretical perspective may not be independent of one another. It is very likely that differences in initial point of view are often responsible for differences in the interpretation of data and for the choice of the object under investigation, with the establishment of a self-reinforcing feedback cycle. It occurs to me that there are three major origins for the problems facing ecology as a scientific discipline:

- Fragmentation renders the field conservative and prevents progress in pressing questions. It has prevented ecology from defining its own identity as a science. A fruitful discussion is often prevented by the lack of conceptual guidelines. This promotes the vagueness of many concepts and results in controversies being often more semantic than substantial.
- Most traditional concepts in ecology emphasize equilibrium and stability. They have arisen in the mould of the mechanistic world view prevalent in classical physics. Development and spread of genuine ecological ideas has been slow.
- Epistemological problems arise when dealing with complex systems. The methods that

currently prevail in scientific argumentation and proof have been developed by the traditional "hard" sciences for mechanistic systems exhibiting well-defined causalities. Despite empirical evidence for the inadequacy of such thinking in the realm of complex systems, old habits still prevail and new ideas have yet to be introduced and/or developed.

The combined effect of these interrelated aspects seems to impede the tackling of important questions, the answers to which should be central to all of ecology. In addition, most researchers do not devote much attention to the basic foundations and implicit assumptions underlying their research. In my opinion the problems prevalent in ecology are expressions of central questions that should be formulated thus:

- What is the "balance of nature"? Is it tied to stability and equilibrium or does not it to the contrary derive from continuous change?
- Are general laws and principles meaningful in the light of the complexity of and the uncertainties associated with ecological phenomena?
- What is genuine ecological knowledge?
- What is the meaning of evolution, and how has it changed with the advent of humankind?

Tackling these questions would require a common effort of all ecological disciplines in collaboration with evolutionary biologists, philosophers and social scientists. Nevertheless, I make here an attempt to contribute towards providing some answers or at least a direction how to proceed. The order in which these questions are listed indicates my confidence that I can do so. Correspondingly, I put the main emphasis on developing a new concept of ecosystems as self-organizing systems operating far from a stable equilibrium point. To do so requires investigating the functioning of ecological interaction networks as a whole in relation to the nature of their internal organization across time and space. I will be rather generous in adopting concepts from population, community and ecosystem ecology to merge them in a more coherent framework and to give them a new meaning in the light of a dynamic perspective. A key aspect will be the meaning of context, the significance of the environment for the identity of a species and/or organism.

Since the notion of an ecosystem is central to the work to be presented, it is worth deviating here for a moment to devote some considerations to its meaning. The question was raised repeatedly whether ecosystems represented entities that have real counterparts in nature or whether they reflected only abstractions of the human mind constituting in reality a random assemblage of individual organisms, subject to change without notice (e.g. Engelberg and Boyarsky, 1979; Ricklefs, 1986; Simberloff, 1980). Obviously any theoretical term, any description of nature, must involve abstractions. However, some abstractions are more consistent with "reality" than others, where reality refers to what we conceive of as being important. There are no criteria to judge what is "real" apart from how our best theories describe and organize the world for us. Hence, as fundamental as it is, the often posed question about ecosystems being real may not be appropriate. A similar point of view has been expounded by Golley (1994) in his excellent account of the historical development of the ecosystem concept in ecology. Let us nevertheless consider some points of critique and some arguments in favour of this concept being reasonable and helpful.

Admittedly, there is a problem in the intrinsic vagueness of the definition of an ecosystem. Odum (1983) called any organized system of land, water, mineral cycles, living organisms, and their programmatic behaviourial control mechanisms, an ecosystem. In this view, the concept of an ecosystem applies to a microcosm just as it applies to a lake or a forest. Such constitutes a possible and valid perspective since, apart from certain microorganisms, no organism can exist in isolation in a purely abiotic environment. To survive, organisms have to be integrated within a functional web. This is obvious for heterotrophic consumers, but also autotrophic plants require for their growth at least the presence of certain microorganisms. To delineate the nature of a species' integration within a functional web, to show that its identity is inextricably linked to its environment, will still be central themes in the discussions to follow. In the context of the work to be presented here, an ecosystem denotes functionally distinct units where biological organization interacts with the abiotic environment to produce a characteristic network of energy and matter flows. Characteristic is not meant to imply a deterministic behaviour and rigid rules for community composition. Rather it refers to overall patterns of organization leaving ample degrees of freedom how such patterns may be realized. Ecosystem boundaries may sometimes seem fuzzy to us, but they can clearly be delineated. The distinction between a large lake with a steep and rocky shore and its terrestrial surroundings is self-evident. In the case of a shallow pond the distinction becomes less clear. However, the distinctions between systems such as a savanna or a forest are also useful and natural ones. Boundaries may be discerned, even when they are not sharp. Due to a subtle combination of physical and biological factors, ecosystems are often separated by broad transition zones, sometimes referred to as ecotones. Boundary zones can be identified by a high migration rate of organisms, by the exchanges of energy and/or matter across the zone exceeding the exchanges within. The converse is true for ecosystems. Most species are more likely to be found within the system than moving across the boundaries. The exchanges of energy and matter by both biotic and abiotic processes are higher within the system than across its boundaries.

The definition given is sufficient for accepting the concept of an ecosystem and for using it in theoretical considerations. Any experimental study of ecological interactions should be based on a more explicit operational definition of spatial and/or temporal boundaries, as will be emphasized in Chapter Six. To summarize, I take the view that the concept of an ecosystem is important and useful to denote an organized network of both living organisms and abiotic components, the development and evolution of which must be looked at as a whole.

After this interlude I return now to the central theme of this chapter, the current state of ecology. Although the diagnosis presented is definitely shaped by subjective judgement, as any such diagnosis must be, it is also based on a considerable body of evidence which has accumulated over a period of decades.

1.2 FRAGMENTATION IN ECOLOGY

Loehle (1988) stressed a crucial problem by noting that many debates in ecology have been more about words than about facts. Often the points at issue seem not to be identical to the real scientific problems. I am going to illustrate this by an example of a long-lasting controversy.

As far as ecological theory is concerned, perhaps the most involved, persistent, and least conclusive argument is that between traditional holists and reductionists (Lidicker, 1988;

Redfield, 1988; Wiegert, 1988; Wilson, 1988). In a coarse approximation one may state that holists believe in emergence, that is, at each level of complexity in the hierarchy of complex systems (e.g. cell – organism – population – community) new qualities emerge that are not only absent, but are clearly meaningless at lower levels. Reductionists argue that all properties of a system can be explained and understood at the level of the constituent parts. A comparison among different meanings associated with holism in ecology suggests that the distinction between holism and reductionism may sometimes be rather futile. Holism has been referred to as (modified from Loehle, 1988):

1. The view that ecosystems are integrated, interconnected systems with their own laws and organizational principles.
2. The practice of embedding a problem in a larger context.
3. A black-box approach that includes questions such as: What is the nutrient loss response of a whole watershed to acid rain? This empirical approach does not necessarily seek for general laws and principles.
4. Detailed systems analysis such as ecosystem models that are mechanism oriented are claimed holistic because they include all components and processes.

Obviously, approach four could equally well be called reductionist. Nevertheless there has been a segregation between these two schools in ecology. Population ecologists have perceived themselves as more reductionist whereas ecosystem ecologists have perceived themselves as more holistic. The former followed Hutchinson's student Robert McArthur and the traditions of Lotka, Volterra, Gause, Grinnell and Lack. This school concentrated on the ecological and evolutionary interactions among species. Being concerned directly with understanding community organization and species diversity, population ecologists developed the coherent body of competition theory. More recently they have been succeeded by a flourishing theoretical community ecology (reviews by Cohen *et al.*, 1990; Pimm *et al.*, 1991).

Systems ecologists, led by E.P. and H.T. Odum, followed the intellectual traditions of Clements, Tansley, Shelford, Lindeman and Elton, and viewed the ecosystem, if not as a superorganism in the strict Clementsian sense, at least as a complete holistic system with interesting and important emerging properties (e.g. Golley, 1994; Odum and Odum, 1959; Odum, 1983). Recognizing the need for an entity that blended biota with the physical environment, the British plant ecologist Tansley (1935) introduced the notion of an ecosystem:

> The whole system (in the sense of physics) including not only the organism-complex, but also the whole complex of physical factors forming what we call the environment of the biome – the habitat factors in the widest sense.

Subsequently, a systems ecology developed being concerned with "functional" attributes of the ecosystem as a whole, or its input–output relations based on matter and/or energy. Species were largely neglected by replacing them with major trophic categories of producer (autotrophs), consumer (heterotrophs), and decomposer (saprophage) organisms. In contrast, population and community ecologists have been concerned with specific populations, their dynamics, genetics, life histories and evolution. The inclusion of abiotic factors such as nutrients as integral part of an ecological entity has been rejected. Instead, abiotic compartments have been perceived as external environment of the biota. Hence, it is not too astonishing that data concerning the relationship between functional properties of ecosystems such as nutrient recycling and structural attributes such as species diversity are rare. This became depressingly evident from a recent collection of essays on these topics (Schulze and Mooney, 1993a).

I should mention at this stage that there are definitely signs of change. The participants of the Fifth Cary Conference on the theme of linking species and ecosystems came to the conclusion that the biggest challenge in this area was how to aggregate species in sensible ways to yield tractable models of ecosystem processes, while still incorporating critical features of species biologies (Lawton and Jones, 1993). The same view has been endorsed by the recent research agenda of the Ecological Society of America (Lubchenco *et al.*, 1991). However, a recent exchange of arguments about the adequacy of either perspective shows also that the debate is still quite alive (Allen and Hoekstra, 1989; Carney, 1989, 1990; Rowe, 1992). Hopefully, this book may contribute to some progress by offering a framework wherein the claimed goal of linking ecosystems and populations can indeed be achieved.

It is important to note that the two schools of "ecosystem ecology" on the one hand and of "population ecology" on the other hand have shared some common practice despite the difference in perspective and the sometimes fierce antagonism: they both have emphasized equilibrium and stability; they both have largely neglected that ecological systems comprise a hierarchy of spatial and temporal scales. With the advent of hierarchy theory and with the increased recognition of spatial and temporal variability, these deeply rooted habits have been questioned (e.g. Levin, 1989, 1992; O'Neill *et al.*, 1986). This casts doubt on the relevance of major traditional concepts such as, for example, the competitive exclusion principle or the relation between diversity (complexity) and stability that were derived from theoretical considerations based on equilibrium assumptions. Both had a far reaching impact on ecological thought and research in general. Systems with a stable equilibrium state are very attractive at first sight because they allow clear and general statements ideally suited for experimental investigation. If, however, more realistic features such as, for example, habitat heterogeneity are included into the considerations, conclusions based on assumptions of equilibrium and homogeneity become obsolete. A more detailed discussion about these topics is deferred to the next chapter.

The apparent failure of simplifying theoretical approaches has only fostered the long-lasting controversy between two other camps, the ones of theoretical and empirical ecologists. Many an empirical ecologist suspects that theoretical ecology is not coping with the real complexity of nature, whereas theoretical ecologists believe that emphasis on complexity impedes progress. A claim by Dayton (1980) is revealing:

> Ecology often seems dominated by theoretical bandwagons driven by charismatic mathematicians, lost to the realization that good ecology rests on foundations of natural history and progresses by use of proper scientific methods.

Ecological theory has to cope with the problem that general laws and principles are scarce. Hence, it was (and I have to admit myself it still is) tempting to take refuge in simple concepts in a simplified world instead of being drowned by the complexity of the subject. The question arises whether the absence of a unifying theory must be attributed to the fact that ecological science has yet to mature and general rules wait to be uncovered. We could visualize ecology as following in the mould of theoretical science as represented by physics. Complex and notably heterogeneous ecological systems would be treated by a theoretical ecology, or even biology based on a body of general laws, principles, constants and mathematical models. I have strong doubts that this is more than wishful thinking. The absence of a unifying theory may as well be simply inherent in the subject – at least when we stick to the rather narrow definition of what theory means in the physical sciences. However, such a definition may not be appropriate for the generation and valuation of ecological knowledge.

1.3   EPISTEMOLOGICAL QUESTIONS IN ECOLOGY

Ecology has frequently been derogated as being immature or nascent because it has not approached the idealized concept of mathematization, axiomatization, or methodology that some philosophers have suggested a science should possess. Supporters of an autonomy of the biological sciences claim that attributes of classical science commonly urged upon biology in general and upon ecology in particular may not be entirely suitable. It would be beyond the scope of this work to give a comprehensive account of this long-lasting dispute. Excellent reviews can be found in Kingsland (1985), Mayr, (1982), McIntosh (1985) and Rosenberg (1985).

But what are the characteristics of a "mature" science? Judging from the ideal of physics, we need a theoretical foundation comprising a coherent body of general laws. Stated provocatively, we could contrast the situation in biology as a collection of facts grouped vaguely around a set of rather inconclusive principles. Biology seems to be characterized by the absence of general laws. At least, general principles become rather futile in practical situations. However, we have also to note that the laws in physics hardly ever work in practice, where empirical corrections have to be introduced (e.g. Cartwright, 1983). In biology empirical corrections abound due to the complexity of the subject matter. It seems to be a combination of the lack of general laws and the production of a qualitatively different type of knowledge that causes biology to be less accepted. Let me take these arguments in turn.

In our positivist conception of knowledge a favourite recipe for the proper way to do science has been the "hypothetico-deductive philosophy". Debates about its appropriateness for, and putative neglect by, ecologists have been common themes (e.g. McIntosh, 1985; Murray, 1986). Figure 1.1 illustrates the typical process of hypotheses generation.

Observations motivate the generation of hypotheses concerning causal relationships. It is not my intention to give here an argument in favour of the inductive method where hypotheses are derived as logical consequences from collections of empirical facts. I assume hypotheses to be generated in a creative process of abstraction and innovation

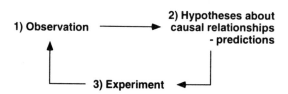

Figure 1.1   Steps required by the hypothetico-deductive method

being nevertheless tied to prior experience. Proper hypotheses should lead to the formulation of experiments, the results of which render hypothesis testing possible. The resulting exchange of arguments and experimental results fosters scientific discussion. Progress in knowledge and understanding is achieved in a continuous spiral of observation – hypotheses generation and improvement – experiment.

In biological systems problems arise at every step of this procedure. Starting with observations (1), in biology it is far more difficult than, for example, in physics to choose appropriate variables to be monitored. What is monitored guides the possibilities for an explanation. Often the limited information available determines the variables the system is reduced to. Due to the nonlinear character of interactions and synergistic effects it is not straightforward to decide whether some variables can be neglected. Hypotheses (2) have often to be formulated in a conditional way referring to specific circumstances of the biotic and/or abiotic boundary conditions. This way hypotheses can hardly be falsified because deviations in experimental data can always be attributed to a difference in boundary conditions, to constraints operating at levels both lower and higher than the one under consideration. When we desire, for example, to conduct investigations at the population level, our results are influenced both by the variations of the individual organisms and by the community of which a population is a part. If we attempt at all to talk about causality in the sense of driving forces as causes and reactions as effects, ecological phenomena are multicausal. Hence, they may often be given different equally valid interpretations from the restricted perspective of a monocausal mechanistic explanation. The situation is aggravated by the fact that ecological experiments are often hardly controlled and hence difficult to reproduce. To the same extent as experiments are simplified to render them better controllable, they lose significance for reality.

An illustrative example is given by the discussions related to the controversy about the dominance of top-down versus bottom-up forces in population and community ecology. For years, ecologists have been debating whether either resources (bottom-up) or predators (top-down) control and limit organisms (e.g. special feature by Matson and Hunter, 1992). More recently the perspective has changed towards a more synthetic, flexible view of the varying roles of top-down and bottom-up forces. Both seem to act on populations and communities simultaneously, albeit with varying degree. There is a shift from discussing which occurs, to what controls the strength and relative importance of the various forces under varying conditions. The conditions may involve a host of factors such as physical environment, species richness, primary productivity etc. As a consequence of the many factors that impinge on the interactions, explaining food web regulation becomes a rather intricate business. Taking this into consideration it is not too astonishing that specific manipulations of food webs have not always met with success. An example is given by lake biomanipulation, which is rather controversial as a management tool.

Figure 1.2 shows an aquatic food chain with the corresponding interactions between trophic levels. The shaded arrows delineate the major routes for controlling excessive phytoplankton growth in aquatic systems. The traditional approach has been to fight the actual cause of eutrophication by limiting the nutrient load. However, this may not always be feasible due to political constraints and/or it may take a long time for measures to show effects. Therefore the application of biomanipulation has enjoyed an increasing popularity. The concept of lake biomanipulation is based on a top-down view of cascading interactions along a linear food chain. It is predicted that increased piscivore abundance

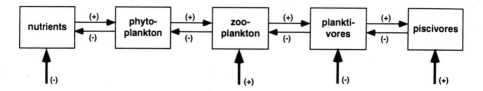

Figure 1.2   Conceptual scheme of cascading trophic interactions in an aquatic food chain. The shaded arrows indicate possibilities of eutrophication control by management

results in a decreased planktivore abundance. A decreased planktivore abundance results in increased zooplankton abundance, and increased zooplankton grazing pressure leads to a reduction in phytoplankton abundance. The most oft employed management tool is to increase stocking of fish which at the same time is beneficial to the fisheries. DeMelo *et al.* (1992) argue that experimental evidence in favour of the success of biomanipulation is equivocal at best, that alternative explanations cannot be excluded because a variety of factors were neglected. Proponents reject this criticism and blame misunderstandings and misinterpretation for causing the controversy (Carpenter and Kitchell, 1992). They attribute failures of biomanipulation to the fact that some factors may have been neglected.

The idealized picture of defined pathways of interaction between defined trophic levels may be an oversimplification. It seems to apply predominantly to rather species-poor and small systems (cf. review by Reynolds, 1994). Strong (1992) compared the trophic interactions in most terrestrial and in species-rich aquatic food webs to trophic trickles rather than trophic cascades. This agrees with a continuous picture of food web regulation to be discussed at a later stage in this book. To say the least, I would describe cause–effect relationships in such food webs as diffuse and strongly dependent on context. Systems of such type seem to escape attempts of external control. Correspondingly, predictions of specific effects are associated with a high degree of uncertainty. The nature of ecological systems organization also provides the reason why ecological knowledge seems to be of little scientific value. It is not amenable to the instrumental approach pervading most of the natural and technical sciences. I try to depict the logical framework in Figure 1.3.

The success of a scientific discovery is often valued by its potential to devise schemes of control and manipulation. It is not by accident that molecular biology nowadays enjoys a high reputation. Finally it seems that even biological knowledge can be converted into something "useful", where useful is identified with the ability to perform specific manipulations. Due to the reasons outlined, ecological knowledge hardly ever falls in this category.

Based on these considerations, I already draw here some conclusions regarding the

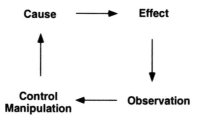

Figure 1.3   Instrumental approach in the natural and technical sciences

nature of theory in ecological research. Instead of being a coherent explanatory framework in terms of cause–effect relationships, I rather view theories as maps guiding our perception of the world, assisting our discerning patterns and context. For a map to be of any use, the purpose the map was designed for and the intention of its user must coincide. Correspondingly, we can assess the value of a map by its potential to assist us in dealing with certain problems, in coming to a required decision. A map can also prove to be wrong when it depicts patterns that have obviously no counterpart in the real world. Here enters also the recognition that science is a social acivity where "truth" is defined in a dialogue in the scientific community. Theory should assist the organization and interpretation of experimental evidence by providing guidelines and a framework for scientific communication. A diversity of approaches and perspectives may assist in shaping a clearer picture of and improving our intuitive understanding of complex systems. In ecology general concepts may rather be found for pattern and characteristics of organization than for mechanisms and cause–effect relationships. It must also be the task of a theoretical ecology to reflect on the nature of questions to be asked, to develop tools on how to value, to deal with, and to communicate knowledge being associated with uncertainties.

1.4 HOW THIS BOOK IS ORGANIZED

This book contains essentially four parts:

- A historical account to elucidate from where my own work started, what it embraces, and where it attempts to make improvements (Chapters One to Three).
- Introduction of a new theoretical framework to describe the organization of ecological networks across time and space (Chapters Four to Six).
- Synthesis and speculations (Chapter Seven).
- Appendices summarizing details of the mathematics involved and of further technical applications (Appendices One and Two).

In the following I should like to sketch briefly the main contents of each chapter.

Chapter Two contrasts traditional approaches emphasizing invariance and stability with recent developments in the science of complex nonlinear systems. In ecology, powerful schools of thought were based on the idea of stability and equilibrium being paramount properties of natural systems. Representative examples are given by the relationship between stability and complexity and by the principle of competitive exclusion. These traditional concepts are challenged by the increased awareness of the importance of spatio-temporal variability in natural systems and by the progress made in the complex dynamics of nonlinear systems.

Chapter Three establishes the link to evolutionary biology where we encounter a similar emphasis on stability and equilibrium concepts. I review the historical development of the niche concept as a major bridge between thinking in ecological and evolutionary dimensions. We will note that evolutionary considerations are largely absent in ecosystem ecology.

In Chapter Four I introduce spatio-temporal organization in dynamic nonequilibrium systems as an alternative view in between stability and equilibrium on the one hand and stochasticity and randomness on the other hand. The approach chosen here, that emphasizes context and relations, is contrasted with one based on a mechanistic conception of natural systems. Spatio-temporal organization is given an operational definition in networks of energy and matter flows. Its meaning is explained in detail both in qualitative and in quantitative terms.

Chapter Five illustrates the functional importance of spatio-temporal organization by results from different multi-species models. They are discussed with special emphasis on the mutual relationship between the level of the species and the level of the system as a whole.

Chapter Six deals with the problem that the network structure of ecosystems is not self-evident but depends on the perspective of the investigator. I introduce a general concept to structure ecological networks. This concept can be seen as a logical consequence of the previous considerations regarding the dynamic, hierarchical nature of ecosystems. Such a coherent description is vital to investigate the mutual relationships among functional, dynamic and taxonomic diversity and functional properties at the level of the ecosystem as a whole.

Chapter Seven is an attempt at a synthesis. The results obtained are integrated within a larger framework. A changed perspective on the dynamic, relational nature of ecological systems may have far-reaching consequences for our attitude towards management, for our dealing with uncertainties, and for our understanding of biological and cultural evolution.

This work is addressed to both a theoretically oriented and an empirically based readership. Hence I decided to choose a form of presentation rendering it accessible and of interest to all of them. Those who feel uneasy when confronted with mathematical equations can simply follow the main text. The conceptual ideas and the typical properties of the mathematical methods and models are described verbally. Those interested in the mathematical foundations will find the basic theoretical and mathematical derivations summarized in "boxes" throughout the main text. Those more interested in the details of the mathematics involved and technical aspects of further applications should consult the two chapters of the appendix.

Chapter Two

Ideas of stability and equilibrium in ecological thought

The scope of this chapter is to review the origins and the influence of concepts based on stability and equilibrium prevailing in traditional ecological theory. The traditional view is contrasted with recent developments in the theory of complex nonlinear systems. It may be useful to clarify at the very beginning the distinction between the notions of equilibrium and stability. Equilibrium refers to a state that is invariant. Invariance may imply that variables do not exhibit any temporal and/or spatial variations. It may equally well imply the constancy of averages over time and/or space. Stability refers to a dynamic behaviour, whether a system returns to its equilibrium state after being perturbed from it. The inconsistent use of both expressions has added to the confusion surrounding their meaning.

2.1 HISTORICAL BACKGROUND

Ecology as a scientific discipline has largely developed in the last century. Correspondingly it has been influenced by traditions from other disciplines. Regarding concepts emphasizing equilibrium and stability we may discern three major sources of influential thought:

- Darwin's theory of evolution and the principle of natural selection
- The balance of nature concept of natural history
- Classical physics and its emphasis on stability and equilibrium

It is not a contradiction to invoke a theory of evolution in supporting equilibrium concepts. Even when Darwin's principle of natural selection explains change on evolutionary time scales, does it require equilibrium on the ecological time scales we are interested in here. They are also the typical time scales of human experience. The ubiquitous influence of Darwinian thought is reflected in the emphasis on competition shaping population and community structure. Darwin's concept of natural selection fostered a flourishing evolutionary ecology searching the explanations for species properties in the optimization of life strategies to gain maximum fitness. Natural selection and maximum fitness are closely tied to the concept of a stable equilibrium point. The latter is required for competitive exclusion to become effective and for optimization strategies to be useful. Despite progress in physics, in other fields of science, and in certain areas of theoretical ecology, the traditional notions of equilibrium and stability seem to prevail in dealing with both ecological and evolutionary phenomena. To develop an understanding of the attractiveness and the persistence of the stability concept it is useful to consider in more detail its historical roots and its applications in ecology.

McIntosh (1985) noted that traditional natural history had as a central theme the concept of the balance of nature, of things being interconnected to preserve an order, which for centuries had been described in Western thought as divinely ordained. Randomness and extinctions weren't usually considered possibilities and variation and change were largely ignored. The "balance of nature" was described by Egerton (1973, 1976) as the "oldest ecological theory". He reviewed its historical origins and influence in 20th-century ecology tracing its transformation from a divinely ordered nature to an order generated by nature itself via evolution. The concept of nature as essentially stable, or balanced, and of species populations as forming assemblages or communities that functioned as orderly integrated units to establish and maintain equilibrium was nearly universal in the 19th century.

Clements (1936) changed perspectives to some extent by introducing a dynamic ecology. He urged for not just descriptions of vegetation, its distribution and classification, but studies of changes in vegetation and explanations of these as a result of a dynamic process. Despite the emphasis on dynamics the processes of succession were assumed to proceed in a regular and predictable order. Clements' concept of the climax posited the development of a stable self-perpetuating association in which the dominants, at least, were in equilibrium. If the climax was not actually present, all sites in an area were assumed to tend towards it, and if disturbed, succession started the trend towards the climax again. In his conception of an ecosystem Tansley (1935) claimed as well a universal tendency towards equilibrium:

> In an ecosystem the organism and the inorganic factors alike are components which are in a relatively stable dynamic equilibrium. Succession and development are instances of the universal processes tending toward the creation of such equilibrated systems.

Starting with the success of the Lotka–Volterra competition models in the 1930s, equilibrium theories of populations and communities flourished as well in theoretical ecology. The subsequent development of the field was centred around the following questions (Lewin, 1983; McIntosh, 1985, Simberloff, 1980):

- Are communities assemblages of populations at or near a stable equilibrium state so that community properties are similarly stable?
- If disturbed by changing conditions, do populations or communities return to, or at least tend toward, the equilibrium state?

In these discussions, an equilibrium state was identified with a stable equilibrium point. Extensive modelling efforts proved at first sight to provide deceptively simple solutions to the problems stated. Equilibrium and stability were emphasized in modelling and support was sought in experimental data. However, to date those questions have remained largely without conclusive answers. There are a variety of reasons for this situation. We have to be aware that any statement about a system's stability or equilibrium properties depends on the space and time scales and the specific variables under consideration as well as the concept we refer to. A major structural problem hindering progress is given by the fact that although the notions of stability and equilibrium are central to ecology there is no unanimity about their definition. They may refer to an intuitive understanding or to more or less rigorous mathematical definitions.

2.2 DIVERSITY OF STABILITY CONCEPTS IN ECOLOGY

Pimm (1984, 1991) must be attributed great credit for having organized the different meanings associated with the concept of stability. He identified five concepts that have been referred to as stability by theoretical and empirical ecologists (again equilibrium refers to a stable equilibrium point where all variables are time invariant):

- In the mathematical sense a system is considered to be stable if and only if the variables return to equilibrium conditions after displacement from them.
- Resilience corresponds to how fast variables return to their equilibrium after having been displaced from it. The higher the resilience the faster the recovery following a perturbation. Systems with high resilience may be said to be stable.
- Persistence is how long a variable lasts before it is changed to a new value. Systems with high persistence may be said to be stable.
- Resistance measures the degree to which a variable is changed, following a perturbation. Systems with high resistance may be said to be stable.
- Variability is the degree to which a variable varies over time. Systems exhibiting little variability may be said to be stable.

There does not yet exist a general agreement on the consistent use of the different terms. In addition, having a concept does not relieve us from having to define the systems to which a concept is meant to refer. Depending on our research interests we may choose for example individual species abundances, species composition or trophic level abundance as variables of interest.

Conclusions with respect to a system being stable and being in an equilibrium state depend as well on the temporal and spatial scales of observation (see also Allen and Starr, 1982; O'Neill *et al.*, 1986; Williamson, 1989). In a forest, for example, little change is noticed, when focusing on the dynamics of single tree stands over the period of a year, which is short in comparison with a tree's lifetime. If the observation period is extended to cover one or more generation times, what may be equivalent to several centuries, the system may appear as highly variable. However, when the spatial observation limits are extended to cover a large ensemble of individual trees, the tree composition may remain rather stationary even on a century time scale. Another example is given by phytoplankton communities. Here one year is a long time period in comparison with the generation time of phytoplankton species. Whereas the species composition exhibits seemingly random fluctuations over an annual cycle, regularities and repetitive patterns may be detected in a comparative analysis of several years.

Hence, when talking about stability and equilibrium we should always be explicit regarding the variables, the spatio-temporal scales, and the concept we refer to. We have also to be aware that the spatio-temporal scales for data resolution and the measuring periods should be chosen according to the variables and the concept we are interested in. In the past, all these aspects were largely neglected thus adding much to the ambiguity associated with the use of the notion of stability in ecology.

Judging whether a system is stable is further impeded because we have to distinguish between exogenous and endogenous sources for variation. A system may be stable with respect to its internal dynamics but it may continuously be prevented from achieving an equilibrium state due to changes in its physical environment. In such cases it is essentially impossible to determine whether the internal dynamics would lead to a stable equilibrium point in the absence of external perturbations. I should like to emphasize that nowadays researchers have recognized the importance of external variations driving populations away from an equilibrium state. However, it is still a vexed question whether the dynamics of species interactions may be the source of instability as well (e.g. Pimm, 1991).

In the context of this book, the notion of stability refers to the mathematical concept stating that a dynamic system is stable when it returns to its equilibrium conditions after displacement from them. It is quite popular to visualize the essentials of the underlying idea in a pictorial representation of the type depicted in Figure 2.1 where the equilibrium state is equivalent to a time-invariant equilibrium point.

The ball symbolizes the state of a dynamic system which at any point in time can be characterized by the magnitudes of the relevant state variables. As the spatial coordinates define the position of the ball in the landscape, the magnitude of the state variables determine the position of a dynamic system in what is called a system's phase space. The phase space embraces as dimensions all the state variables relevant for system dynamics. In a competition model, for example, with three species competing for a limiting resource, the phase space includes four dimensions – the abundance of each species and of the resource. We can now describe a system's dynamics as movement in this phase space. The ball in the physical model in Figure 2.1 moves in the landscape driven by the forces of gravitation. Similarly, we can imagine the state of a dynamic system to move in phase space driven by forces arising from interactions among state variables. In both Figures 2.1a and b the ball is in an equilibrium state where it does not move. The stability of the equilibrium state depends on the surrounding landscape. Whereas in (a) the ball returns to

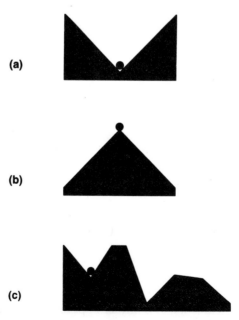

Figure 2.1   Pictorial representation of the stability properties of dynamic systems. Systems with (a) a stable equilibrium point, (b) an unstable equilibrium point, (c) several equilibrium points both stable and unstable

the equilibrium state after having been displaced from it this is not the case in (b). Hence, the equilibrium state in (a) is characterized as stable and the one in (b) as unstable. For a dynamic system a corresponding equilibrium state is given when the magnitudes of the state variables are constant. The nature of the landscape in phase space depends on the nature of the interactions among state variables. An equilibrium point of a biological community is stable when the interactions among the populations counteract any deviations in population densities from their equilibrium values.

We can further characterize an equilibrium point by what is called its domain of attraction. To illustrate this, I extended the ball example by introducing the rugged landscape in Figure 2.1c. We can now distinguish several locally stable equilibrium points. The domain of attraction of each can be identified with the extension of the surrounding valley. The equilibrium points differ in their recovery time defined as the time the system needs to return to the equilibrium point subsequent to a perturbation. In our model the recovery time is inversely related to the slope of the surrounding valley. The steeper the slope, the shorter the time for the ball to return to its equilibrium state. For a dynamic system the slopes are again determined by the nature of the interactions among state variables. They can be determined by the so-called eigenvalue analysis of a system of differential equations linearized around its equilibrium point (e.g. Luenberger, 1979). The eigenvalues correspond to the slopes in our model. Their magnitude tells us something about the rate of change whereas their signs tell us the direction of change regarding the deviation of a state variable from its equilibrium value. Linear models or models linearized around the equilibrium point have the advantage of allowing a rigorous definition of stability and powerful quantitative statements are possible (for a system to be stable all eigenvalues must be negative). This explains also the popularity of this method and the emphasis on linear or linearized systems in modelling as, for example, in investigations of the relationship between stability and diversity.

Such methods rely on an equilibrium state being equivalent to an equilibrium point. However, we may also talk about equilibrium systems that possess or approach a state where the long-term averages of variables such as the number of species, the biomass and the amount of resources remain constant. This is self-evident for systems with a stable equilibrium point. It may also apply to systems that show a high degree of variability on short time scales. Expressed in the language of dynamic systems, we may talk of a dynamic equilibrium state if the system moves on a bounded attractor in phase space. A system's attractor denotes the ensemble of all observed system states. Chaotic systems may also possess a dynamic equilibrium state despite the irregular nature of individual trajectories. Trajectory refers to the path in phase space that is followed by a system as a function of time. It comprises the variations of all the individual state variables. A system can be called a real nonequilibrium system when the long-term averages of the variables change as well, either caused by external physical disturbance or by internal processes of evolution.

Based on these considerations I derived the definitions given in Box 2.1 of equilibrium and stability properties of dynamic systems. These are used throughout this book.

In most discussions, stability and equilibrium have referred exclusively to the notion of a static equilibrium point. To avoid confusion I will in such cases explicitly refer to point-stability. As defined in Box 2.1, stability and equilibrium cover a broader range of system behaviour. The classification of any real system depends on the temporal and spatial scales of observation and the monitoring variables chosen. In systems with irregular population dynamics, for example, the mean of a species abundance may

---

Box 2.1   Definitions of equilibrium and stability properties in dynamic ecosystems

Equilibrium system
Dynamic system possessing at least one defined equilibrium state.
- Static equilibrium point: The basin of attraction is equivalent to a point equilibrium where all state variables are time invariant.
- Dynamic equilibrium state: The basin of attraction is equivalent to a periodic orbit or even a chaotic attractor where state variables fluctuate periodically or even chaotically. The long-term averages of state variables are time invariant.
- Stable system: The system returns to its equilibrium state (static or dynamic) subsequent to perturbations.

Nonequilibrium system
Dynamic system where the phase space itself as well as the attractor change over time due to a combination of exogenous and endogenous influences. It is per definition always unstable. The long-term averages of state variables vary over time. The variability of state variables increases rather than decreases with increasing length of the observation period.

---

converge to some finite value. However, the time for doing so may exceed the generation time by orders of magnitude. Hence, for practical purposes such a system has to be considered a nonequilibrium system, in particular when we take into account that the environment varies as well over time.

Let us now return to our general discussion about the application of the stability concept. We have already noted that theoretical considerations focused mainly on investigations of linearized mathematical models. The search for general principles in linearized models, discarding thereby variations as noise obfuscating the essential, had its origin in classical physics. Such a type of analysis is associated with a certain perspective on the natural world.

2.3   LINEAR WORLD VIEW IN CLASSICAL PHYSICS

The increased recognition of the ubiquitous presence of complex dynamics in physical systems seems to have inspired several researchers to some critical reflections on the pervasive influence of simple dynamics dominating traditional physical concepts. Ruelle (1991), for example, gave an illuminating account of the importance of stability and linearity in classical physics. West (1985) wrote an excellent review about the implications of a linear world view pervading classical physics and its subsequent spreading into most other areas of natural science.

One implication of a linear world view with far-reaching consequences is the principle of superposition. It qualitatively states that a complex event can be segmented into a number of separate simple components. These components can then be recombined back into an organized whole that can be understood in terms of the properties of the components. The origin of superposition may be traced back to Newton's treatment of the propagation of sound in air. Newton argued that a standing column of air could be modeled as a linear harmonic chain of equal mass oscillators. The linear character of the model allowed a simple treatment of the problem. The success of the application of a harmonic chain to the description of complicated physical processes established a precedent that has evolved into the backbone of modelling in physics. The principle of superposition to be formulated subsequently states that the most general motion of a vibrating system is given by a linear superposition of its characteristic (eigen, proper) modes. The solution to sets of complex partial differential equations may be represented as a linear superposition of the characteristic or eigen-motions of the physical problem thus rendering it accessible to rigorous mathematical treatment. Such an approach was applied in almost every area in physics including diffusion, heat propagation and quantum mechanics.

West (1985) notes that any deviation from linearity in the interacting motions was looked upon as a perturbation from the equilibrium point and was assumed to be treatable by perturbation theory. West further points to the fact that the transition from a linear to a nonlinear world view has been forced on the physical sciences, a transition that is still in progress and that cannot hide the linear world view presently dominating the physical sciences. One may argue that only the lack of computing facilities prevented the proper treatment of nonlinear phenomena, the importance of which was always recognized. This argument cannot be simply dismissed. If we consider, however, the slow acceptance of and the fierce antagonism against chaotic phenomena we may conclude that the belief in stability and linearity must be deeply rooted.

More recently, chaos theory has become quite popular among physicists. Nevertheless, the original paradigm of linearity and stability seems to be still omnipresent in physical concepts and has infiltrated into other fields of science. Consequences are given by the firm belief in the predictability of natural systems, or by the habit of decomposing complex systems into their component parts and of investigating the latter in isolation. Even when the assumption of linearity was relaxed at an early stage, by including for example effects of saturation, the idea that natural systems must possess a stable equilibrium point has prevailed. Internal forces have been assumed to drive every natural system back to its equilibrium point. In ecosystems the internal forces were identified with the interactions among species. It suggested the need to start searching for relationships between a system's properties (e.g. diversity) and the stability of its equilibrium point. Methods of linearized stability analysis were finally applied in the attempt to solve the controversy about the relationship between diversity and point-stability, a controversy with a long history.

2.4 THE EQUIVOCAL RELATION BETWEEN DIVERSITY AND STABILITY

The following brief summary emphasizes the points relevant to the considerations in this book. More comprehensive reviews of the topic were given by Goodman (1975) and Pimm (1991).

The predisposition to expect increasing stability with increasing diversity of ecosystems was derived from weak arguments only. At its origin were observations that species diversity increased from the pole to the equator. Tropical rain forests were perceived as the prime example for a direct relationship between high species diversity and stability where stability was rather vaguely defined. The conceptual model relating complexity and stability was thought to be supported by the following lines of fragmentary evidence (Elton, 1958):

- Simple one-predator/one-prey systems undergo violent population oscillations and go extinct rather quickly. This applies to both experimental systems and mathematical models.

- Man-made monoculture is vulnerable to pest outbreaks.
- Island biota are vulnerable to invasion.
- The relatively depauperate arctic and boreal faunas are subject to obvious population fluctuations, whereas the rich and complex tropical rain forests look stable.

Based on these observations it was implicitly assumed that an increase in complexity would foster point-stability even when it was never shown to be true in the case of mathematical models. The idea that there might be a simple relationship between such gross variables as stability and complexity was intriguing, in a field where generalizations are scarce. Note that basic to these derivations was the underlying idea that natural systems must necessarily possess an equilibrium point and that the latter must be stable. However, neither stability nor complexity were well defined at this early stage of the considerations (e.g. Pimm, 1984). Subsequent attempts to improve this situation and to cast both stability and complexity in more rigorous mathematical terms were fraught with difficulties.

Following an essentially intuitive argument, MacArthur (1955) suggested that a measure of trophic diversity reflecting complexity could be used to describe community stability. The proposed function was derived from information theory:

$$S = -\sum_i p_i \log p_i \qquad (2.1)$$

where p_i denotes the proportion of a community's food energy passing through the ith pathway in the food web. This diversity index increases with the logarithm of the number of distinct pathways in a food web. It is highest for an equal distribution of the energy over all flows resulting in a minimization of the variation among flow magnitudes. MacArthur laid here the early foundations for the development of food web theory that carried the diversity side of the diversity–stability relationship one further step by including interactions or connectance among the species, and not just counting the species. Hence, taxonomic diversity was replaced by trophic complexity.

Subsequent attempts to provide evidence in favour of the hypothesis that trophic complexity confers a degree of community stability largely failed. Both empirical studies and numerical simulations of model food webs seemed to contradict the expectations of the theory of diversity–stability relationships. The most influential studies were performed by May (1973) who approached the problem using a general set of equations for a community with n species.

As outlined in Box 2.2, the nonlinear equations may be linearized around the equilibrium point to yield the community matrix, A. The community matrix is a matrix of dimensions $n \times n$ whose elements a_{ij} describe the effects of species j upon species i near the equilibrium point. May neglected details of the original equations and focused on the stability properties of the community matrix A which can be used to construct a landscape in a phase space comprising all component populations. As outlined in the previous paragraph, such a landscape can be used to analyse the stability of equilibrium points. Subsequently, the community matrix developed into the backbone of investigations of community stability. The equilibrium point is stable relative to small perturbations if all eigenvalues of A are negative.

Box 2.2   Linearized approximation of system dynamics around an equilibrium point

May (1973) used a general set of equations for a community with n species:

$$\frac{dN_i}{dt} = F_i(N_1(t), N_2(t), \ldots, N_n(t)) \tag{2.2}$$

He defined as a measure for the deviation of a population's size, N_i, from its size at the equilibrium point, N^*_i:

$$x_i = N_i - N^*_i \tag{2.3}$$

to obtain the set of linearized equations:

$$\frac{dx_i}{dt} = \sum_{j=1}^{n} a_{ij} x_j \tag{2.4}$$

with $a_{ij} = \left(\frac{\partial F_i}{\partial N_j}\right)^*$ where $*$ refers to the equilibrium point. In matrix notation:

$$\frac{dx}{dt} = A\, x(t) \tag{2.5}$$

where x is then $n \times 1$ column matrix. The $n \times n$ Jacobi matrix A was called the community matrix.

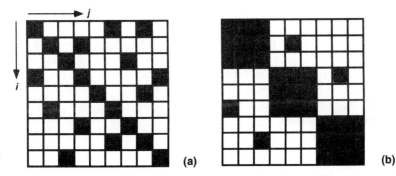

Figure 2.2   Schematic representation of community matrices. Each field denotes an element a_{ij} describing the effect of species j on species i. The shading denotes the intensity of an interaction. (a) Interspecific interactions are distributed at random. (b) Interspecific interactions are stronger within subgroups of species

May used the community matrix to perform systematic investigations of the relationship between complexity and stability. He identified complexity with the degree of connectance in a community. Connectance refers to the fraction of possible links that are realized in a given community. Figure 2.2a sketches the type of matrices chosen by May. The shading denotes the intensity of an interaction. May assumed the presence of intraspecific competition ($a_{ii} = -1$, $i = 1,2 \ldots, n$) as indicated by the dark shading along the main diagonal. He further allowed a fraction C of the interaction coefficients a_{ij} to be determined by random drawing from a symmetrical distribution of values with a mean of zero and a variance of σ. May showed that the equilibrium point will almost certainly be locally stable if the condition stated in Eqn (2.6) is fulfilled and almost certainly unstable otherwise.

$$nC < \frac{1}{\sigma^2} \tag{2.6}$$

where

n is the number of species,
C is the average connectance, $C \leq 1$,
σ is the average interaction strength.

Thus, May concluded:

> This ensemble of very general mathematical models of multispecies communities, in which the population of each species would itself be stable, displays the property that too rich a web connectance (too large a C) or too large an average interaction strength (too large a σ) leads to instability. The larger the number of species, the more pronounced the effect is.

May's results contradicted the diversity–stability hypothesis. His work impressed by its methodological elegance, by the clear definitions and by the logic of argumentation. He phrased his interpretation with caution, being aware of the model's underlying simplifications. It is only valid for a stability concept referring to a localized linear approximation in the close vicinity of an equilibrium point. Diversity is equivalent to the number of interactions. Nevertheless, May's results were readily assumed as being of general relevance. In addition, the statements in terms of food web connectance were clear and amenable to experimental investigation.

Much effort has subsequently been devoted to find experimental evidence supporting relation (2.6). The whole field of food web research was to a large extent motivated by the prediction that patterns observed in nature ought to correspond to the patterns that were shown to yield stable equilibrium points in model investigations. The finding of an inverse relationship between the number of species and food web connectance C has been taken as support for the theoretically derived claims. Deterministic models, mainly of the Lotka–Volterra type, were used to derive further properties to be expected in systems displaying point-stability, as the absence of omnivory or the absence of feeding loops. The lack of finding these properties in data sets from natural food webs was taken as further evidence for natural systems exhibiting configurations assuring point-stability (Cohen *et al.*, 1990; Lawton and Warren, 1988; Pimm *et al.*, 1991). However, we should be aware that analyses of the stability properties may lead to completely different results when, for example, hierarchical interaction structures and/or spatial patterns are included in a food web model (e.g. Hastings, 1988; Hogg *et al.*, 1989). An example for a hierarchically structured community matrix is provided in Figure 2.2b. As Gardner and Ashby (1970) noted earlier, the stability is enhanced when a community is organized into guilds of strongly interacting populations that interact only weakly among each other. It was only recently that these observations were cast into more formal terms (Hogg *et al.*, 1989).

Recent years have witnessed a heated debate on the validity of food web theory (e.g., Hastings, 1988; Paine, 1988). It may be seriously compromised by ignoring largely spatial and temporal scales, by the weak data base exhibiting, for example, a highly heterogeneous level of aggregation (taxonomic and trophic species, functional groups) and by the absence of criteria to record a link or not. The diversity (complexity)–stability discussion has always suffered from the lack of coherent definitions for the two properties and the resulting heterogeneity and incompleteness of both theoretical arguments and field studies. Detailed food web studies may lead to results differing largely from the ones derived from simplified webs. A comprehensive discussion of the points at issue is deferred to Chapter Six.

Despite the condensed style of presentation, the foregoing discussion should have conferred some feeling for the problems associated with stating general principles and laws in ecology. A similar situation is encountered in another important area of ecological theory, competition theory, and for its most important component, the competitive exclusion principle.

2.5 THE COMPETITIVE EXCLUSION PRINCIPLE

Key assumptions of classical competition theory involve the following notions of equilibrium and stability (for review see Chesson and Case, 1986):

1. The life history characteristics of a species can be adequately summarized by the population's per capita growth rate.
2. Deterministic equations can be used to model population growth; in particular, environmental fluctuations can be ignored.
3. The environment is spatially homogeneous, and migration is unimportant.
4. Competition is the only important biological interaction.
5. Coexistence requires a stable equilibrium point.

An important prediction of the theory is the competitive exclusion principle (CEP) stating that at least *n* limiting resources are required for the coexistence of *n* species. In such a case the *n* fittest species dominate and the others become extinct. The CEP was derived from the Lotka–Volterra competition equations as summarized in Box 2.3 for two competiting species. These equations predict that two species competing for one resource cannot coexist when they exploit their environment in exactly the same way. As with other ecological principles of great generality, the application and testing of the CEP has been

---

Box 2.3   Lotka-Volterra competition equations for two species

The dynamics of two competing species are described by a pair of differential logistic equations extended with a competition term:

$$\frac{dN_1}{dt} = r_1 N_1 \left(1 - \frac{N_1 + a_{12}N_2}{K_1} \right) \tag{2.7}$$

$$\frac{dN_2}{dt} = r_2 N_2 \left(1 - \frac{N_2 + a_{21}N_1}{K_2} \right) \tag{2.8}$$

K_1 and K_2 refer to the carrying capacities of species 1 and 2 in the absence of one another. a_{12} and a_{21} are competition coefficients expressing the degree of competitive inhibition of one species on the other. Depending on the parameters, the equations predict either coexistence or competitive exclusion. Coexistence is impossible if the species use the environment in the same way which means that $K_1 = K_2$ and $a_{12} = a_{21}$.

associated with considerable ambiguities. Competition is notoriously difficult to demonstrate in natural communities. Generally, support is sought indirectly by invoking competition as a driving force for an observed outcome. In ecological practice, the CEP has also been stated in rather inconclusive ways, as can be seen from the following selection of citations (Murray, 1986):

- Two or more resource-limited species having identical patterns of resource utilizations cannot continue to coexist in a stable environment.
- Complete competitors cannot coexist indefinitely.
- Two or more species with identical ecological niches cannot occupy the same environment.

However, identical patterns will hardly ever be found, thus limiting the application of the principle to rather unlikely cases. Indefinitely is a long and undefined period of time. And, like stability, ecological niche and environment are only useful terms when they are explicitly defined.

We encounter here again several of the major problems when stating any ecological "law". If formulated in a generalized form, claims for ecological regularities are not very enlightening and they are hardly ever applicable to a real situation. Modified to fit the needs of practical problems, "laws" risk becoming vague and confusing. The counterparts of the constants and laws which are characteristic of physics seem to be difficult to identify in ecology. Critical voices may reason that ecological theory only lacks rigour and the appropriate methodology. Murray (1986), for example, argued that the CEP might acquire the status of an ecological "law". In his opinion, it was the vagueness of how the CEP was stated not the contents of the CEP *per se* that caused the problems. He further argued that the success of physics in contrast to ecology lies in the explicitness of the physicists' theories, the rigorousness of their evaluations, and their understanding that general theory does not and cannot usually account for all empirical data. He suggested that ecological theory could profit considerably if it would adopt some rules from physical theory.

I agree with Murray's claim that ecological theory needs a more coherent body of terminology and concepts to facilitate unambiguous communication. His point is also well taken that physical laws are of less general applicability than we naively may believe. However, I question the usefulness of his promoting the CEP to an ecological law. There is no doubt that the CEP applies once the restrictive assumptions required are fulfilled. Sommer (1984) observed competitive exclusion to take place in laboratory algal communities when the laboratory conditions and the nutrient supply were held constant. It took, however, several weeks – or 10 to 20 generation times – for inferior species to become extinct. When the nutrient was added in pulses, coexistence was greatly enhanced. Such observations make us suspicious whether the conditions required for the CEP to apply will ever be encountered in a natural environment.

We have further to be aware that relaxing some key assumptions changes the conclusions of the CEP not only quantitatively but qualitatively. Similar species may coexist indefinitely in simulation models once the restriction of the equilibrium state being equivalent to a point equilibrium is abandoned. Such is the case if predation is introduced, leading in most cases to a periodic equilibrium state with predator–prey cycles. In a patchy environment favouring different species in different patches, it is possible for n species to coexist on one limiting resource in a system consisting of at least n different patches. Despite the spatial heterogeneity, this is also an equilibrium situation with *n* different equilibria for each of the *n* patches. Both situations can still be accommodated within an extended, but increasingly more complex and specific, competition theory. In the case of predation the competitive abilities of species vary in time; in the case of spatial patchiness they vary in space. These modifications of the original theory still have the properties of conservation, recovery, assembly, and irrelevance of history that equilibrium theories possess, albeit these properties refer to temporal and/or spatial averages rather than to a single moment and/or patch.

The situation becomes completely different if we consider nonequilibrium systems that are in a continuous process of change either caused by a continuously changing environment (e.g. climate) and/or by a continuous process of evolution of the biological community itself. Here history and chance events come to play a major role. Currently, the type of dynamic models matching most closely these properties seem to be given by systems with chaotic dynamics and stochastic elements. Even when there is as yet a lack of more systematic investigations, we can note that the interest in, and the awareness of the possibility of ecological systems exhibiting complex dynamics has increased during recent years. This was not in the least due to the general progress in the field of nonlinear dynamic systems.

2.6 CHAOTIC DYNAMICS AND SELF-ORGANIZATION OF NONLINEAR SYSTEMS

In recent years we could observe a burst of interest in the dynamics of nonlinear systems. The flourishing of this new branch of science has been mediated by the advent of new computer technologies rendering possible detailed investigations of complex systems behaviour. Contrary to prior experience with simple linear systems, nonlinear systems may exhibit irregular time evolution and a sensitive dependence on initial conditions. What was previously dismissed as perturbation has meanwhile become fashionable under the name of chaos, being promoted to the dignity of nonlinear science, and various institutes were created to study it. New scientific journals continue to appear, entirely dedicated to Nonlinear Science. The success story of chaos and fractals took on the dimensions of a media event. Fractal images have a lot of visual appeal. It is intriguing that seemingly complex patterns arise from simple rules. Self-organization, chaos and fractals encountered in physics, climate, biology, medicine and economics are all sold under the same label of a new theory on complex nonlinear systems (e.g. Anon., 1989).

Chaos seems to have swept the community of science journalists and popularizers even more than it has the scientific establishment. Up to now the whole turmoil has hardly affected the everyday life of most scientists. Scientists are basically conservative, this being, to some extent, self-protection. Despite the extreme popularity of chaos, fractals and self-organization, the real consequences for scientific practice have remained largely inconclusive. To accept that irregular, chaotic behaviour may be the rule and not the exception should have serious consequences for our attitude towards natural systems.

Let us have a closer look at some characteristics of chaotic systems illustrated with the famous Lorenz model that is the prototype of a model displaying chaotic behaviour (see Box 2.4). It was developed by Lorenz (1963) to describe weather phenomena with a simplified nonlinear model of fluid convection.

Box 2.4   The Lorenz model

$$\frac{dX}{dt} = \sigma(Y - X) \tag{2.9a}$$

$$\frac{dY}{dt} = rX - Y - XZ \tag{2.9b}$$

$$\frac{dZ}{dt} = XY - bZ \tag{2.9c}$$

Roughly speaking, the variable X measures the rate of convective overturning, Y the horizontal temperature variation, and Z measures the vertical temperature variation. The three parameters depend on physical properties of the region under consideration and are hence liable to change Comprehensive reviews about the dynamic characteristics were given by Sparrow (1982, 1986).

Figure 2.3a illustrates the characteristic dynamics of the Lorenz model for the X variable. We observe irregular fluctuations, and X switches unpredictably from positive to negative values. What happens becomes clearer when we focus on the representation of X versus Z represented in Figure 2.3b. Despite the irregular appearance of the time course of single variables the behaviour of the system as a whole is confined to a bounded attractor. If the behaviour was completely at random we would expect a uniform covering of the X,Z-plane. The two lobes of the attractor correspond to positive and negative values for X. The system moves on one lobe and then switches unpredictably to the other lobe where it may stay for some time. These lobes may be thought of as representing entirely different regimes of weather patterns. We could also conceive of a similar behaviour for ecosystems where the system switches between two different states. Such a switching between alternate states is currently under discussion to explain the highly unpredictable effects of biomanipulation on lakes (Ulrich Sommer, pers. comm).

A striking property of chaotic systems is their extreme sensitivity to small changes in initial conditions. This is demonstrated for the Lorenz model in Figure 2.4 showing the result of two simulation runs that were performed under exactly the same conditions except for a small difference in the initial conditions. The temporal variations of X obtained in the two simulations (solid and dashed lines) are represented in (a). We note a visible difference after about 13 time steps. A more sensitive measure is given by the Euclidian distance between the two trajectories on the attractor that is calculated as: distance $= \sqrt{\{(X_1 - X_2)^2 + (Y_1 - Y_2)^2 + (Z_1 - Z_2)^2\}}$. As shown in (b), this distance increases exponentially from its initial value of 10^{-6} to reach a plateau where it fluctuates around the average distance between any two points on the attractor. This means two initial states that differ by a minor fluctuation diverge exponentially. After approximately 13 time steps a trajectory has lost any memory of its previous fate.

Another possibility to evaluate the "temporal memory" of a system is given by autocorrelation analysis. The autocorrelation coefficient measures the degree of dependence between two points of a time series as a function of their temporal distance denoted as the lag time. In a periodic pattern the correlation coefficient attains its maximum value of one at a lag time equal to the oscillation period. Figure 2.4c shows that the autocorrelation function obtained for the Lorenz model decays to zero within

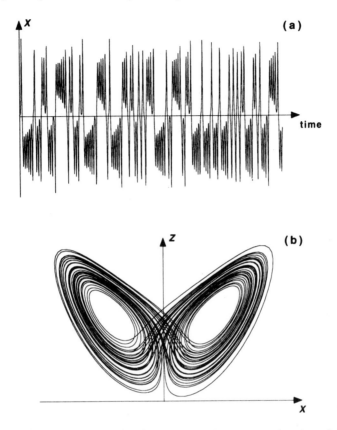

Figure 2.3   Results from simulations with the Lorenz model (a) X as a function of time, (b) phase space representation of the $X-Z$-plane. Parameter values: $\sigma = 10.0$; $b = 8/3$; $r = 28$

approximately one time step. This period corresponds to the average length of one oscillation. Such a finding implies that successive oscillations are essentially independent of each other.

Ubiquitous measurement errors, noise and computer round-off severely limit the time over which one can predict a chaotic system's evolution from a defined initial state. Let us now assume that the time series obtained from simulations with the Lorenz model represented a set of experimental data. If we had only statistical methods available the range of predictability would be limited to essentially one time step (with sophisticated methods to a few time steps with rapidly decreasing reliability). If we were fortunate enough to know the equations governing system behaviour the range of predictability would still be limited to about 10 time steps due to the intrinsic chaotic nature of the system. Such behaviour is not incredibly welcome when designing models for the purposes of forecast and prediction. Thus it should not be too surprising that acceptance of the relevance of chaotic dynamics has been slow and that it has preferably been avoided in environmental models. Even when Poincaré (1894) had already discovered that dynamical systems can exhibit chaotic behaviour, it took decades for these ideas to be taken seriously by the scientific community at all. Ruelle (1991) gave an illuminating account of their historical development in physics. When Ruelle and Takens (1971) first

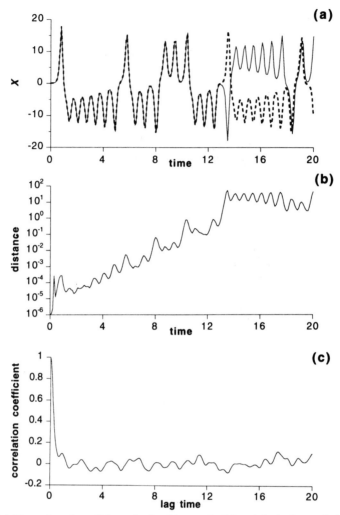

Figure 2.4   (a) X as a function of time obtained in two (solid and dashed curve) simulation runs with the Lorenz model under exactly the same conditions except for the starting values of X that differed by 10^{-6}. (b) The Euclidian distance between the two trajectories as a function of time. (c) Autocorrelation coefficient as a function of the lag time obtained for a time series of X values

tried to publish their idea that turbulent flow was not described by the traditional superposition of many independent oscillators but can be explained by a single chaotic attractor, their paper was rejected. They also claimed that sudden changes in behaviour had to be expected, contrasting with changes occurring more smoothly in the traditional picture. Now it is accepted that turbulence is a case of spatio-temporal chaos, albeit a very complicated one (Ottino *et al.*, 1992).

It is worth mentioning one episode from these early days of chaos. Ruelle (1991) reports that he explained to a group of chemists studying periodic chemical reactions how they might expect to see nonperiodic "turbulent" oscillations as well as periodic ones. Discussions with one of the chemists involved revealed that reactions had indeed been observed to become "turbulent". However, they had been declared as "failures" and had been discarded. At least here, the impact of "chaos" has already changed scientific practice. As a consequence of having a method, we have become increasingly aware of the ubiquity and importance of irregular evolutions in space and time. Instead of being neglected they are studied with care. This episode shows how preconceived expectations shape our perception. New theoretical discoveries may foster scientific progress by promoting new perspectives on seemingly well-known phenomena. This applies as well to certain fields of medicine and physiology where the application of chaos has provided significant new insights.

The healthy individual displays a mosaic of rhythms in the various body systems. These rhythms rarely display absolute periodicity. Indeed, recent interest in the measurements of such rhythms as heart rate and respiration, has frequently revealed much greater fluctuations in these systems than might naively be expected (review by Glass and Mackey, 1988). Goldberger (1990) has even suggested that the normal healthy dynamics are "chaotic" and disease is associated with periodic behaviour. One prominent example is given by epileptic seizures which are frequently characterized by an EEG (ElectroEn-cephaloGram) displaying uniform and periodic electrical activity over a large part of the brain, whereas the EEG of a healthy patient shows irregular patterns. Whether normal brain function is chaotic is still a matter of dispute. Even when it seems to be too early for generalizations regarding the impact of chaos in these fields, one consequence is the increasing awareness of diseases characterized by abnormal temporal organization. Glass and Mackey (1988) proposed to call these pathological states dynamical diseases.

One may wonder why just chaotic dynamics should be characteristic for the normal state of the organization of the brain or the body. However, self-organizing systems may exhibit both chaos and order. Whereas individual trajectories may be inherently unpredictable, regularities and order become evident if we focus on the characteristics of a system's attractor that can be identified within the envelope of all possible behaviours of a system. It seems that owing to self-organization the macroscopic behaviour of complex systems is characterized by a bounded attractor with limited dimensionality, whereas the behaviour of specific traits may be highly erratic. The very essence of self-organization relies on the close interaction and mutual shaping of a system's constituents. This contrasts with the mechanistic conception of nature where it is still a favourite practice to analyse a complex system by breaking it down into its constituent parts, by studying these in isolation, and by deriving a description of the system as a whole from a synthesis of the parts. Self-organization is a property of the system as a whole, it is meaningless at the level of the parts. Ideas of this type have already exerted a considerable influence on our perception of brain function. Previously the brain was visualized as a mechanistic network of interconnected neurons where function was associated with the specificity of wiring in certain areas of the brain. If our brain functioned this way we would have some trouble in carrying around the huge mass of brain tissue thus necessary to account for the accomplishment of any complex task. Such rigid a picture has started to be replaced by a dynamic perspective where memory and brain function are associated with changes in ensembles of cooperating neurons (Braitenberg and Schütz, 1989; Freeman, 1991; Haken and Stadler, 1990; Shephard, 1990). We may thus conclude that function is stored in and is related to patterns of interaction. The combination of chaotic dynamics at the microscopic level, the constraints at the level of the attractor, and the change in functional associations may give the brain its unique capabilities.

Translated into ecological terms this could, for example, imply that the fate of a single species exhibits a highly irregular behaviour. However, functional properties at the level of the system as a whole remain within certain bounds and exhibit less variability. The attractor can be identified as characterizing the macroscopic system behaviour preserving order and function of the ecosystem (whole) at a higher level despite the many degrees of freedom at the level of the species (parts). We will still find out that it is just the degrees of freedom at the level of the species that mediates integrity of function at the level of the system as a whole. Such a statement reminds us of the famous relationship between stability and diversity. However, diversity will have to be defined quite differently from an enumeration of taxonomic species and/or trophic interactions. Further, we have to refrain from identifying the integrity of a system's function with stability of the component populations.

In general, chaotic dynamics or self-organization have not yet found their way into ecology in spite of the early contributions in the field of population dynamics. May (1976) noticed already in the 1970s that simple discrete time models of one or two species may exhibit complicated dynamics. Initially, ecologists assumed that chaotic behaviour was confined to some special cases of simple models only and that an increase in model complexity would be associated with an increase in stability. However, research in recent years has demonstrated that complex and even chaotic dynamics can arise in model systems (discrete and continuous in time) with three or more species as well (e.g. Guckenheimer and Holmes, 1983; Pahl-Wostl, 1993c; Schaffer, 1985). In multi-species models chaos seems even to be the rule rather than the exception. Studies with models of three-level food chains revealed that chaotic dynamics result for biological reasonable choices of parameters (Abrams and Roth, 1994; Hastings and Powell, 1991). The authors question the relevance of conclusions drawn from stability analyses of models linearized around the equilibrium point. We should recall that such analyses are central to food web theory.

Unfortunately, it is not easy to find experimental support for or against these arguments. Unlike in medicine or physiology, ecologists rarely have access to long time series of high quality data. It is not too astonishing that most investigations of chaotic behaviour in population dynamics were performed in the field of epidemics, where detailed records are available (review by Schaffer, 1985). Meanwhile, further support has been provided in favour of chaotic dynamics in population data from other sources (Turchin and Taylor, 1992; but c.f. also Perry *et al.*, 1993). There is also increasing theoretical evidence that systems with chaotic dynamics have an even higher potential for adapting to changing environmental conditions and for resisting perturbations than systems with a stable equilibrium point (Allen *et al.*, 1993; Hastings, 1993; Pahl-Wostl, 1993c; Wilson, 1992). Models based on equilibrium assumptions have little potential for adaptation, or evolution. Hence they lack properties essential for and characteristic of biological systems. Here, models displaying chaotic dynamics could become potent tools for conception of systems having the potential to change, and to address topics such as the importance and evolution of functional diversity.

2.7 SUMMARY AND CONCLUSIONS

Traditional concepts in ecological theory, such as the competitive exclusion principle, were derived under the premises of a stable, homogenous environment and the presence of a time-invariant equilibrium state. Real systems hardly ever yield to these demands. Correspondingly, the relevance of the conclusions derived under such assumptions must be questioned. Nowadays, it is readily accepted that ecosystems may not be in an equilibrium state because of fluctuations in the physical environment. Acknowledging the importance of exogenous sources of variability does not mean that we have to give up the idea that the endogenous dynamics of ecosystems are characterized by a stable equilibrium point. Whether complex or even chaotic dynamics may arise endogenously from the interactions among the component species is still controversial.

Recent progress in the theory of nonlinear dynamic systems suggests strongly that systems with a stable equilibrium point are rare in nature. Effects of history and change are important in systems with positive feedback. Such systems often exhibit chaotic dynamics and must be considered nonequilibrium systems if the time scale of interest covers only a small segment of a system's time evolution. In ecology, positive feedback (mutualism, symbiosis) and time lags have for a long time been neglected and dismissed as inherently destabilizing features. Indeed, it is not very attractive to abandon the idea of a "balance of nature" that has guided ecological thought for such a long time. At first sight this seems to be the consequence, when chaotic, irregular dynamics come to play a major role. However, stability and equilibrium must not be replaced by randomness and stochasticity. I am going to point out regularities in the hierarchical pattern of spatial and/or temporal variations in ecosystems, the elucidation of which constitutes one of the major goals pursued in this book. Finally, we may end up with a familiar statement that diversity generates the constancy of whole system function. Before doing so we still have to develop new concepts of what diversity and constancy mean in the light of spatio-temporal organization in ecosystems operating far from a static equilibrium point.

Chapter Three

Evolution and ecology

Why a chapter on evolution and ecology, when the impatient reader finally expects to learn more about the mysterious spatio-temporal organization I continue talking about? To understand the present state of ecological systems and to reflect on how they might develop in future requires a long-term perspective. Introducing a concept of spatio-temporal organization in ecosystems without the goal to give it as well an interpretation within an evolutionary context does not seem sensible. In this respect it might be useful to clarify my points of view on certain attitudes that seem to prevail in present thinking about evolution.

Before starting to do so it is worth reflecting a second on the meanings associated with the notions evolution and ecology. Evolution is a phenomenon – rather narrowly focused it is identified with the processes responsible for the formation of biological species. Whereas evolutionary biology is a subdiscipline of biology, ecology has developed into a separate scientific discipline studying the mutual relationships between organisms and their natural environment. Is such a comparison only an overemphasis of minor semantic differences? I do not think so. I am convinced that it indicates a real and important difference. Evolutionary biology is entirely centred on organisms and populations. At these levels we find also strong links to ecology. What seems to be missing is an appreciation of the environment, of the mutual shaping of the organism–environment context.

3.1 FOCUS OF THINKING IN EVOLUTIONARY BIOLOGY

All evolution occurs within an ecological context given by a reticulate web of ecological interactions. At present, however, this context is not very well represented in evolutionary theory. In a current textbook about evolutionary ecology by Pianka (1988) I discovered one single page on community evolution. The author states:

> Can selection operate between natural communities? . . . Most important, selection acts only by differential reproduction, and it is most difficult to envision reproduction by a community or an ecosystem (Pianka, 1988).

These statements reveal that the meaning of biological evolution is essentially identified with the action of the Darwinian principle of natural selection. The questions evolutionary theory has been concerned with focus on:

- What is it that evolves; what is (are) the unit(s) of selection?
- How is evolution effected; what are the mechanisms, the "driving forces"?

At the risk of being accused of gross oversimplification, I summarize the answers given thus: Species evolve, whereas organisms are the units of selection. The major driving forces are competitive interactions and a tendency of species to optimize their life strategy and hence their reproductive success. To invoke any higher levels of biological organization has been perceived as a type of heresy testifying an adherence to some obscure superorganism concept. Such rejection reveals the mechanistic focus of evolutionary thinking. In order to be relevant in the evolutionary process any unit must be able to compete with others engaging thus in the process of natural selection. At the community level it would require a rigid association of co-evolving species that move as a tightly coupled unit through evolutionary time. Such a narrow perspective and the lack of evolutionary considerations at the community level indicate further the lack of a theoretical understanding of community organization – an understanding that exceeds the simple meaning of a community as an assemblage of species that just happen to occur together. Some people may not perceive any lack because they see no need for considering any evolutionary processes at the level of the community and/or ecosystem. I claim, however, that an integration of these higher levels of ecological organization should lead to a fruitful and essential extension of current evolutionary concepts.

The main theoretical framework of contemporary evolutionary theory comprises what is called the "Modern Synthesis". In a nutshell, it may be characterized as the combination of Darwin's principle of natural selection with concepts of classical genetics and population genetics. According to the Synthesis, evolution proceeds as given below (Maynard Smith, 1975; Mayr, 1970, 1982, 1991; Wuketits, 1988).

The gene is the unit of heredity. At this level primary evolutionary factors, genetic recombination and mutations, generate random variations in the offspring. The individual is the medium of natural selection. The degree of an individual's being adapted to its environment determines its fitness, its reproductive success. By increasing specialization organisms differentiate and occupy separate niches, thereby avoiding competition. As a result one observes a continuing adaptation of organisms to their environment. The biological species is then the final result of the evolutionary processes of genetic variation, selection, adaptation, genetic drift, and reproductive isolation.

The details of processes are still a matter of debate. Regarding the final process of speciation for example, proponents of the gene flow hypothesis argue that species exist and persist primarily due to their comprising an ensemble of cross-breeding individuals. Here the main driving force would be reproductive isolation deriving from a spatial separation of populations. The proponents of a theory of adaptation by natural selection argue that species are the sole result of the selection of a set of adaptive traits by the environment. The arguments aren't mutually exclusive, and presumably the future will witness some compromise. Despite these divergences there is a general agreement on the main elements of the Synthesis as summarized above.

Hence, the mainstream of thinking in contemporary evolutionary theory can be characterized by the following essential elements: A mechanistic conception of the evolutionary process (= speciation) prevails. The focus is on explaining how we can understand evolution to occur. Fitness is defined at the level of individual organisms with a rather static conception of the environment. What are missing are attempts to explain the qualitative nature of how evolution proceeds, to develop also some understanding of a direction. Such may be due to the touch of teleological thought associated with the notion of a direction or progress, to racist abuses and the possible identification of progress with increasing quality in the evolutionary process. We shall still see that there is no need to invoke any teleological reasoning for introducing directionality in the evolutionary process. It requires an explicit consideration of ecological context departing thereby from emphasizing the individual only.

Now we are going to discuss another issue where the practice in evolutionary theory resembles to a large extent the one encountered in ecology already. Natural selection and optimal fitness are based on an equilibrium concept – it is assumed that the environment is sufficiently stable for natural selection to become effective. At first sight the combination of stability and evolution seems to be a contradiction in terms. However, stability is a question of time scales. Most models in evolutionary theory assume that some equilibrium is attained over the time scale for natural selection to become effective.

3.2 EQUILIBRIUM CONCEPTS IN EVOLUTIONARY THEORY

Numerous mathematical models were developed in population genetics to derive predictions for the consequences of natural selection (review by Pianka, 1988, Roughgarden, 1979). One of the most elementary and basic results of these theoretical considerations is the Hardy–Weinberg "law". It is based on the assumption that a population contains genetic variation at one locus with two alleles (the classical situation of Mendel's experiment with smooth and wrinkled peas). The Hardy–Weinberg law states that in the absence of external forces (natural selection, mutation, genetic drift) any variation initially present in the gene pool will persist without change. An equilibrium is predicted for the genotypic frequencies, and the approach to equilibrium is very rapid, it is attained after only one generation. The degree of deviation from equilibrium may then be interpreted as a measure of the forces of selection actively favouring one gene over the other.

To account for more complex situations, theorists have developed a great variety of sophisticated models considering, for example, natural selection for the case of multiple alleles at one locus, or for multiple loci. The models of population genetics have been combined with models of population ecology to predict fate and distribution of genetic traits in populations under particular environmental circumstances. The potential for different combinations of models is inexhaustible, and it has been exploited extensively. All these models are based on a mechanistic conception of genetic variation and of natural selection. As with most modelling efforts in ecology, conclusions are derived on the premises that the populations approach a stable equilibrium point.

A popular approach to analysing evolutionary strategies uses optimization models. In fact models of optimization and evolutionary games, originally imported from economics, have developed into one of the backbones of evolutionary theory (review by Parker and Maynard Smith, 1990). The rationale behind using optimization models is that adaptation towards maximum fitness is a pervasive feature of living organisms, and that it is to be explained by natural selection. These assumptions are tested by comparing predictions derived from models based on these premises with experimental observations. The procedure for doing so is best understood by following in more detail the steps involved in the construction of an optimization model.

The optimality approach involves the construction of a model about adaptation. First, it is necessary to ask an explicit biological question, for example, why is the sex ratio often unity, or a more specific one, for example, why are male spiders smaller than females?

Next, a range of alternative actions or strategies relating to the question is defined. The strategy set simply specifies the plausible alternatives, given what we consider it possible for evolution to achieve. It is usually assumed that some direct or indirect measure of fitness is to be maximized. Once payoffs to the strategies have been stated, the optimal solution(s) are deduced by an appropriate analytical technique. The final step in the optimality approach is to test the predictions, quantitatively or qualitatively, against the observations. The quality of the fit is then assumed to be a measure for the appropriateness of the assumptions.

We have to be aware that any model of optimal strategies is reasonable only if we assume that a biological community approaches a stable equilibrium point. In nonequilibrium systems where effects of chance and history at the population level come to play a significant role, such models lose their significance. In such systems both fitness and optimality have to be defined differently. Rather than referring to a fixed character set, fitness could involve strategies on how to deal within environmental variability. Optimization must be seen relative to an environmental context – not absolute for the isolated individual.

The examples support our first impression that the neo-Darwinian synthesis relies heavily on mechanistic thinking and on ideas of linear propagation of causality. It is based on a definite assumption about how evolution occurs. The evolutionary process is seen primarily as the adaptive transformation of phenotypic attributes through time mediated by the driving forces of random genetic variations and of natural selection. Organisms are regarded as being decomposable into distinct traits, each of which must necessarily confer some selective advantage, and most of which are studied and modelled in isolation. Thus it is largely neglected that evolution is a process embedded in lower- and upper-level constraints as sketched in Figure 3.1.

Several people suggested a hierarchical approach by viewing evolution as a multi-level phenomenon. Wuketits (1988) suggested an extended Synthesis where the simultaneous action of internal and external selection factors at the level of the organism drive evolution. It may be visualized as the external world posing problems to the organism; the internal possibilities determine the type of strategies an organism can pursue. Wuketits acknowledged the possibility that organisms may modify their environment to some

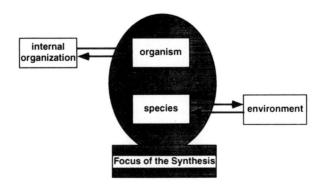

Figure 3.1   Current evolutionary theory focuses on the levels of organisms and taxonomic species, thereby largely neglecting the mutual interaction with the internal organization of organisms and the external organization of the environment

extent. Salthe (1985) introduced what he called the basic triadic structure. It means that any level of biological organization must be viewed as being embedded within lower-level and higher-level constraints. In case of the organism the lower-level constraints are inherent in the morphological constraints. The higher-level constraints are given by the environment. Despite such awareness, what still seems to be missing is a theoretical framework explicitly taking account of the mutual relationship between levels, in particular between organisms and their environment.

Let us now consider some of these constraints. It would be beyond the scope of this work to extend the considerations of the evolutionary process to the internal organization and/or the genome of organisms. Therefore I will only briefly mention some approaches for doing so, and then direct the focus again on the mutual relationship between organisms and their environment.

3.3   THE INTERNAL ORGANIZATION OF ORGANISMS AFFECTS EVOLUTION

Over recent years molecular genetics has been revolutionized. It becomes increasingly clear that simple models of linear causality are inappropriate at the level of genetic information transfer as well. The genome is more fluid than previously anticipated as illustrated by the ubiquitous presence of movable genetic elements, so-called transposons. Nevertheless, most models in evolutionary population biology continue to rely on mechanistic conceptions of rigid structures and isolated traits at the molecular level. Certain attempts have been made to develop models of genetic networks based on the notion of genetic landscapes to study evolution and co-evolution (Kauffman, 1993). These modelling approaches emphasize characteristic patterns of self-organization at the level of the genome as a whole ignoring thereby particular details of single genes. Yet traditional evolutionary theory still assumes a reluctant attitude with respect to such unconventional approaches.

Pre-Darwinian morphologists clearly perceived a deep connection between the systematics of natural forms and the organization of developmental processes by which forms are realized (Goodwin *et al.*, 1989). Since Darwin, this connection has been interpreted purely in terms of descent with modifications. Organisms are seen to be the sum of past random variations accumulated by natural selection and preserved by heredity. Opponents point out that organisms are not preformed in the germ, they take shape epigenetically, in the course of development. Molecular genetic mechanisms do not provide an explanation whatsoever for the development of complex organisms.

Structuralism assumes that there is a logical order to the biological realm and that organisms are generated according to rational dynamic principles. Goodwin (1990) contrasts the structuralist approach with the current practice of assuming that biology is a historical science, species morphology being the result of random variation and natural selection of functionally adapted forms. He argues that these forms are therefore not intelligible in terms of generative principles. Structuralists use very sophisticated mechanistic models (invoking morphogenetic fields and complex regulatory sequences) to show that only a small subset of biological forms can be realized due to constraints imposed by physical and chemical processes. Structuralism represents a recent, albeit somewhat extreme, expression of early claims for internal mechanisms at the level of the organism influencing and directing the evolutionary process. Obviously life and evolution of organisms must also be determined by their own body plan specifying morphological characteristics and being constrained by rules of proportion.

The structuralist approach emphasizes structure and neglects function. Function cannot be explained by generative principles at the level of the organism. I support a point of view invoking both function and structure. Organisms are constrained by the integrity at the organismal level itself being thereby coordinated harmonized wholes that are not just free to develop and evolve in an arbitrary direction. However, organisms cannot be viewed isolated from their environment. To the contrary, I conjecture here that organisms internalize their environment and vice versa due to the mutual dependence between them.

3.4 DOES ORGANIZATION AT THE COMMUNITY/ECOSYSTEM LEVEL AFFECT EVOLUTION?

Numerous approaches exist to explain or describe properties of specific communities or ecosystems. I mentioned before that there is a lack of a coherent theoretical framework – in particular regarding the link between species properties and ecosystem function. Such deficiency is fostered by the segregation between bottom-up approaches emphasizing properties of communities viewed as collections of species and top-down approaches emphasizing properties of ecosystems viewed as functional entities of energy and matter flows. An attempt to combine environmental elements and species properties is given by the concept of the ecological niche. Therefore it might be useful to review the current state of niche theory with regard to its consideration of the mutual relationship between organisms and environment.

3.4.1 The niche – conceptual tool to link organisms and environment

The notion of the niche pervades most of ecological thinking. As with stability the meanings are manifold. In its common meaning, the niche is still closely related to the balance of nature concept as a vague notion of a species' place in an orderly natural environment. In modern niche theory, that has, nowadays, become more or less synonymous with competition theory, the niche concept has been formalized and been delimited in scope. The original concepts developed by Elton (1927) and Grinnell (1924) were broader but also less tangible. Both Elton and Grinnell identified the niche as the place/role a species happens to occupy in its environment comprising biotic and abiotic factors. Grinnell emphasized abiotic habitat factors, whereas Elton characterized the niche primarily in terms of functional roles in a food web. Both also shared the attitude of remaining rather vague about the real content of the niche concept. It was only with the advent of modern niche theory that the niche concept received a formal treatment. Hutchinson (1957, 1978) laid the foundations for the subsequent perception of the niche as a property of the occupying species and not of a species' environment. Instead of being a description of a state, the niche was derived as the result of competitive interactions based on the following sequence of arguments. Natural selection favours survival of those species that exploit resources most efficiently – remember that the principle of competitive exclusion predicts extinction of the competitively inferior species. Competition is the ecological mechanism for natural selection to become effective. In agreement with the principle of competitive exclusion it is assumed that, to coexist, species must differ in their niche characteristics specifying resource utilization and environmental requirements. As shown in Figure 3.2, the niche of a species is then defined by a fitness function along an environmental gradient. Fitness is identified with reproductive success. Environmental gradients may, for example, be given by resource availability or physical variables such as temperature.

Niche theory assumes that species reduce competition by separating along environmental gradients with respect to their optima in fitness. The niche overlap that is defined as the degree of utilizing a resource in common then provides a direct measure for the degree of competitive interactions between species. In Figure 3.2, species (2) experiences the highest competitive pressure, whereas species (1) experiences the least. The overlap can further be reduced when species separate along further dimensions within an n-dimensional niche space. The overlap integrals among the niches of several species can be used to quantify the coefficients of interspecific competition in the Lotka–Volterra equations (cf. Box 2.3). An n-dimensional niche space in combination with the Lotka–Volterra competition equations thus provides a comprehensive base for profound theoretical treatments. Although such an n-dimensional hypervolume model of the niche is extremely attractive conceptually, it is too abstract to be of much practical value. Theory predicts that niche overlap is a direct measure of competition. However, in experimental observations one can never know whether all relevant factors impinging upon organisms have really been taken into account. Due to the fading popularity of competition theory, the concept of the niche has as well experienced a certain decline in use over recent years (Schoener, 1989a).

Does contemporary niche theory establish a link between species properties and community organization? Competition and niche theory have proved to be a considerable disappointment in explaining patterns of species distribution and diversity. To the contrary, the coexistence of many species in seemingly homogenous environments

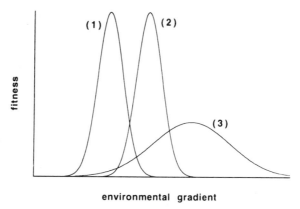

environmental gradient

Figure 3.2   The niches of three species segregated along a one-dimensional niche space. We observe a trade-off between specialists (species 1 and 2) having narrow tolerance limits but using the resource space very efficiently and generalists (species 3) having broad tolerance limits but using the resource space less efficiently. This way one accounts to some extent for internal restrictions of organisms having a limited energy budget only. Different strategies may pay off under different environmental conditions. Related to community organization, specialists are more likely to occur in species-rich, mature ecosystems

contradicts theoretical predictions. Due to the failure of mechanistic explanations based on competitive interactions, phenomenological approaches have been chosen to explain species diversity by assuming partitioning rules in an abstract niche space. Sugihara (1980), for example, developed a model to explain the log-normal distribution of species abundances. He assumed a one- or multi-dimensional niche space to be partitioned among an ensemble of species. Over evolutionary time scales, fractionation arose by "sequential breakage" of the niche axes with the entry of each new species into a community. Such attempts to link species pattern with niche space without invoking mechanistic details may prove to be more fruitful in revealing relationships between species properties and community organization.

In summary, we can conclude that present niche theory does not prove to be of particular help in understanding the mutual relationship between species properties and ecosystem organization. Population biology in general and niche theory in particular put an overwhelming emphasis on competition. The argument goes that species occupy different niches to avoid competition, thus preventing competitive exclusion. However, we should also be aware that in order to specialize, a species depends on other species to be specialized as well. Let us imagine, for example, a field where two guilds of plant species can be discerned. One guild comprises shallow-rooted species having their maximum activity in spring. The other guild comprises deep-rooted species having their maximum activity in summer (examples by Fitter, 1986; McKane *et al.*, 1990). Obviously we could interpret this pattern as resulting from competitive interactions only. However, we could also argue that the species guilds depend on each other. Due to their simultaneous presence the nutrient is retained in the soil over the whole year and the soil is stabilized. Similar conclusions can be drawn from comparisons of a natural ecosystem with an agroecosystem. In the latter competition is excluded to maximize the production from a single species. As a result the state of the whole system degrades due to a loss in soil fertility, increased nutrient leaching, run-off etc. In a natural community where species differ in their tolerances and requirements for abiotic factors, a balance is maintained. To then invoke an argument that the species collaborate intentionally to maintain system function would be rather weird indeed. There is also no need to introduce again a type of super-organism concept. We can derive an understanding of such phenomena by adopting a multi-level approach in paying more attention to the mutual dependence between species characteristics and ecosystem constraints and function. To do so we will need to develop a dynamic perspective on ecological organization where the niche is seen within a dynamic environmental context.

3.4.2 Concepts of evolution at the ecosystem level

Loehle and Pechmann (1988) deplored that evolutionary considerations are virtually missing in the field of ecosystem ecology – evolutionary in the sense of organic evolution manifested in changes at the genetic and species level. Many systems ecologists believe that, given sufficient time, evolution provides mutual adaptation of species to form organized, functionally integrated ecosystems (e.g. Odum and Biever, 1984; Patten and Odum 1981; Ulanowicz, 1986). Hence, system-level constraints and causal feedback loops are often treated as higher-order phenomena that can be studied and modelled at the level of aggregated variables such as biomass without considering the individual organisms and/or species. Ecosystems have been depicted as thermodynamic systems, the macroscopic properties of which can be dealt with on a phenomenological basis without taking into account detailed mechanisms. Based on these premises several attempts were made to derive goal functions and driving forces for ecosystem development and evolution. Biological systems in general and ecosystems in particular have been claimed to maximize power (Odum, 1971; Odum and Pinkerton, 1955), exergy (Jørgensen, 1992), ascendency (Ulanowicz, 1980, 1986), or entropy (Brooks and Wiley, 1988; Wicken, 1980). I will discuss in more detail Odum's and especially Ulanowicz's approaches because they contributed important ideas to the work presented in this book.

One of the earliest claims of a goal function was the maximum power principle. Odum and Pinkerton (1955) suggested that ecosystem developed structures maximizing the flow of useful energy. More recently Odum (1988) included ideas of self-organization and offered an all-embracing explanation for the development of biological systems based on the maximization of emergy. The spelling is not a typing error. EMERGY is a hybrid for EMbodied enERGY. Rather than quantifying the energy content of a flow or biomass in absolute units such as joule or calorie, it is expressed in equivalents of the solar energy input required to maintain a flow or biomass. Figure 3.3 illustrates the principle for a linear food chain. The symbols correspond to Odum's system language where half ovals refer to primary producers, and hexagons refer to consumers (Odum, 1983). The grounding signs denote losses due to energy degradation in respiratory processes. An increasingly smaller percentage of the solar energy entering a system is transferred to the next higher trophic level if one ascends along the food chain. Correspondingly, the energy transformation ratio equivalent to the solar energy embodied in one unit energy increases to the same proportion as the absolute amount of energy transferred decreases. Emergy can thus be interpreted as a value term where the currency is the solar energy entering a system. In analogy to economic principles, emergy reflects the accumulation of capital in terms of solar energy during the processes of producing a flow or biomass.

As indicated in the top of Figure 3.3, an increase in trophic level is associated with a decrease in the number of organisms and an increase in time scales. Such applies in particular to aquatic ecosystems where the energy flow is directed along a gradient of increasing body size. From there one can derive a hierarchy of time scales in parallel to an increase in emergy.

Odum claims that ecosystems develop system designs that reinforce energy use, characteristically with alternate pulsing of production and consumption (Odum, 1983, 1988). He claims further that ecosystems are organized in hierarchies because this design maximizes useful energy processing. He considers emergy to measure the value of a flow or storage to a system in the long-run after self-organization selection processes have been at work. Emergy is thus assumed to indicate the potential of a flow to exert control on the system as a whole. Regarding the approach pursued in this book, Odum's perception of ecosystems comprising a self-organized spatio-temporal hierarchy of pulsing consumer–producer interactions will be of special relevance. In general, I feel that Odum's hypotheses require additional efforts to contrast the general claims with reality, and to translate them into terms that the average biologist can understand.

By claiming that ecosystem design results from maximizing useful energy processing, Odum assumes that a system's emergy captures as a measure both a system's physical size and its structural organization. Ulanowicz (1980, 1986) makes an explicit distinction between extensive (= physical size) and intensive (= structure) properties. He merged the quantitative and qualitative aspects of ecosystem development by scaling the structural

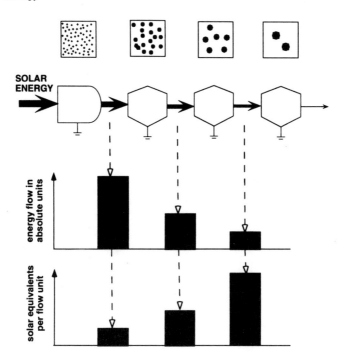

Figure 3.3   Food chain illustrating that, with increasing trophic level, there is a trade-off between energy flow measured in absolute units and the energy transformation ratios = embodied solar equivalents per unit flow (Odum, 1988)

organization of a network (quantified by a measure of organization = I) with the total system throughflow (sum of all flows in an ecological network as a measure of size = T). Ulanowicz called the new measure ascendency, $Asc = T \times I$. A more detailed consideration of the measure used to quantify organization is postponed to the next chapter. Here, I discuss only its qualitative properties and the underlying assumptions.

Ulanowicz (1986) derived his view on organization from cybernetic arguments. As shown in Figure 3.4, any network can be depicted as being composed of linear flows passing through a system and of cyclic flows staying within the system. Ulanowicz stated the hypothesis that the self-reinforcement of autocatalytic feedback cycles drove system development in a mature stage. As a consequence selection would work towards a highly specialized and maximally streamlined network topology with progressive ecosystem development.

To illustrate what such a topology implies, two different network configurations are contrasted in Figure 3.5. The network depicted in (a) is maximally connected. Flows are not specific with respect to where a compartment receives its inputs from and where its outputs are directed to. The network depicted in (b) displays a maximally streamlined network topology with only one link between any two compartments. According to Ulanowicz's definition, the network in (a) displays no organization whereas the one in (b) is highly organized. Ulanowicz assumes such a streamlined topology to result from effects of positive feedback favouring the dominance of selected pathways with increasing self-organization and autonomy of a system.

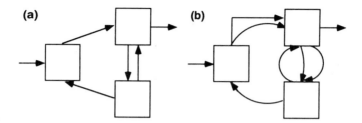

Figure 3.4   A network that includes recycling, like the one depicted in (a), can be looked upon as being composed of linear flows and of cycles as shown in (b)

Contrasting with Odum's emphasis on pulsing and dynamics, Ulanowicz limits his attention to static network descriptions where flows are averaged over time and space. His statement can be identified with a type of competitive exclusion principle where the competing units are identified with positive feedback cycles. As discussed in detail in the previous chapter, any such argumentation is valid for systems with a stable equilibrium point only. In nonequilibrium systems and/or in systems with a dynamic equilibrium state we should expect the network structure to exhibit a high dynamic diversity. As we observe the coexistence of species in such systems, we should also expect the coexistence of feedback cycles. To accurately describe organization we have to include the spatio-temporal dynamics in the description of a network's structure and organization. Instead of focusing on static network descriptions where the flows are averaged over time and space, the dynamics of the flows have to be resolved along the temporal and spatial dimensions. This is a central issue of the present work.

Coming back to the general question of goal functions for ecosystem development, we should be aware that a source for misunderstandings and problems lies in the fact that questions of spatial and temporal scales have largely been neglected in the phenomenological approaches discussed. Even when Odum emphasized the dynamic nature of ecosystems, he remained rather vague on how to explicitly account for it in an ecosystem's description. What are the appropriate temporal and spatial scales for an ecosystem to maximize its emergy or ascendency? How should the internal structure of an ecosystem be resolved with respect to temporal and spatial scales and aggregation into functional compartments? A consideration of these questions is crucial for any of these approaches to be of general use. At present, there is no experimental evidence supporting the view that any of the goal functions mentioned is maximized during ecosystem development and/or evolution. However, macroscopic system descriptors (I prefer this term to the expression goal function) may prove helpful to organize observations and data and to establish links between different levels of organization. Maximization principles may form a background against which to contrast the reality in ecosystems.

3.4.3   Present situation

Obviously we have the odd situation that on the one hand evolutionary ecologists are entangled in a jumble of extremely detailed sophisticated models maximizing fitness at the level of the organism and the species, largely ignoring higher levels of biological organization. On the other hand systems ecologists focus their attention on energy and matter flows and on optimization criteria at the system level, thereby ignoring species and

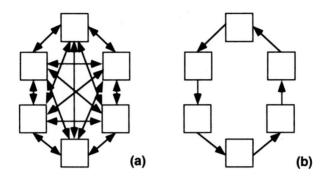

Figure 3.5   Two extreme network structures representing the maximally (a) and the minimally (b) connected configurations. According to Ulanowicz (1986) the network in (a) displays no organization whereas the one in (b) is highly organized

organisms. This is yet another facet of the splitting between ecosystem and population ecology that does not foster progress in bridging these levels.

3.5   OBJECTIVES OF THIS WORK

Contrasting with the traditional practices of emphasizing stability concepts and of separating considerations about populations from those about ecosystems, I posit the claims that:

1. Ecosystems are characterized by a dynamic order crossing a wide range of spatial and temporal scales. This dynamic order renders possible the combination of a highly variable behaviour and an integrity of function.
2. Evolution proceeds by the dualistic interplay and the mutual dependence between the level of the parts (organisms) and the whole (ecosystem). Constraints operate at the levels of both the organisms and the ecosystem. Characteristics are revealed by considering an intermediate level of functional aggregations.

The main objectives of the present work may be summarized thus:

1. Derive an operational definition of parts and whole in ecosystems.
2. Develop means to describe the spatio-temporal organization of ecosystems that arises from the mutual interaction between parts and whole.
3. Derive conclusions with respect to the predictability and evolution of ecosystems.
4. Outline the role of theory and mathematical models in ecology (and evolution).

The last point adds a new aspect and deserves further attention. It may be viewed as a logical consequence of the critical statements made before with respect to the scope of theory and mathematical models in ecology and/or evolution. In addition, I will make extensive use of simulation models to develop and to illustrate my arguments. To clarify the scope of the application of models in the present work, it is worth focusing on the logic behind some mathematical models applied in ecology.

In contrast to physical systems, driving forces remain elusive in ecosystem dynamics. As outlined above, several attempts have been made to derive goal functions for ecosystem development in close analogy to theromodynamics. The validity of such approaches remains to be shown. It is more common to quantify ecological interactions with mechanistic models as, for example, the Lotka-Volterra predator–prey model. In this model, the feeding rate of a predator is calculated as $k \times P \times B$ where B is prey biomass, P is predator biomass, and k is a biomass-specific rate constant. The idea behind this formulation may be derived from reaction kinetics. For the predator to capture a prey an encounter between the two must take place the probability of which is assumed to be proportional to the product of prey and predator biomass. The rate constant k includes all environmental influences, predator-specific behaviourial characteristics, efficiency terms etc. It is not too difficult to imagine that such a coarse description is hardly suitable for reliable quantification. We have to remember that the same description is chosen for the reaction between molecules and the feeding relationship between organisms. Being mainly determined by physical and chemical factors the parameters in a chemical reaction are reasonably well behaved. The parameters in ecological models, however, are influenced by a large variety of factors. Correspondingly, parameters in ecological models are not constant but are subject to continuous change. One hope in applying goal functions was just that they could assist in the derivation of flexible models where parameter adjustment is guided by a maximum principle at the level of the system as a whole (Salomonsen, 1992).

From this discussion we may already suspect that mathematical models in ecology may serve different purposes to models used, for example, in physics and/or engineering. In many cases models may be of an illustrative character, assisting our intuition which is quite limited when dealing with complex systems. Models may provide qualitative results making some dynamic behaviour such as oscillations of population densities more intelligible. Even when theoretical ecologists like Lotka, Volterra or MacArthur may have been aware of the qualitative, intuitive nature of their blending theory and empirical predictions, it was also their stated goal to improve theory making it a quantitative, predictive tool. It seems to me that the distinction between what had been achieved and what was yet intended became progressively blurred. Conclusions derived from simplified models have started to lead a life of their own. Restrictive assumptions tend to become neglected and the validity is taken far beyond the scope the models were initially designed for. This has, for example, been the case for the complexity–stability relationship and to some extent also for the competitive exclusion principle. The disappointment of empirical ecologists with what general theory can accomplish may not in the least derive from such misunderstandings.

Models that were designed for illustrative purposes should not be judged by their deriving quantitative predictions. An example of the confusion that may arise is the never-ending discussion about the Daisyworld model by Lovelock (1990). Lovelock designed a model to visualize how biological diversity might foster a decrease in the variability in the physical environment (temperature). Despite the model's illustrative character it was criticized that the simulated effect on temperature would be about 20% less if more realistic assumptions for the physical environment were made (Lovelock, 1990). Even if one is not favourable to Lovelock's Gaian hypothesis, such a type of criticism signifies a complete misunderstanding of the purpose the model was designed for. It illustrates also a danger of using such models. The results of heuristic models like Daisyworld cannot be directly tested one to one against real-world data. Such models can be used to derive and illuminate the implications of certain hypotheses, but not to establish their absolute validity. Authors must be cautious to state the scope and the limitations of their models to avoid misleading discussions. The crucial question arises whether models of this type can be of any use in shaping scientific opinion, in decision making and planning, especially in fields where unequivocal evidence cannot be obtained.

To avoid committing the same mistakes just criticized, I should like to state clearly that the models to be used in this work are of an illustrative character. Their formulation was motivated by phenomena observed in real ecosystems. However, none of them is meant to portray reality in a quantitative fashion. Nevertheless, I will introduce methods on how to quantify organizational properties of ecological systems. Even when I do not adhere to numbers as the only way to find access to nature, I consider that clear definitions and quantitative methods assist us in shaping our understanding and conferring ideas to other people. We must, however, be aware of and point out simplifications and assumptions made to avoid a possible confusion of models and the numbers produced with reality. Any number must be embedded in a context. The danger lies in isolating mechanisms and quantitative results.

3.6 SUMMARY

In a critical review I noted that present evolutionary biology relies heavily on a mechanistic conception of nature. Natural selection is viewed as the dominant evolutionary process. Attention is focused on the organisms as evolutionary units of selection and on the species as the units preserving evolutionary relevant information. Higher levels of organization are largely neglected. Such a restricted perspective derives also from the lack of a coherent theoretical framework linking species with higher-level organization. Phenomenological approaches to derive goal functions directing evolution at the ecosystem level largely neglect properties of the component species. What is required is a multi-level approach taking into account the mutual relationship between species and environmental contexts. A tool to foster progress in this direction could be provided by the niche concept. However, the current emphasis on competition has rendered it to a rather static and narrow description. Niches are predominantly perceived as outcomes of competitive interactions. The niche concept needs a redefinition to make it a versatile tool in a description of ecosystems as self-organizing dynamic entities.

Evolution is viewed here as a hierarchical process that involves, as essential ingredients, phenomena at the level of the ecological network. Ecosystems as evolving entities are not evolving entities in the Darwinian sense. Evolution at this level has to be defined with respect to irreversible changes in the patterns of interactions, with the nature and the quality of interactions between species and their environment.

Chapter Four

Spatio-temporal organization

In this chapter I introduce spatio-temporal organization in dynamic ecological networks as a novel perspective in between stability and static point-equilibrium on the one hand and stochasticity and randomness on the other hand. First, I attempt to convey an intuitive understanding of what the notion of spatio-temporal organization is meant to imply and of the general perception of the natural system in which it is embedded. Subsequently, I give spatio-temporal organization an operational definition in ecological networks of energy and matter flows. Its meaning is explained in detail both in qualitative and in quantitative terms.

4.1   A RELATIONAL PERSPECTIVE AS COUNTERPART OF THE MECHANISTIC PARADIGM

A mechanistic perspective dominates theoretical concepts and mathematical models in both ecology and evolutionary biology. The common recipe for describing a system comprises the identification of the relevant variables (e.g. nutrient pools, functional groups, species) and the processes (e.g. predation, nutrient uptake) linking them. In Figure 4.1a, I tried to sketch schematically the major elements of a mechanistic model thus arising: two entities *A* and *B* interact to produce an effect *C* via a process indicated by the arrow.

To give a more tangible example let us assume that we intend to model plant growth in a community where the main limiting factor is nitrogen. Entity *A* can then be identified with the nitrogen pool, *B* is plant biomass. The environment comprises biotic elements such as herbivores or microorganisms and abiotic variables such as precipitation or temperature. Nitrogen uptake by the plants is the main mechanism and the effect is biomass

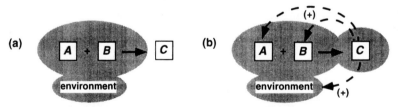

Figure 4.1   (a) Schematic sketch of a mechanistic description where two entities *A* and *B* interact to produce an effect *C*. (b) The presence of feedback renders the situation more complex, in particular when the feedback is positive. Cause–effect relationships and the time sequence become blurred. Further explanation in the text

production. In a first step we have to determine the relevant environmental influences impinging on our entities *A* (e.g. nitrogen loss) and *B* (e.g. herbivory) and on the process (e.g. temperature). Then we can come up with a quantitative expression linking biomass production to the states of nitrogen availability, *A*, and the biomass, *B*, under specific environmental conditions. The model could, for example, be refined by introducing species with specific functional properties that compete for nitrogen.

As depicted in Figure 4.1b, the situation becomes progressively more complex when feedback effects are present, in particular when these are positive. Recently, Hobbie (1992) summarized evidence that plant species create positive feedbacks to patterns of nutrient availability and cycling in natural ecosystems. In nutrient-poor ecosystems cycling is slowed down by plants growing slowly, using nutrients efficiently and producing poor-quality litter that decomposes slowly and deters herbivores. In contrast, in nutrient-rich ecosystems cycling is further enhanced by plants growing rapidly, producing readily degradable litter, and sustaining high rates of herbivory. Hobbie continues to note that correlative evidence as obtained from field-studies cannot distinguish cause from effect with respect to whether certain species thrive under certain regimes of nutrient availability or whether they actually generate them. The question arises whether it would be at all reasonable to attempt such a distinction. Hobbie gives a clue to an answer by remarking that the distributions of plant species are both a cause and an effect of nutrient cycling. If cause and effect cannot really be separated, what is then the meaningfulness of any mechanistic description?

We have to be aware that any mechanistic description follows a sequential resoning in terms of cause and effect such as: given $A(t)$ and $B(t)$ follow $C(t + \Delta t)$. To describe any real situation with circular feedback it may be more appropriate to use a parallel reasoning of mutual causation such as: given $A(t)$, $B(t)$, and $C(t)$ follow $A(t + \Delta t)$, $B(t + \Delta t)$ and $C(t + \Delta t)$. You may now argue that establishing mutual dependencies is just what systems theory is all about. True – essentially we could take such mutual dependencies into account by including more and more feedback effects into our model. However, doing so makes an explanation in terms of a mechanism obsolete – namely explaining the existence of *C* by the fact *A* and *B* interact to produce *C*. Further we have to be aware that $A(t_1)$ or $B(t_1)$ cannot be reduced to mere quantities characterizing the states of *A* or *B* at a specific time t_1 – they represent *A* and *B* embedded in a qualitatively specific time-dependent context. I argue that we continue running into trouble in our understanding of and interacting with ecological systems when we keep to our habit of focusing on single elements and processes and of investigating them in isolation.

Sceptics may argue that such a situation arises only because of the complexity of the subject matter, not because of fundamental differences in cause–effect relationships between, for example, physical and ecological systems. They may continue to point out that an individual person, a bird, a tree, all are well-defined entities with characteristic properties of their own. Since it is valid and reasonable to describe the interaction between defined entities as a process, there exist no strong arguments against inferring mechanistic cause–effect relationships. That is true – to a very limited extent. Individual persons can be looked upon as isolated biological entities. But what distinguishes the specific human, the human personality, is just the being embedded in a social network. Consciousness depends on the presence of an environment. Similarly, the essence of any living organism is its ability to engage in specific relationships with its environment that have meaning to the organism, and that are an integral part of its identity.

To give an example, let us imagine that we intend to make a model of communication among persons. We could give a description in an entirely mechanistic fashion by focusing on individuals, their specific properties, and their ability to communicate with other individuals via the tool of a common language. Once an individual generates a message there is a certain probability that this message will be received, understood, and will generate a response. We may also come up with a quantitative estimate of the probability that two persons meeting each other engage in a process of communication. The frequency of "successful encounters" depends on the probability that the two persons know each other, that they share a common interest, that they share a common language etc. Even when such a description cannot be said to be entirely wrong, it does not do justice to the significance of language and communication. Language is obviously meaningless at the level of the individual. The process of communication must be looked upon as being embedded in and even more as being entirely dependent on a social context. Persons represent a social environment, they have certain social functions that influence their behaviour. These functions are defined and are meaningful in a social context only. Hence, to understand the processes of communication we have to investigate the requirements for successful communication to take place, where successful refers to meaningful information being conveyed. Instead of asking why did persons A and B engage in a communication, and what is the probability that they engage in a communication the next time they meet, we should rather pose questions of the kind: What are the requirements that allow persons A and B to engage in a meaningful communication? How do the means of communication utilized depend on the social context? What is the effect of an exchange on their social situation, on the social network as a whole?

At first, the example chosen may seem to be rather peculiar to human societies. We may, however, draw analogies to biological systems. Consider, for example, a bee colony. Again we could attempt to explain the organization of a colony entirely in terms of the characteristics of the individual bees and their interactions. But what would be the significance of a queen bee or of the famous bee dance in the absence of the rest of the colony and the social context?

When we go down to the level of an organism the relation becomes even more obvious. I do so, even when I am aware of the prevailing prejudice against invoking arguments at the organismal level in discussions of ecological topics. The function of an organ like the liver makes sense within the context of the organism as a whole only. Nevertheless, many investigations in medical research focus on functional disturbances at the level of a single organ and consequently on the effect of specific drugs on single organs. However, many diseases may be far better characterized by subtle disturbances in relations and context at the level of the organism as a whole, rather than by targeting a single organ. The problem is that disturbances in context tend to escape defined countermeasures and strategies of treatment. Therefore it is understandable that the search for defined cause–effect relationships prevails. Being honest we have to admit that we adopt a similar strategy in dealing with environmental problems as well.

The objection may be raised that the emphasis on relations is already common ecological practice. Isn't it included in and isn't it even the essential core of notions such as primary producers and herbivores, making explicit reference to functional properties in an ecosystem? That is true – to some extent. One specific property, a rigid and frozen context, is assigned to a group of organisms. Subsequently their properties are described in a mechanistic fashion. Disturbances and change are in general attributed to single

Box 4.1 Comparison between a mechanistic and a relational approach

	Mechanistic	Relational
Question:	What are the causes for an even to happen?	What are the circumstances rendering possible a pattern of interactions?
Goal:	Derive causal mechanistic explanations for system dynamics	Find relationships between structural and functional properties
Method:	Identify and isolate entities and processes	Identify patterns of interaction and their requirements
Theory:	Models that predict events	Models that make patterns intelligible
	Rules how processes act on entities to produce events	Rules on how to proceed in detecting and characterizing relational patterns

processes and/or properties of entities, but not to changes in context. Therefore I should like to contrast a mechanistic perspective with what I call a relational perspective that emphasizes context. To clarify the distinction between the two approaches I portrayed both in their extremes in Box 4.1.

The comparison shows that the approaches are not mutually exclusive but complementary. However, I claim that in order to understand the nature of ecological systems, we have to adopt a relational perspective. The problem with the mechanistic approach is its focus on a single level of a system's description. Thus it is rather insensitive to account for function and constraints arising at a higher level of organization. Using only mechanistic models leaves us with a rather fragmented and distorted perspective on nature.

To avoid any misunderstandings let me state the following. I do not argue in favour of an all-embracing complex interconnected system where everything interacts with everything else, which due to its overwhelming complexity nobody can understand. Actually, such a perception would be entirely compatible with a mechanistic perspective. However, I strongly object against an isolationist view separating organisms from their environment that is an integral part of their identity. If the relational perspective is taken seriously, part and whole cannot be separated. A part gains its identity only in the context of the whole. The whole is continuously regenerated by its parts. Because these issues are of utmost importance, I will not stop repeating them throughout this work. If we adhere to mechanistic monocausal explanations in feedback systems where cause and effect cannot be separated, and where situations are often historical and unique, we remain entangled in contradictions and in circular reasoning.

4.2 ECOSYSTEMS VIEWED FROM A RELATIONAL PERSPECTIVE

How does adopting a relational perspective change our perception of ecosystems? It is rather trivial to conclude that we must focus on contexts rather than on single objects. One way of doing so lies in looking simultaneously at different levels of organization – at the interacting components and the system as a whole – and their mutual dependence. We have to account for the fact that species generate, modify and respond to their environment. The environment that signifies both the biotic surroundings consisting of other species and the physical abiotic environment provides both opportunities and constraints acting back on the organisms. We cannot say that species show a specific behavioural repertoire because of the characteristics in their environment, nor can we say that the environment has specific traits because of the characteristics of the species. Both are mutually dependent and may reinforce each other, as in the patterns of nutrient cycling discussed in the previous section. What evolves is a context.

To have a conceptual base, I describe an ecosystem as an interaction network consisting of an ensemble of compartments that are linked by interactions. For the time being, I assume the compartments as given, representing some aggregation of organisms. The derivation of proper guidelines for such aggregations will prove to be a major task to be accomplished at a later stage of this book. A compartment is embedded in a network context where it interferes with its environment and where it experiences feedback effects related to its own past actions. Feedback effects are diffuse due to their being propagated along an intricate network of interactions – intricate both with respect to the pathways of interaction and their spatio-temporal dynamics. Feedbacks are in most cases experienced as system-level constraints. To develop some intuitive understanding of the diffuse nature of feedbacks it might be helpful to make reference to an example from daily experience. As members of social networks we are continuously confronted with events and phenomena in our social surroundings. As active members we take part in shaping these social surroundings, as for example in the workplace or by contributing to political decisions. Even when it is in general impossible to trace back the influence of us as single persons or of single social groups, we have to be aware that we experience surroundings that are generated by us. The lack of such an awareness mediated by the vagueness of cause–effect relationships is actually central to most environmental problems facing us today. The situation is aggravated by the mismatch in temporal and spatial scales. We act locally and experience adverse effects at a national or even global scale. It is rather difficult to establish links between local actions and global threats. Rather than tracing single processes in detail, we need to develop an improved overall understanding of the dynamic nature and organization of such systems.

Can we already conceive of any consequences for ecosystem dynamics? The behavioural repertoires of organisms are evidently more restricted than those of human beings. Nevertheless, organisms in an ecosystem shape their environment by acting due to past experience, information about their environment, and anticipation of the future. I conjecture that there is a continuous imbalance between the information received and the actions propagated along the network that results in a continuous driving force for change. The effect of an input depends to a large extent on the context it is embedded in and on the recipient's previous history. The information propagated along the ecological network and the memory of the constituent parts never match and the reciprocal shaping prevents equilibrium and fosters evolution.

Let us again consider an example from a human system by having a closer look at a stock market. People base their decisions to buy or sell stocks on the current market prices, on the general situation on the stock market, and on the general economic situation. Not in the least, their decisions depend on their anticipation of the future, an anticipation that is largely determined by their past experience. For obvious reasons, many attempts have been made to predict and profit from future trends. However, by acting one interferes with and changes the whole situation on the market. It is in the nature of this self-organizing highly dynamic system to be inherently unpredictable and to remain far away from a stable equilibrium point. The interaction of many individual agents embedded in a network of positive and negative feedback keeps the system within bounds – at least in most cases. Spectacular crashes of the stock market give evidence that such dynamics comprise as well the possibility for sudden major changes. The example of the stock market is illustrative because the market is embedded in a complex web of economic, political and social interactions and because at least the immediate feedback effects become rapidly visible.

The situation in ecosystems may be perceived along similar lines. Organisms are selective with respect to the type and the meaning of environmental signals received, they have certain degrees of freedom on how to react. The infinite number of possible combinations of internal and external states renders situations unique and prevents the approach to a stable equilibrium point. The question arises, how systems are organized that combine a high potential for change with the maintenance of function. Such self-organizing systems with a distributed and flexible "control" are in sharp contrast to most technical systems where a centralized control optimizes a desired function.

4.3 SELF-ORGANIZATION – AN EMPTY PHRASE?

Like stability or complexity the notion of organization embraces a wide range of intuitive understandings of its meaning. This diminishes its usefulness for scientific discussions unless one attempts a more rigorous definition. We typically talk about societies being organized, about the organization of a living organism. In most cases organization is perceived as a static structural framework being associated with order and complexity – contrasting with chaos and randomness. The structural organization may change but the change itself is not perceived as part of the organization. In this respect our perception resonates with a type of organization governed by what may be called a centralized control. I refer to such organization as imposed organization because it is in general imposed onto a system by a rigid top-down control and/or external influence. I consider imposed organization to be typical for designed systems, but to be atypical for ecosystems that arise from processes of dynamic self-organization. The notion "dynamic" expresses that such organization must comprise change as essential element. Box 4.2 contrasts some properties that can be conceived as being characteristic for the two types of organization.

Examples for imposed organization are typically given by systems that are designed to serve a specific function. In an army a centralized top-down control exercises power onto the members, the interactions among which are determined by strict rigid rules. The organization of an army is obviously designed to make it a controllable instrument that resembles in this respect a machine operating in a rather rigid mechanical fashion. Such systems have the advantage that they can be optimized to fulfil a specific single task. They are, however, rather inflexible and unable to respond and adapt to changing circumstances.

Liberal, democratic systems, where rules and decisions emerge continuously from the interactions among the members, agree more with the notion of self-organization. Political structures exist in any social system, but in a real democracy these structures are subject to change according to the internal dynamics of the society. Systems of this type are characterized by what I refer to as distributed control. Whole system function is generated by interactions in a relational network. The degrees of freedom of the network

4.2 Types of organization	
Imposed organization	Self-organization
externally imposed	endogenously generated
centralized control	distributed control
rigid networks of interaction	flexible networks of interaction
little potential to adapt and change	high potential to adapt and change
predictable	unpredictable
Examples	
machine	organism
army	democratic society
agroecosystem	natural ecosystem

members are restricted by their being embedded in a network context. Distributed control results from this dualistic interplay.

The distinction between centralized and decentralized control has become a major topic in computer science during recent years (e.g. Hogg, 1990; Ramadge and Wonham, 1989). In large heterogenous computer networks the limited resources must be managed to benefit a wide range of tasks. In this field centralized control has been the reigning paradigm implying that an omniscient central controller monitors activities and allocates resources to the various tasks and users. Such a central control requires that all the interacting tasks can at any time be controlled by the central unit which must have a complete knowledge of every single event. Even when such structures can in principle be realized by appropriate design, computer scientists have become increasingly aware that they are slow, inflexible and liable to disturbance. A new approach to system design has been referred to as computational ecosystems (reviews in Huberman, 1988). This approach is based on the insight that natural ecosystems are examples for distributed systems managing large ensembles of more or less independent units each of which makes decisions based on its limited and imperfect view of the world. Even when there exists neither a central control nor a universally recognized overall goal, ecosystems do operate reliably and produce effective solutions to problems of resource allocation.

It is as yet a largely unresolved question as to what, in detail, are the essential properties of an ecosystem and its constituent components for achieving the combination of effective resource utilization and adaptability. Attempts to design systems show that it does not suffice to simply conceive of such systems as random ensembles of competing units. Of major importance seem to be elements such as cooperation, uncertainties in decision making, anticipation of the future in a given context, the limited amount of information available to each unit, or time lags in response (Hogg, 1990; Huberman, 1990; Huberman and Hogg, 1988; Kephart *et al.*, 1990). As ecologists, we have to admit to not being able to contribute much knowledge in this respect. Intuitively we may be aware of the dencentralized nature of ecosystems, of the fact that some type of organization must be generated within the system. However, any such insights have not yet been incorporated into our theoretical concepts nor in our approaches towards management. Notions such as top-down or bottom-up control resonate much more with centralized than with distributed control. I argue that our current perception of natural systems from organisms to ecosystems resembles much more the idea of machines than of genuine living systems. It is a major goal of this book to assist in changing perspective by elucidating and characterizing patterns of self-organization in ecosystems.

4.4 SPATIO-TEMPORAL ORGANIZATION IN ECOLOGICAL NETWORKS

Given an interaction network we may wonder what should be considered a system that is both complex and organized. We may start with the intuitive guess that the most complex system we could think of would be a maximally connected network where every compartment interacts with every other. However, such a network could hardly be considered to be very organized. It may not even be considered as complex because it is based on a rather simple rule. A system that is both complex and organized should be characterized by a diversity of well-defined communication pathways. Many interactions among compartments may be present but they all have to be meaningful.

As an illustrative example let us consider the interactions among people in a large company. The organization may be said to be high when any information exchange results in useful action. For a person it is presumably not very useful and effective to communicate with all people in the company. The other extreme of receiving the information from one person only and of transferring it to only one other person might not be very effective either. Our intuition tells us that an optimum should be somewhere in between. To become more explicit, each person may communicate with a variety of other people, however, not necessarily at the same time and at the same location. You may object that the advent of modern means of communication give the advantage that you can communicate with several people at the same time at different locations. That is true – to some extent. Spreading some pieces of information over an ensemble of people is useful only when the people concerned are selected. An even distribution of all information to all people is inefficient and confusing. This also comes closer to our understanding of complexity and organization because neither random interaction, nor interacting with one single partner, ask for complex decisions. Choosing, however, one interaction pattern among several possibilities requires a sophisticated exchange with the environment. The organization of the people in a company may now be said to increase with increasing diversity of the number of temporally distinct communication partners, where a partner may also represent a whole group of people sharing the same interest. This intuitive argument links in with the concept of spatio-temporal organization that will be at the centre of the considerations to follow.

We obviously cannot identify interactions in an ecosystem with the conscious acts of communication among human beings. However, ecological interactions are not occurring at random either. In analogy to the above example, the organization of ensembles of organisms in an ecosystem may now be said to increase with increasing diversity of the number of spatio-temporally distinct interaction partners, where a partner may also represent a group of organisms sharing the same function.

To develop now a formal definition of such organization we have to make a decision on a certain conceptualization of an ecosystem, always being aware that any such endeavour cannot cover all aspects of organization. In Figure 4.2 I sketch schematically three levels that can be distinguished in an ecological network: the microscopic level of the organisms,

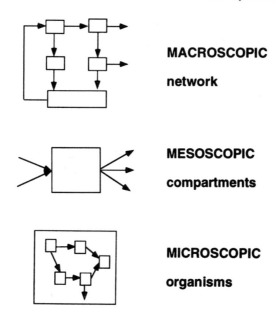

MACROSCOPIC

network

MESOSCOPIC

compartments

MICROSCOPIC

organisms

Figure 4.2   Levels in an ecological network

the mesoscopic level of the compartments representing aggregations of organisms, and the macroscopic level of the network itself.

The level of the organisms is the most tangible since it consists of well-defined entities. The level of the system as a whole can be defined reasonably well by aggregated functional properties such as primary production, biomass or nutrient recycling. The level of the network compartments is the most ambiguous. Talking about compartments requires an ecological network to be delineated first. Especially when we are interested in describing organizational characteristics we have to exercise care on how to define the network. We must be aware that any type of network description imposed onto a system reflects the perspective of the investigator. The traditional assignment of organisms to trophic levels as primary producers, herbivores and carnivores is ambiguous in any real ecosystem because ecological networks resemble food webs more than food chains, where trophic levels are well defined. An assignment in terms of taxonomic species may not be very desirable even when we have the information required to do so. We risk losing structural information by the overwhelming complexity of the network thus arising. Further, we have to be aware that the criteria on how to define a taxonomic species are not the same over the whole hierarchy of biological genera.

Arguments have been put forward in favour of individual-based descriptions as the most promising and appropriate concept for modelling populations, communities and ecosystems (e.g. Huston *et al.*, 1988; Villa, 1992). The individual organism obviously has the advantage of being a well-defined unit, whereas any higher level of aggregation necessarily includes a certain degree of arbitrariness. However, function in an ecological network appears at the level of organism ensembles rather than at the level of the individual. A proper description of a network's structure may thus reveal important characteristics of a system's functional organization. We will still find out that the network structure can be perceived as deriving from spatio-temporal self-organization within the constraints arising from the level of the organisms and the level of network as a whole. We may thus identify at least two major tasks in pursuing the approach chosen:

1. Set up rules on how to define a network and its internal structure.
2. Develop methods to describe and quantify a network's spatio-temporal organization.

Even when it may not seem to be the logical way to proceed, I will deal with the second point first. The network concept to be introduced in Chapter Six has been developed by viewing ecosystems as characterized by a dynamic organization that extends over a wide range of spatial and temporal scales. Hence, following the rationale for the network concept is facilitated by an improved understanding of the nature of a network's spatio-temporal organization. For illustrative purposes, I start with extremely simple configurations that at first sight might contradict any idea of distributed control and diffuse interactions. The examples chosen will become progressively more complex when we deepen our understanding and improve our models. I will now move on to the introduction of the formalized concept of spatio-temporal organization in ecological networks and the derivation of measures for its quantification. A more rigorous treatment of the mathematical derivations is given in the technical appendix.

4.5   DERIVATION OF MEASURES FOR (SPATIO-TEMPORAL) ORGANIZATION

To define organization from a network perspective means that the focus is on the interactions (energy, matter, information) between the compartments instead of being on the compartments (biomass, internal structural organization) themselves. In a straightforward and operational approach I identify such interactions with material exchanges. This is based on the reasoning that natural systems are dissipative systems the existence of which depends on the permanent exchange of energy and matter with their surroundings. A description based solely on material flows may be questioned because a variety of interactions (e.g. social) cannot be identified directly with and quantified by material exchanges. Nevertheless, I assume the characteristics of energy and matter networks to integrate and reflect the consequences arising from various types of interaction. Energy and matter flows are the necessary material basis for life to exist. They are amenable to experimental investigation, whereas it is far more difficult and ambiguous to investigate or even quantify other types of interaction.

Measures derived from information theory have been applied to quantify structural properties of ecosystems. Most readers may be familiar with the entropy measure introduced by Shannon (1948). The Shannon entropy measure, which should not be confused with entropy as defined in thermodynamics, has found widespread application in ecology as a diversity index. In his search for a tangible expression for the vague notion of ecosystem complexity, MacArthur (1955) was the first to apply the Shannon entropy measure in ecology (cf. section 2.4). He suggested the following measure to quantify trophic diversity that he assumed reflected ecosystem complexity:

$$S = - \sum_i p(x_i) \log p(x_i) \qquad (4.1a)$$

where $p(x_i)$ denotes the proportion of a community's food energy passing through the ith pathway in the food web. This diversity index increases with the logarithm of the number of distinct pathways in a food web. It attains its maximum in a network where all compartments are connected and where all flows are of equal magnitude. If we recall Figure 3.5, the diversity of the network depicted in (a) is higher than the one of the network in (b). If we assume all flow magnitudes to be equal, the network in (a) has a diversity index of log(30) whereas the one in (b) has a diversity index of log(6). Obviously, this diversity measure does not account for structure or flow specificities. To characterize organization it is not sufficient to simply enumerate flows. We must take account of the compartments embedded in a network context. Following the environ concept of Patten (1981), such an embedding can be described in terms of an input- and an output-environment as represented in Figure 4.3a.

The compartment depicted in Figure 4.3a exhibits a high degree of redundancy. Redundancy refers to a multiplicity of flows performing the same function in a network. In case of the network element depicted in Figure 4.3a, a flow's function is given by its serving as input or output to this very compartment. We may also say that the compartment has on average five degrees of freedom with respect to where it receives its inputs from or where its outputs are directed to. A flow serves to link two compartments. Hence, in order to describe a flow's function in the network, we need to take into account that a flow emanating from one compartment serves as input to another compartment, as shown in Figure 4.3b. What type of configuration would we now expect to find in a network displaying a high degree of organization?

Recall that we discussed in section 3.4.2. Ulanowicz's concept of ecosystem organization based on energy networks where the flows are given by long-term averages over time and space. Ulanowicz (1986) identified an increase in a network's organization with a decrease in the overall redundancy. He argued that out of the numerous pathways serving as inputs or outputs to a compartment, single pathways are selected and optimized over evolutionary time scales. Due to its redundancy, the network configuration depicted in

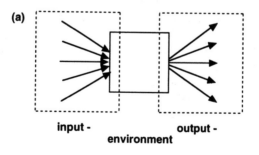

(a)

input -   output -
environment

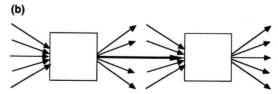

(b)

Figure 4.3   (a) A compartment's embeddeding within a network can be characterized by an input- and an output-environment that include all links averaged over time and space. (b) Link between two compartments indicating the complexity of a compartment's embedding in the network as a whole

Figure 4.3b would thus be identified with a low level of organization. The situation depicted could for example represent a predator–prey relationship where the prey is consumed by a number of predators and the predator feeds on a number of prey species. A high organization, as defined by Ulanowicz, would be given if the predator–prey relationship was highly specific in that the prey was exclusively consumed by this single predator of whom it comprised the sole food item.

Such a streamlined network topology contradicts the idea of spatio-temporal organization (cf. also section 3.4.2). In dynamic networks we expect an array of functionally similar network configurations to coexist. Coexistence does not necessarily imply a simultaneous presence. It may also be realized by a temporal succession and/or a spatial segregation. Correspondingly, we expect a high redundancy, as for the compartment represented in Figure 4.3a, if the flow patterns are averaged over temporal and spatial extensions that are large in comparison to the typical spatio-temporal scales of the internal dynamics (e.g. an annual average of the phytoplankton species that have generation times of the order of hours to a few days). We expect configurations as represented in Figure 4.4 for short observation periods, short when compared to the typical time scales characterizing the organisms involved. At a specific moment in time a single interaction dominates. The pattern of dominance may, however, be highly variable over time and space, this being characteristic of what I call spatio-temporal organization. The compartments should not be identified with taxonomic species. Instead the compartments represent ensembles of organisms characterized by the same function in a network context. Guidelines on how to delineate such ensembles will be given in Chapter Six.

In ecosystems with high spatio-temporal organization we thus expect a major decrease in redundancy when the temporal and spatial flow patterns are resolved compared to a network configuration where the flows are averaged over time and space. As an example we could conceive of algal species that are functionally redundant with respect to their being primary producers. Let us further assume that they differ in their growth optima with respect to light intensities and temperature. As a consequence, different guilds are active during different phases of a seasonal cycle (e.g. Reynolds, 1984b). Only the temporal resolution of the growth patterns gives us the full information about the functional importance of the different species that would simply be functionally redundant in an annually averaged description.

To now derive a measure for such spatio-temporal organization I proceed as follows:

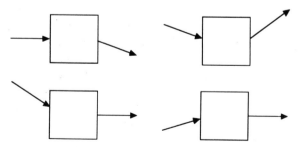

Figure 4.4   Temporal snapshots of a compartment's time-averaged network environment depicted in Figure 4.3a. Contrasting with the time-averaged representation the input- and the output-environments comprise only a single interaction at a certain moment in time

The network averaged over time and space serves as reference state. In a first step I will introduce measures quantifying the redundancy and organization in networks averaged over time and space. Subsequently, I will derive measures quantifying the increase in organization equal to a decrease in redundancy that results from accounting for the temporal and spatial flow patterns.

4.5.1 Measures of organization and redundancy in networks averaged over time and space

Rutledge *et al.* (1976) used the concept of average mutual information (AMI) for deriving a measure quantifying the redundancy of ecological networks. Originally, the concept of AMI was introduced in the context of an information channel to measure the amount of information transferred per signal (e.g. Abramson, 1963). An information channel consists of a signal source and a receiver. The AMI is zero when the response generated in the receiver is completely independent from the signal generated by the source. In this case no information is transmitted. The AMI is maximal if each signal generated in the receiver can be traced back unequivocally to one specific signal generated in the source. The AMI is a contingent measure that takes into account the functioning of and the relationship between source and receiver. It is high when meaning and hence information are transferred unambiguously. In contrast, the Shannon entropy of the source would simply increase with the number of different signals produced irrespective of the response generated in the receiver. We could as well derive the Shannon entropy for the ensemble of all possible events corresponding to all possible combinations of signals generated and signals received. This measure would be maximal when the AMI is zero, namely when the number of possible combinations is maximal and when the signals received are independent of the signals generated.

In a more general sense, the AMI between two sets of events is zero when they are statistically independent. The AMI increases with degree of dependence of the events upon each other. How can we make use of this property to derive a measure of redundancy in ecological networks? To do so we identify the two sets of events with the compartmental inputs and outputs, respectively. One event can thus be characterized by $p(a_j)$, the probability of one unit of energy and/or matter leaving a compartment j. The other event is given by $p(b_i)$, the probability of one unit of energy and/or matter entering a compartment i. The dependence between inputs and outputs is low in a network with high redundancy where each compartment receives its inputs from many other compartments and distributes its output to many other compartments. Correspondingly, the AMI between inputs and outputs is high if each input into a compartment can be traced back to one specific output from another compartment. In this case $p(b_i/a_j)$, the conditional probability of an energy unit entering compartment i when it leaves compartment j, equals one. Instead of simply enumerating flows, as done in diversity measures, we focus on the whole context of the output environment of the donor compartment and the input environment of the receiving compartment, and their relation to the network as a whole. The AMI between inputs and outputs constitutes a measure for a network's organization as defined by Ulanowicz (1986).

To describe a flow network some conventions are introduced as depicted schematically in Figure 4.5. The total number of functionally distinct compartments in a network is referred to as n. Hypothetical compartments are introduced to account for the flows

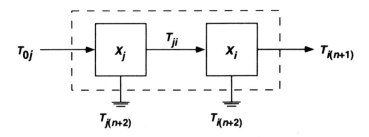

Figure 4.5 Basic elements of an energy network. A flow from a compartment X_j (e.g. primary producer) to another compartment X_i (e.g. herbivore) is referred to by T_{ji}. $T_{j(n+2)}$ and $T_{i(n+2)}$ denote respiratory losses. T_{0j} refers to imports (e.g. solar energy) and $T_{i(n+1)}$ refers to exports (e.g. migration)

crossing system boundaries (indicated by the dashed line). Compartment 0 serves as source for all exogenous inputs into a system such as solar energy or nutrient inputs. Compartment $n+1$ serves as sink to accept exogenous exports such as losses due to sedimentation or nutrient losses by the watershed. Compartment $n+2$ is important for energy flows only to serve as a sink to accept respiratory losses. From thermodynamic considerations it is reasonable to make the distinction between useful exports where the energy quality is maintained and respiratory losses where the energy is degraded into heat. T_{ji} denotes a flow from compartment j to compartment i. T_{0j} denotes an import from the environment into compartment j, $T_{i(n+1)}$ denotes an export to the environment from compartment i, $T_{j(n+2)}$ and $T_{i(n+2)}$ denote respiratory losses from compartments j and i, respectively.

Using these conventions, the AMI between inputs and outputs can be quantified in terms of flows, as summarized in Box 4.3. A more detailed derivation of the mathematical expressions is given in the technical appendix.

Box 4.3 Measure of organization in networks averaged over space and time (Pahl-Wostl, 1992a; Ulanowicz, 1986)

The average mutual information, I, measures the reduction in the mean uncertainty about the occurrence of an event from the joint set AB that is provided by a knowledge of all events from the single sets A and B. I is zero when the two sets of events are statistically independent. I achieves its maximum whenever knowledge of the events from the singular sets A and B leaves no uncertainty regarding the corresponding events from the joint set AB.

The average mutual information $I(A;B)$ of two sets of events $\{A\}$ and $\{B\}$ may be expressed as:

$$I(A;B) = H(A,B) - H(A/B) - H(B/A) \qquad (4.2)$$

where

$$H(A,B) = -\sum_j\sum_i p(a_j,b_i)\log p(a_j,b_i) = \text{Shannon entropy of the joint set } AB$$

and

$$H(A/B) = -\sum_j\sum_i p(a_j,b_i)\log p(a_j/b_i) = \text{conditional entropy of the set } A \text{ given set } B$$

(Continued overleaf)

Box 4.3 *(continued)*

and

$$H(B/A) = -\sum_j \sum_i p(a_j,b_i)\log p(b_i/a_j) = \text{conditional entropy of the set } B \text{ given set } A$$

$p(a_j,b_i)$ refers to the joint probability that the events a_j and b_i occur simultaneously. $p(a_j/b_i)$ refers to the conditional probability for the event a_j to occur given b_i has occurred.

To derive a measure of organization in a network with n compartments, the joint set AB is identified with the ensemble of all intercompartmental flows $\{T_{ji}\}, j,i = 1,2\ldots,n$, the set A to the ensemble of all compartmental outputs $\{T_{j.}\}$ and the set B to the ensemble of all compartmental inputs $\{T_{.i}\}, i = 1,2,\ldots,n$. The symbol T_{ji} denotes a flow from compartment j to compartment i, $T_{j.}$ the total output from compartment j and $T_{.i}$ the total input into compartment i. A point as an index means summation over the corresponding dimension in the flow matrix.

The calculation of the information-theoretic measures is based on the probability for events to occur. The probability for an energy unit to be transferred via a certain flow can be estimated as the flow magnitude normalized by the total system throughput, T, the sum of all flows in the system:

$$T = \sum_{j=0}^{n} \sum_{i=1}^{n+2} T_{ji} \tag{4.3}$$

Normalization according to Eqn (4.3) allows the treatment of systems that are not in steady state with respect to inputs and outputs. The measure of organization, I, expressed in terms of flows, yields:

$$I = \sum_{j=0}^{n} \sum_{i=1}^{n+2} \frac{T_{ji}}{T} \log \frac{T_{ji}T}{T_{j.}T_{.i}} \tag{4.4}$$

Basic results from information theory require that I is bounded by:

$$H(A,B) \geq I \geq 0 \tag{4.5}$$

The system's redundancy, R, is defined as the deviation of I from this upper bound:

$$R = H(A,B) - I = H(A/B) + H(B/A) \tag{4.6}$$

The term information should not be taken literally when using measures based on the mutual information in the context of quantifying a network's organization. It might be misleading if these quantities were identified with any type of information transferred in the system. In the first place the term was derived from system's dynamics and describes a type of network structure. This does not exclude that we may still draw parallels between an increase in spatio-temporal organization and the change in the patterns of information transfer. To prevent misunderstandings I am therefore going to avoid the term mutual information in connection with the measures defined and use the term organization instead. The measure of organization *sensu* Ulanowicz in the time- and space-averaged network yields (expressed in terms of flows):

$$I = \sum_j \sum_i p(a_j,b_i)\log\frac{p(a_j,b_i)}{p(a_j)p(b_i)} = \sum_{j=0}^{n} \sum_{i=1}^{n+2} \frac{T_{ji}}{T} \log \frac{T_{ji}T}{T_{j.}T_{.i}} \tag{4.4}$$

A point denotes summation over the corresponding index, e.g. $T_{j.}$ is the total output from compartment j summed over all compartments in the system. T is the total system throughput, the sum of all flows in the system. The ratio of T_{ji} to T corresponds to the probability $p(a_j,b_i)$ that one unit of matter travelling in the network is found in the flow from compartment j to compartment i. Each flow contributes then to the overall measure of organization with the product of its quantitative weight as expressed by $p(a_j,b_i)$ and a logarithmic structural term accounting for its specificity. To emphasize the difference between this measure and the Shannon entropy defined in Eqn (4.1a) I express the latter in the same notation:

$$S = -\sum_j \sum_i p(a_j,b_i)\log p(a_j,b_i) = \sum_{j=0}^{n}\sum_{i=1}^{n+2}\frac{T_{ji}}{T}\log\frac{T}{T_{ji}} \qquad (4.1b)$$

Whereas the Shannon entropy increases simply with the number of flows present in a network, the measure of organization takes the network configuration into account. The maximum, I_{max}, is obtained in a streamlined network where each compartment has one input and one output only (strictly speaking all flows must have the same magnitude for I to become maximal). In this case $T_j = T_{\cdot i} = T_{ji}$ and the redundancy equals zero. The redundancy of a network, R, can be quantified by the difference between this maximum and the realized configuration: $R = I_{max} - I$. R corresponds to the logarithm of the average number of links in the input- and output-environments of the compartments. If we choose logarithms to the base 2, a decrease of I by one unit corresponds to a doubling of the average number of flows entering or leaving a compartment. To give an example let us assume that we have a network with 16 compartments. I attains its maximum equal to 4 in a network where the input- and output-environments of each compartment comprise on average one link only. A further decrease in I corresponds to an increase in redundancy according to:

I ($\log_2$)	4	3	2	1	0
R ($\log_2$)	0	1	2	3	4
Average number of links in input- and output-environments	1	2	4	8	16

The redundancy in the network averaged over time and space is an essential requirement for an ecosystem's spatio-temporal organization. A comparison of Figures 4.3 and 4.4 reminds us that we need a multiplicity of flows leaving and entering a compartment in order to realize different network configurations as a function of time and space. The latter process is exactly what I have called spatio-temporal organization, a quantitative measure of which is derived next.

4.5.2 Measures of organization in networks resolved in time and/or space

The increase in organization associated with a temporal and/or spatial flow pattern is identified with the reduction in redundancy that is achieved upon resolution of the network along the dimensions of time and/or space. A quantitative measure can be derived by extending the measure of organization introduced in the previous section to include the temporal and/or spatial dimensions (Pahl-Wostl, 1990, 1992a). The mathematical expressions are summarized in Box 4.4, and a more detailed derivation is given in the technical appendix.

The measure of temporal organization yields (expressed in terms of flows) is as follows:

$$I_t = \sum_{j=0}^{n}\sum_{i=1}^{n+2}\sum_{k=1}^{r}\frac{T_{jik}}{T}\log\frac{T_{jik}T_{jik}T}{T_{ji\cdot}T_{\cdot ik}T_{j\cdot k}} \qquad (4.8)$$

Box 4.4   Measures of organization in networks resolved along the dimensions of time and/or space

The dimension of time is introduced as a new set C comprising the ensemble of all time intervals $\{\Delta t_k\}$. Based on $H(A,B)$ as the upper bound, the logical extension of I (Eqn 4.2) to another dimension, C, is defined as:

$$I(A,B;A,C;B,C) = H(A,B) - H(A/B,C) - H(B/A,C) \qquad (4.7)$$
$$= I(A;B) + I(A;C/B) + I(B;C/A)$$

A formal derivation is given in the appendix. The measure of temporal organization, I_t, as defined in Eqn (4.7) yields the following expressed in terms of flows:

$$I_t = \sum_{j=o}^{n} \sum_{i=1}^{n+2} \sum_{k=1}^{r} \frac{T_{jik}}{T} \log \frac{T_{jik}T_{jik}T}{T_{ji.}T_{.ik}T_{j.k}} \qquad (4.8)$$

The total observation period is resolved into r time intervals of equal duration. T_{jik} denotes a flow from compartment j to compartment i during a time interval t_k. If r is chosen to be equal to 1, Eqn (4.8) becomes equivalent to Eqn (4.4) derived for the time-averaged network.

Basic results from information theory require that $H(A/B,C) \leq H(A/B)$ and $H(B/A,C) \leq H(B/A)$. Hence, I_t is bounded by:

$$H(A,B) \geq I_t \geq I \geq 0 \qquad (4.9)$$

R_t, the redundancy in the time-resolved network, yields:

$$R_t = H(A,B) - I_t = H(A/B,C) + H(B/A,C) \leq R \qquad (4.10)$$

The measure of temporal organization can be extended to further dimensions. A spatial dimension may be introduced as a new set D of all spatial intervals $\{\Delta s_l\}$. The measure for spatio-temporal organization is then defined as:

$$I(A,B;A,C,D;B,C,D) = H(A,B) - H(A/B,C,D) - H(B/A,C,D) \qquad (4.11)$$
$$= I(A,B;A,C;B,C) + I(A;D/B,C) + I(B;D/A,C)$$

Expressed in terms of flows Eqn (4.11) yields:

$$I_{ts} = \sum_{j=0}^{n} \sum_{i=1}^{n+2} \sum_{t=1}^{r} \sum_{l=1}^{q} \frac{T_{jikl}}{T} \log \frac{T_{jikl}T_{jikl}T}{T_{ji..}T_{.ikl}T_{j.kl}} \qquad (4.12)$$

The spatial dimension is resolved into q equal intervals and denoted by the index l. T_{jikl} represents a flow from j to i during a certain time interval t_k at a spatial location s_l.

I_{ts} is bounded by:

$$H(A,B) \geq I_{ts} \geq I_t \geq I \geq 0 \qquad (4.13)$$

The maximum possible organization always remains bounded by the capacity $H(A,B)$ inherent in the network averaged over time and space. A generalized form of multi-dimensional measures of organization is given in appendix one. There exist other possibilities on how to introduce the dimensions of time and space in a network description. The different measures that might be defined and their meaning are summarized in appendix one. However, the measures introduced here proved to serve best the purposes of quantifying spatio-temporal organization as defined in the context of this work.

T_{jik} refers to the flow from compartment j to compartment i during a specific time interval t_k. Again a point denotes summation over the corresponding index, e.g. $T_{j.k}$ is the total ouput from compartment j during the time interval t_k summed over all compartments in the system. A comparison with Eqn (4.4) reveals that the maximum of temporal organization is obtained when $T_{j.k} = T_{.ik} = T_{jik}$, when during a specific time interval t_k the flow T_{jik} constitutes the only output of compartment j and the only input to compartment i, respectively. Hence, the measure of organization increases upon resolution of the temporal flow pattern $(I_t > I)$ if redundant interactions are confined to different time intervals. As an example we might conceive of guilds of plant species that are active during different phases of a seasonal cycle.

Similarly, spatial organization refers to networks where redundant interactions are confined to different spatial locations. As an example we might conceive of a spatial mosaic of different plant species. Spatio-temporal organization refers to networks where redundant interactions are confined to different spatial locations during different time intervals. As an example imagine predator species whose foraging area changes over a seasonal cycle. The corresponding measure of spatio-temporal organization is obtained by replacing T_{jik} in Eqn (4.8) by T_{jikl}, which refers to the flow from compartment j to compartment i during a specific time interval t_k at a specific location s_l. Mathematical details are summarized in Box 4.4.

When we investigate a network we can now quantify the temporal and/or spatial organization inherent in the temporal and/or spatial variations of the flow pattern according to:

$$\Delta I_t = I_t - I \qquad \Delta I_s = I_s - I \qquad \Delta I_{ts} = I_{ts} - I \qquad (4.14)$$

The reduction in redundancy, equivalent to an increase in the measure of organization, is expressed as deviation from I, the measure of organization obtained for the flow pattern averaged over time and space, ΔI_t and ΔI_s correspond to the increase obtained for a resolution along time or space, respectively. ΔI_{ts} corresponds to the increase obtained for a resolution along both time and space.

Accounting for spatial and temporal flow patterns can only increase the measure of organization or leave it unchanged (cf. Eqn 4.13). We cannot increase the overall redundancy of interactions by taking into account temporal and/or spatial variations. The measures remain unchanged if the network configuration is invariant in time and/or space. This does not imply the absence of variations at the level of the system as a whole. Let us again use the example of algal species. Imagine a situation where the total activity of a species ensemble changes over a seasonal cycle whereas the relative contribution of each species remains constant. In this case we cannot talk about a temporal organization of the algal community. The redundancy remains constant and correspondingly the measure of organization does not increase when we account for the seasonal variation.

Regarding the appropriate resolution in time and space we have to be aware of the following. The resolution must be fine enough, to detect meaningful differences in activities. The resolution must be coarse enough to average over random short-time fluctuations. (It is unlikely that we ever face the latter problem in a real ecological data set.) The resolution must be coherent over the wide range of spatial and temporal scales in an ecosystem. What is fine and coarse and what is coherent must be judged relative to the internal dynamics of the populations under consideration. I will still deal with this topic in some detail in Chapter Six.

We now have a tool to evaluate the functional contribution of the temporal, spatial and spatio-temporal variations of flow patterns in an ecological network. To foster an understanding of the meaning of the measures defined and to avoid confusion with other information-theoretic measures I will illustrate the characteristic behaviour with a variety of sample examples.

4.6 SPATIO-TEMPORAL ORGANIZATION IN SIMPLE MODELS

I will discuss two major sources for spatio-temporal organization in ecological networks. The first concerns spatio-temporal resource partitioning in ensembles of species sharing a common resource. The second is given by pulsed consumption in producer–consumer interactions. I will sketch the characteristic patterns of such interactions with simple conceptual models that should be looked upon as representing elements of more complex networks. The description of these isolated network elements is reduced to the utmost necessary for demonstrating some characteristic spatio-temporal flow patterns and the associated changes in the measures of spatio-temporal organization. Even when they can give us only a limited understanding of the nature and significance of spatio-temporal organization at the ecosystem level, this simplification facilitates the understanding of the behaviour of the measures introduced here. Due to working on a phenomenological base, these measures do not reflect how a certain pattern is generated. However, I claim that a network's spatio-temporal organization as quantified by these measures is indicative of a system's state of self-organization. Support for this statement will be given in Chapter Five by results obtained with a variety of multi-species models.

4.6.1 Spatio-temporal organization resulting from resource partitioning

Imagine terrestrial plant or phytoplankton species that share a common pool of a limiting resource and that differ in their period of activity over a seasonal cycle. The corresponding pattern of exchanges with the resource pool is depicted for two species in Figure 4.6. In the time-averaged representation shown in Figure 4.6a the two species are functionally redundant regarding their exchanges with the resource pool.

The situation changes when we account for the temporal activity pattern as shown in Figure 4.6b. The two species engage in temporal resource partitioning. The inputs into each compartment are confined to different time intervals, and the original redundancy is resolved. What appears as two separate redundant feedback cycles in the time-averaged representation reveals itself to be a feedback spiral linking the two species across time. The time-averaged annual network would not adequately reflect the real dynamic relationship between the species during the various seasons of a year. Figure 4.6c indicates that this temporal shift between resource utilization and recycling results in a positive feedback. Instead of competing for the resource the species assist each other in growth. One could obviously argue that they can only coexist because of avoiding competition by temporal resource partitioning. However, the presence of both may be vital for each of them to persist, for maintaining the functional state of the system. I will give ample examples for such mutual dependencies.

Experimental evidence for the pronounced seasonal pattern of phytoplankton species was summarized and discussed in detail by Reynolds (1984a, b). In recent publications Fitter (1986) and McKane *et al.* (1990) investigated the spatial and temporal patterns of root activity in species-rich nitrogen-limited grasslands. They discovered major spatio-temporal differences in the uptake of nitrogen. In both studies two main guilds of plants could be distinguished: early-active, shallow-rooted species, and late-active, deeper-rooted species. Figure 4.7, which was derived from McKane *et al.* (1990), shows the uptake of nitrogen of the two dominant species, *Poa pratensis* and *Schizachyrium scoparium*, as a function of depth and season. The nitrogen uptake is given as a percentage of the total uptake averaged over depth and season. *Poa* is mainly active in spring and in a

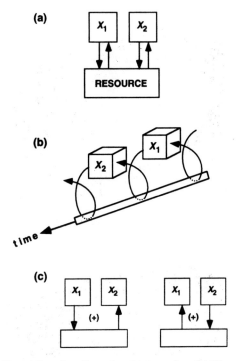

Figure 4.6   (a) X_1 and X_2 are functionally redundant species. (b) The activity of each species is confined to a different time interval, and the redundancy is resolved. The rectangle along the time axis denotes the resource pool. What appears as two separate feedback cycles in the time-averaged representation reveals itself to be a feedback spiral linking the two species across time. (c) The temporal shift in activities results in a positive feedback

horizon close to the surface. *Schizachyrium* has its main activity period in summer and prefers deeper horizons. McKane *et al.* (1990) showed with statistical methods that the seasonal differences were mainly responsible for resource partitioning and that the spatial differences were of less importance. These effects showed up clearly in the relative contributions of resource partitioning to the temporal and spatial organization which were quantified with the measures introduced above (Pahl-Wostl, 1991).

I use a simple conceptual model focusing on the resource utilization of two species ensembles to illustrate the behaviour of the measures of organizations. The dynamics of the interactions responsible for the generation of spatio-temporal variations are not described explicitly. Instead, I assume the nutrient inputs to vary periodically around a seasonal mean as represented in Figure 4.8a. Spatial variations are neglected for the moment.

The phases of temporal overlap in the activity of the two ensembles are shaded. The phaseshift φ determines the temporal overlap and hence the degree of functional redundancy with respect to the output-environment of the resource pool. The mathematical expressions describing this model are summarized in Box 4.5.

If we average the pattern depicted in Figure 4.8a over time, both compartments are functionally redundant and have the same share of the resource. Let us assume that we had recorded this temporal pattern by making repeated measurement every time step Δt

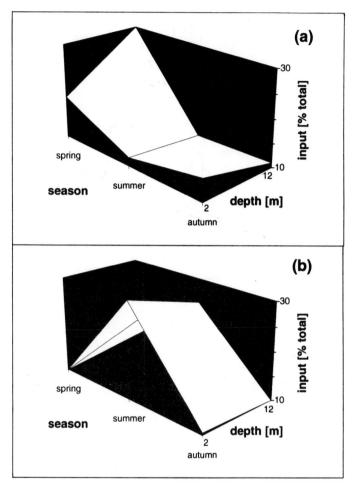

Figure 4.7   Input of nitrogen in percentage of the total input as a function of depth and season for (a) *Poa pratensis* and (b) *Schizachyrium scoparium*, the two dominant species in an old-field plant community. Data from McKane *et al.* (1990)

that we express as a fraction of the oscillation period. Figure 4.9 shows the resolution of the temporal pattern for two different sampling strategies: (a) coarse grained with a time step $\Delta t = 0.25$, and (b) fine grained with a time step $\Delta t = 0.0625$. Note that the temporal average is retained in both cases. More and more structure of the temporal pattern is revealed when we increase the resolution along the dimension of time.

The measure of temporal organization, I_t, was then determined for different levels of temporal resolution. In this simple model the maximum of I_t is given by $\log_2 2 = 1$. This maximum is obtained when the temporal overlap between the two flows equals zero. Figure 4.10 shows ΔI_t ($\Delta I_t = I_t - I = I_t$ since here $I = 0$) as a function of the resolution time step. ΔI_t reflects the progressive accounting for the temporal flow pattern with increasing resolution. If the time step chosen equals the length of the observation period the configuration corresponds to the time-averaged network. ΔI_t was calculated for three different phaseshifts given at the corresponding curves. The time step of resolution and the

---

Box 4.5   Sine model with two compartments

(a) Variations of the nutrient inputs in time

$$In_1 = <In>\{1 + \sin(\omega\,t)\} \tag{4.15a}$$

$$In_2 = <In>\{1 + \sin(\omega\,t - a\varphi)\} \tag{4.15b}$$

The temporal overlap of the inputs can be quantified as:

$$OvL_t = \frac{\int_0^{2\pi} In_1 In_2 dt'}{\int_0^{2\pi} In_1^2 dt'} = \frac{2 + \cos\varphi}{3} \tag{4.16}$$

where $t' = t/\omega$, and $<In>$ was set equal to 1.

The overlap is maximal when $\varphi = 0$, and minimal when $\varphi = \pi$.

(b) Variations of the nutrient inputs in time and space

This model may be given a spatial dimension by assuming the inputs to be functions of both time and space:

$$In_1 = <In>\{1 + \sin(\omega_t t)\}\,\{(1 + \sin(\omega_s\,s)\} \tag{4.17a}$$

$$In_2 = <In>\,\{1 + \sin(\omega_t t - a\varphi_t)\}\,\{(1 + \sin(\omega_s\,s - a\varphi_s)\} \tag{4.17b}$$

where s denotes the spatial coordinate. The overlap along the temporal dimension is determined by the phaseshift φ_t, and that along the spatial dimension by the phaseshift φ_s. The total overlap along both dimensions can be quantified as:

$$OvL_{ts} = \frac{\int_0^{2\pi}\int_0^{2\pi} In_1 In_2 dt' ds'}{\int_0^{2\pi}\int_0^{2\pi} In_1^2 dt' ds'} = \frac{2 + \cos\varphi_t)(2 + \cos\varphi_s)}{9} \tag{4.18}$$

where $t' = t/\omega_t$, $s' = s/\omega_s$, and $<In>$ is set equal to 1.
Since $\cos\varphi \le 1$ it follows that $OvL_{ts} \le OvL_t$.

---

phaseshifts are expressed in fractions of the oscillation period. ΔI_t converges to a limit with increasing temporal resolution once the reduction in functional redundancy has been resolved.

A phaseshift of 0.5 corresponds to the situation represented in Figure 4.8a where the temporal overlap and hence the functional redundancy of the two inputs are at minimum. For a phaseshift of zero the two inputs oscillate in phase and the temporal variation of the flows does not lead to a reduction in functional redundancy. Even when the amplitudes of the flows fluctuate over time both flows always have the same magnitude. As a consequence, the output-environment of the resource pool always exhibits the same structural pattern.

Due to the sine shape chosen for the temporal variation of the flows, the temporal overlap between the two inputs cannot become zero. Figure 4.8a depicts the optimal situation with respect to temporal resource partitioning. The residual functional redundancy can be further reduced by a spatial resource partitioning as in the example discussed before where the nitrogen uptake of the plant species differed in season and depth. To demonstrate such an effect let us assume the inputs in our simple model to exhibit additional periodic variations along a dimension of space. The resulting spatio-temporal patterns are represented in Figures 4.8b and c for an equal phaseshift of 0.5 along both time (φ_t) and space (φ_s). The overlap in activities is now determined by the combinations of time (φ_t) and space (φ_s) (cf. Eqn (4.18) in Box 4.5).

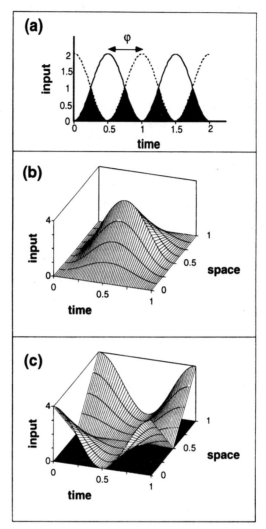

Figure 4.8   Periodic variations of the nutrient inputs. (a) In time for both compartments.
Compartment 1, full curve, compartment 2, dashed curve. Phases of temporal overlap in the activity
of the two compartments are shaded. (b) In space and time for compartment 1. (c) In space and time
for compartment 2. The units refer to fractions of the oscillation period in time and space, respectively

To investigate the combined effect of spatial and temporal variations, I increased the
phaseshifts stepwise from 0 to 0.5 assuming thereby equal phaseshifts along time and
space ($\varphi_t = \varphi_s$). Figure 4.11 shows the results obtained for the measures of organization as
a function of the phaseshift. ΔI_t was obtained for the activity pattern averaged over space
resolved along time, ΔI_s was obtained for the activity pattern averaged over time resolved
along space, and ΔI_{ts} was obtained for the activity pattern resolved along both time and
space (cf. Eqn 4.14). We note that the sum of the increase in organization when accounting
for the spatial and temporal patterns separately ($\Delta I_t + \Delta I_s$) exceeds the increase when they
are accounted for simultaneously (ΔI_{ts}). This result can also be formally derived from the

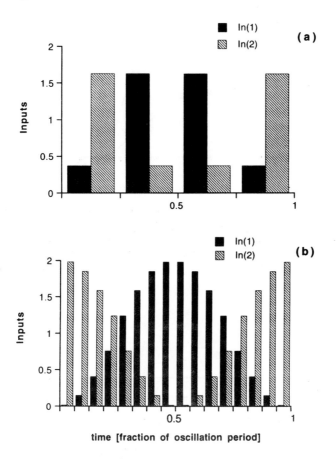

Figure 4.9 Flow pattern depicted in Figure 4.8a sampled with two different time steps of resolution, Δt: (a) coarse grained – $\Delta t = 0.25$ and (b) fine grained – $\Delta t = 0.0625$. Δt is expressed in fraction of the oscillation period. The scheme of resolution was chosen to match with the oscillation periods

mathematical expressions as summarized in Box 4.6.

On the one hand, such behaviour is due to the fact that the temporal and spatial flow patterns were chosen to be equivalent with respect to the change in the functional redundancy. Resource partitioning along an additional dimension reduces the same fraction of the residual functional redundancy. On the other hand the flow patterns along time and space were chosen to be independent. A look at Figures 4.8b and c shows us that the spatial distributions of the compartmental activities are time invariant. The maxima and minima are always at the same spatial location even if their amplitudes fluctuate over time. Such a spatio-temporal pattern may be encountered for sessile plants when we consider periods of observation that are smaller than the typical time scales of successional changes. The situation is different in open water bodies where the plants are not restricted to a fixed spatial location. I will discuss below, in more detail, different possible relationships between the patterns in space and time. Before doing so I am going to introduce pulsed consumption, another major source of spatio-temporal organization.

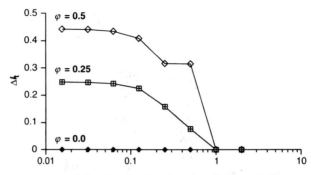

resolution time step [fraction of oscillation period]

Figure 4.10 ΔI_t as a function of the resolution time step obtained for different phaseshifts φ that are given in fraction of the oscillation period. (The irregularity in the curve for $\varphi = 0.5$ when going from a resolution of 0.5 to 0.25 of the period length is due to the fact that not very much structure is delineated in this step when the scheme of resolution is chosen to match with the oscillation periods. In both cases the resolved points are centred close the maxima or minima of the sine curves)

Box 4.6   Reduction in redundancy effected by temporal and/or spatial patterns

A reduction in redundancy leads to an increase in organization, which can be expressed as:

$$\Delta I_t = I_t - I = \frac{1}{H(A,B) - R_t} = \frac{1}{H(A,B) - f_t R} \tag{4.19}$$

$$\Delta I_s = I_s - I = \frac{1}{H(A,B) - R_s} = \frac{1}{H(A,B) - f_s R} \tag{4.20}$$

where $R_x = H(A,B) - I_x$, $x = $ t, s, ts, and f_t, f_s are fractions of the redundancy in the averaged network, R, that remain after resolving flows along the dimension of time or space, respectively.

These fractions depend on the phaseshift between the input functions of the two compartments. Due to the equivalence of the spatial and temporal dimension, each dimension contributes the same to the reduction of the systems overhead, i.e. if $\varphi_t = \varphi_s$ then $f_t = f_s$.

Resolving both dimensions results in:

$$\Delta I_{ts} = I_{ts} - I = \frac{1}{H(A,B) - R_{ts}} = \frac{1}{H(A,B) - f_t f_s R} \tag{4.21}$$

and using Eqns (4.19), (4.20) and (4.21) one can derive:

$$\frac{\Delta I_{ts}}{\Delta I_t + \Delta I_s} = \frac{1 - f_t f_s}{2 - (f_t + f_s)} \leq 1 \tag{4.22}$$

The ratio defined in Eqn (4.22) may take any value between 1 (for $f_t = f_s = 1$, the limiting case where resolution does not provide any reduction in overhead) and 0.5 (for $f_t = f_s = 0$, the limiting case where resolution along either dimension reduces the overhead to zero).

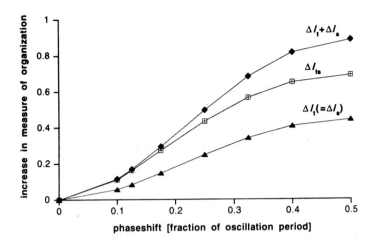

Figure 4.11   Increase in the measures of spatial in the measures of spatial ($\Delta I_{\rm s}$), temporal ($\Delta I_{\rm t}$), and spatio-temporal ($\Delta I_{\rm ts}$) organization as a function of the phaseshift that was chosen to be equal in time and space – $\varphi_{\rm t} = \varphi_{\rm s}$

4.6.2   Spatio-temporal organization resulting from pulsed consumption

In the previous examples, the source for spatio-temporal organization was given by a spatio-temporal sharing of the same functional niche. Ensembles of primary producers were segregated along the dimensions of time and/or space in their utilizing a common resource pool. Having a food web in mind we may refer to such a pattern as horizontal because it involves ensembles of species at a similar trophic position. Spatio-temporal organization may also derive from vertical organization in producer–consumer interactions. In this case the source for organization is given by a switching between functional states, by a spatio-temporal segregation of production and consumption. Figure 4.12 summarizes the essential elements of such a segregation in a model comprising a predator–prey pair and a nutrient pool. It suggests identification of the prey with a primary producer. We could, however, conceive of the "prey" to represent as well a sub-module of a food web comprising an ensemble of species on which a predator at a higher trophic position depends.

Figure 4.12a represents the network of all nutrient flows in a time-invariant steady state situation. However, in natural systems we hardly ever encounter an equilibrium between production and consumption. Recently, Seasteadt and Knapp (1993) summarized experimental evidence for the importance of nonequilibrium resource availability in both terrestrial and aquatic systems. They concluded with the hypothesis that the productivity in many ecosystems was greatly enhanced by their being in nonequilibrium states of transition rather than close to an equilibrium point. Such a point of view was also supported by Ruess and Seagle (1994) in their study of landscape patterns in the Serengeti National Park. They found that herbivores tracked plant growth, which was highly variable both temporally and spatially, and in doing so profoundly impact nutrient cycling processes. Such a pattern corresponds to the situation depicted in Figure 4.12b, which shows the temporal succession of a predator–prey–nutrient cycle resulting from producer–consumer pulses: a buildup of the prey's biomass depleting thereby the nutrient

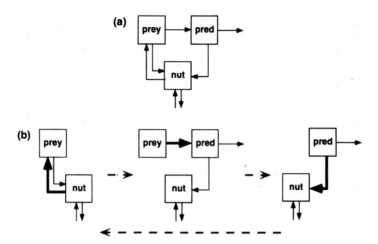

Figure 4.12   Nutrient flows between a predator–prey pair and the nutrient pool. (a) Flow pattern averaged over time and space. (b) Possible temporal succession of different functional states

pool→a buildup of the predator's biomass depleting thereby the prey's biomass→a decay of the predator's biomass and a buildup of the nutrient pool. Such a segregation of functional states may occur along the dimension of time and/or space. Another example for the simultaneous temporal and spatial segregation is given by the diurnal vertical migration of zooplankton in lakes. Geller (1986) showed, for example, that in Lake Constance certain species stay far below the lake surface during the day, whereas at night they migrate closer to the surface to feed on phytoplankton biomass that was built up during the day. To illustrate now the characteristic spatio-temporal patterns and the associated behaviour of the measures of organization, I intend to limit the attention at this first stage to an isolated network element. However, we can do real justice to this phenomenon only when at a later stage we look at it integrated into a network context.

Most ecologists may be familiar with the Lotka–Volterra predator–prey model that is known to generate predator–prey cycles. It is frequently used in spite of its displaying a number of undesirable properties, for example, the shape and amplitude of the cycles depend on the initial conditions chosen. Essentially, the Lotka–Volterra model behaves like an ideal pendulum without friction. Having been pushed into motion such a pendulum oscillates for ever, preserving thereby the energy initially received. Similarly, the Lotka–Volterra model behaves like a closed system where the total biomass that is determined by the initial conditions oscillates between predator and prey. To illustrate pulsed consumption, I will use here another model, the Brusselator, which received its name due to being the favourite model of the Brussel's school centred around Prigogine (c.f. Nicolis and Prigogine, 1977). The Brusselator may be called the "E-coli" of the modelling community, being interested in self-organization where the patterns in time and space are generated by internal system dynamics in a uniform environment. The Brusselator is a prototype for models describing the generation of so-called dissipative structures in space and time. All living systems are dissipative structures that regenerate themselves continuously by degrading high quality energy into heat. In the Brusselator model, frequency, amplitude and wavelength are all determined by the system and its characteristic spatial dimensions, independently of the initial conditions chosen. This is a characteristic of dissipative systems. In this respect the Brusselator model is definitely closer to real ecological processes than the often used Lotka–Volterra equations. The dissipative nature endows the system with remarkable self-regulatory properties that shape its internal spatio-temporal structures, and we are clearly justified in talking about "spatio-temporal organization".

The model that comprises two variables was originally derived as a simplified description of the essential characteristics of a more complex biochemical reaction involving five constituents. The mathematical expressions are summarized in Box 4.7. A diagram that includes all flows is depicted in Figure 4.13a. The model may be looked upon as consisting of a linear throughflow and a positive feedback loop. As shown in Figure 4.14, the source for the spatio-temporal patterns exhibited by the Brusselator are the switching between the functional modes of storing and pulsing.

Richardson and Odum (1981) used models similar to the Brusselator to describe pulsed consumption in ecological systems. Their models were more complicated involving at least a resource, a producer and a consumer. The dynamic behaviour, however, was also characterized by the alternation of the modes of storing and pulsed consumption, as in the case of the Brusselator. To help the imagination the Brusselator may be thought of as an aggregated representation of a three-compartment model where X corresponds to the aggregated compartments of consumer and resource, and Y corresponds to the producer (cf. Figure 4.13b). In comparison with a more complex but ecologically more realistic model, the Brusselator has the major advantage of being one of the best-studied so-called

---

Box 4.7   Equations of the one-dimensional spatial Brusselator model

$$\frac{\partial X}{\partial s^2} = a - (b+1)X + X^2Y + D_1\frac{\partial^2 X}{\partial s^2} \qquad (4.23)$$

$$\frac{\partial Y}{\partial t} = bX - X^2Y + D_2\frac{\partial^2 Y}{\partial s^2} \qquad (0 \le s \le L) \qquad (4.24)$$

The Brusselator admits a uniform steady state solution, $X_0 = a$ and $Y_0 = b/a$. Beyond the instability point of the steady state this reaction–diffusion model gives rise to a variety of spatial and temporal patterns. The particular behaviour encountered depends on the geometry of the system, parameter values and initial and boundary conditions. Here, the focus will be on the functional changes in a one-dimensional system for the boundary conditions:

(a) zero flux at boundaries:

$$\frac{\partial}{\partial s}X(0,t) = \frac{\partial}{\partial s}X(L,t) = \frac{\partial}{\partial s}Y(0,t) = \frac{\partial}{\partial s}Y(L,t) = 0 \qquad (4.25a)$$

(b) Dirichlet $\equiv$ fixed concentrations at boundaries:

$$X(0,t) = X(L,t) = Y(0,t) = Y(L,t) = b/a \qquad (4.25b)$$

Whether a bifurcating solution is periodic or monotonic is dictated by whether the dominant eigenvalue is complex or real, respectively. The characteristics of the spatial pattern depend on the critical wavenumber at which a mode becomes unstable. For details of the mathematical treatment readers are referred to Auchmuty and Nicolis (1976), Herschkowitz-Kaufman (1975) and Nicolis and Prigogine (1977).

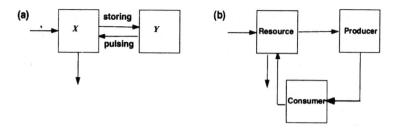

Figure 4.13   (a) Flow diagram of the "Brusselator" model that may be thought of as an aggregated representation of the model in (b) consisting of a producer, a consumer and a resource pool

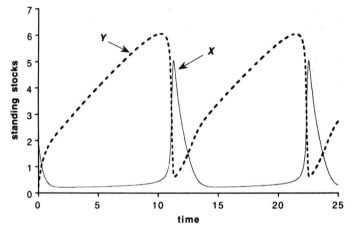

Figure 4.14   Standing stocks in the Brusselator model as a function of time obtained for no flux boundary conditions. Parameter values: $a = 0.75$ and $b = 3$, $D_1 = 0.008$, $D_2 = 0.004$. (Reprinted from Pahl-Wostl, 1992a, with kind permission from Elsevier Science Ltd, The Boulevard, Langford Lane, Kidlington, OX5 1GB, UK)

reaction–diffusion models (Haken, 1983; Nicolis and Prigogine, 1977). Therefore the characteristics of its spatial and temporal behaviour are well documented, making it thus an ideal tool for investigations of the type we are interested in here.

To give the Brusselator model a spatial dimension we assume that X and Y are free to move along one spatial dimension. More specifically we describe the movement as random diffusion, the speed of which is determined by the diffusion constants D_1 and D_2, respectively. In contrast to the previous model of resource partitioning, the spatial pattern is thus not fixed, but depends on the movements of X and Y. Diffusion was originally derived for molecular processes to describe the random movement of molecular particles. Meanwhile diffusion models have also found widespread application in ecology to describe, for example, active random dispersal or the passive displacement of planktonic organisms in turbulent aquatic environments (review by Levin, 1986; Okubo, 1980). Such models consist of a set of reaction–diffusion equations where the state of a species is characterized by its abundance at each spatial location. Population abundance is not simply an average quantity but it has a spatial structure. On the one hand, the change in population abundance at a specific spatial location depends on the local inter- and intraspecific interactions a species engages in. On the other hand, it depends on the population abundances in the neighbourhood that determine the direction and magnitude of spatial exchanges. Hence, the spatio-temporal patterns in such reaction–diffusion models arise from the combination of biological interactions and spatial movement. If the movement is very fast in comparison to the interactions, any spatial pattern is homogenized. If the movement is very slow in comparison to the interactions, the spatial patterns are time invariant. A rich variety of patterns are possible when the time scales of movements and reactions are within a similar range.

Giving a model a spatial dimension implies that we have to define an interval or area of interest that we wish to describe. We must account for the characteristics of spatial movement within this spatial unit as well as for exchanges with the environment across the boundaries. Exchanges across the boundaries are specified by the so-called boundary conditions, the two most common types of which are referred to as no or zero flux and Dirichlet, respectively (cf. Eqn 4.25a and b). No flux boundary conditions describe a system without movement across the spatial boundaries. We may think of an oasis in a desert or an island in the ocean as examples approaching closely such a description. No movement across the boundaries does not imply that such a system is closed with respect to material exchanges with the environment. In the case of an island, for example, we could conceive of an import of nutrients via precipitation that is distributed more or less homogenously over the island's surface. In the case of Dirichlet boundary conditions the model system may be thought of as being embedded in a large environmental reservoir. Hence the population densities and the nutrient concentrations at the spatial boundaries correspond to those in the environment and there may be intense exchange with the environment across the boundaries. We can also conceive of different conditions for different species. In a forest, for example, some species may easily disperse across the boundary, others may be restricted to remain within. In another situation we may have to choose different boundary conditions for different spatial environments. If we make, for example, a model for the euphotic zone of a pelagic system with the spatial dimension equal to depth, the boundary at the surface has definitely different characteristics to the boundary with the deep water body. Combining characteristics of movement, boundary conditions and species interactions, we may come up with a wealth of models displaying a rich variety of behaviour.

After this interlude let us return to the Brusselator model. Using this model we may generate a large variety of different types of spatio-temporal patterns by varying the parameter values and the boundary conditions. I investigated for a representative set of patterns the spatio-temporal organization associated with a temporal and/or spatial segregation of pulsing and consumption. The pulsed consumption represented in Figure 4.14 corresponds to a pattern that varies periodically in time but is homogenous and invariant in space. We observe thus only an organization in time but not in space. In analogy, one may choose different conditions to obtain a time-invariant pattern where the activities of pulsing and consumption are confined to different spatial locations. Thus in this case, we observe an organization only in space but not in time. A more detailed description of these patterns and the associated changes in the measures of spatio-temporal organization is given in Pahl-Wostl (1992a). Let us focus here on the most interesting case when the patterns vary in both time and space. In this case the spatial pattern is time dependent. As depicted in Figure 4.15 we obtain consumer pulses travelling over the spatial dimension. A mathematical explanation for the patterns observed is given in Box 4.8.

What are the relative contributions of the temporal and spatial patterns to

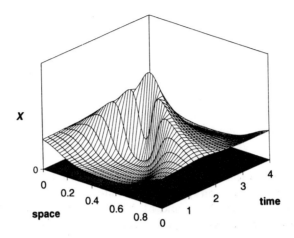

X

space   time

Figure 4.15   A consumer pulse, X, as a function of time and space. Parameter values: $a = 2.0$; $L = 1.0$; $D_1 = 0.008$, $D_2 = 0.004$; $b = 6$; $b_{crit} = 5.1$. The spatial dimension was divided in 100 elements. The initial conditions chosen were $i = 1,2 \ldots , 100$, $Y_i = 1.5$; $X_i = a$. Dirichlet boundary conditions

---

Box 4.8   Explanation of spatio-temporal patterns observed

Under certain conditions the local temporal oscillations become coupled via diffusion and can produce synchronized oscillations. The pattern is reasonably well explained by assuming a superposition of standing waves. The superposition of different oscillations give rise to propagating fronts, which subsist, however, only during part of the overall interval. This front becomes obvious in the expression for the time-periodic solutions for x, the deviation of X from the steady state:

$$x(t, s) = \varepsilon \cos(\omega t)\sin(k\pi r) + \varepsilon^2 \sum_{k=1} \{a_{k+}b_k \cos(\omega t + \psi_k)\}\sin(k\pi r) \qquad (4.26)$$

where $r = s/L$.

Eqn (4.26) was obtained using nonlinear approximations based on perturbation theory. A similar relation holds for y. For details of the derivation of Eqn (4.26) and the meaning of the parameters therein, the reader is referred to Auchmuty and Nicolis (1976).

---

spatio-temporal organization now? To study the relationship between the dimensions of space and time it is quite illustrative to compare what happens when we average over either dimension. Figure 4.16 shows X_t, the spatial average of X, as a function of time, and X_s, the time average of X, as a function of space. Both are given as a fraction of the maximum obtained for X_t. Note the difference in scale between X_t and X_s. Whereas the temporal pulse is clearly maintained in the spatial average, the spatial pattern has largely disappeared in the time average.

This is also reflected in the behaviour of the measures of organization represented in Figure 4.17 as a function of b, the growth rate of the storage. The parameter b has a major influence on the nature of the spatio-temporal pattern. A comparison with Figure 4.11 reveals that b effects the spatio-temporal segregation between production and consumption

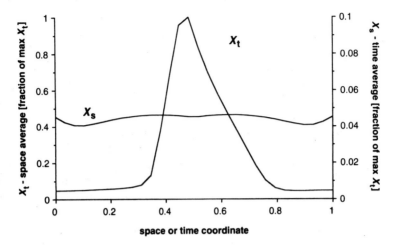

Figure 4.16 $X_t \equiv$ average over space, as a function of the time coordinate t, and $X_s \equiv$ average over time as a function of the spatial coordinate s. t is given in fraction of the oscillation period, s is given in fraction of the length of the total spatial dimension. Parameter values as for Figure 4.15. (Reprinted from Pahl-Wostl, 1992a, with kind permission from Elsevier Science Ltd, The Boulevard, Langford Lane, Kidlington OX5 1GB, UK)

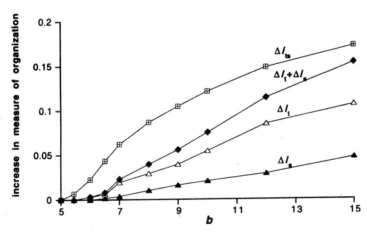

Figure 4.17 Increase in the measures of spatial (ΔI_s), temporal (ΔI_t), and spatio-temporal (ΔI_{ts}) organization as a function of b. Parameter values as for Figure 4.15

in a similar fashion to the way the phaseshift in the previous example determined the spatio-temporal resource partitioning.

The changes in the measures of spatial and/or temporal organization are again expressed as deviation from the measure obtained for the space- and time-averaged system (cf. Eqn 4.14). As expected after inspection of Figure 4.16, the contribution along the dimension of time largely exceeds the contribution along space ($\Delta I_t > > \Delta I_s$). We note that the sum of the increase in organization when accounting for the spatial and temporal patterns separately ($\Delta I_t + \Delta I_s$) is less than the increase when they are accounted for simultaneously (ΔI_{ts}). This means that the contributions of the spatial and temporal dimensions are not simply additive but they are mutually dependent. Such behaviour contrasts with the model for resource partitioning studied in the previous section (cf. Figure 4.11). Owing to the independence of the dimensions of time and space, we observed temporal variations of a fixed spatial pattern. Comparisons of Figure 4.15 with Figure 4.8b and of Eqn (4.26) with Eqn (4.17) reveal the different relationships between the spatial and temporal dimensions in the two situations. Averaging over either dimension in the sine model still preserves the pattern in the other dimension; space and time are equivalent and independent. Figure 4.16 shows that for the Brusselator the temporally averaged deviations in space are only minor corrections to the steady state. The functional significance of the spatial pattern with respect to a segregation between production and consumption depends largely on the variation in time. In such a case neither the temporal pattern nor the spatial pattern investigated in isolation reveals the real functional significance. Only a simultaneous consideration of both time and space gives us a full picture of the spatio-temporal organization. Similar conclusions can be drawn from another model of spatio-temporal resource partitioning where the source of spatio-temporal organization is given by a complex pattern of variation in the physical environment.

4.7   SPATIO-TEMPORAL ORGANIZATION MEDIATED BY ENVIRONMENTAL VARIABLES

The next model illustrates how variations in the physical environment can foster the spatio-temporal organization of a species community. Contrasting with the fixed spatial pattern of vegetation in terrestrial systems, all organisms including primary producers are subject to spatial transport in open water bodies. Results from a model of a pelagic system demonstrate that complex spatio-temporal patterns may easily arise in a seemingly simple system. The patterns are generated by the combined effect of spatio-temporal variations in light intensity, spatial movement, and the growth characteristics of phytoplankton species. The model comprises three phytoplankton species and one herbivorous consumer preying on all three species indiscriminately. Hence, the phytoplankton species are functionally redundant both with respect to their role as primary producers and with respect to their serving as prey for the same consumer. The model includes a spatial dimension equivalent to depth that is resolved into discrete layers. Movement may be present due to an exchange of biomass between adjacent depth layers, the intensity of which is determined by the exchange rate γ. We account thus only for passive transport by physical mixing processes. Model equations and patermeters are listed in Box 4.9.

The intensity of spatial movement determines whether spatial organization is possible or not. Figure 4.18 contrasts two extreme situations for the euphotic zone of a pelagic system that extends from the surface to the depth where growth is prevented by light deficiency. (a) depicts the situation for a stably stratified water column whereas (b) represents the situation for full turbulent mixing. The profiles for light and temperature indicate what we would measure as a function of depth. In both cases the light intensity decays exponentially with depth. In (a) the temperature gradient shows us that there is little exchange of water in the vertical dimension. Since the phytoplankton species may remain at a fixed depth we may actually talk about the presence of a spatial dimension

---

Box 4.9   Equations and parameters for the simulation model

All light intensities are expressed in fractions of the annual mean light intensity at the surface, time is expressed in fractions of the inverse of the maximal growth rate equivalent to the minimal turnover time of the phytoplankton species, biomass is expressed in fractions of the half saturation rate of grazing.

Change of the biomass of phytoplankton species i in depth z, $P_i(z)$:

$$\frac{\partial P_i(z)}{\partial t} = (g_i(z) - r_i(z))P_i(z) - G_i(z) - D\frac{\partial^2 P_i(z)}{\partial z^2} \qquad (4.27)$$

where

z = depth incremented into discrete layers of extension $L = 1.1$ m

$i = 1, 2, 3$ index of phytoplankton species

$g_i(z) = Y\, e^{\,[-\varepsilon z - Ye^{-\varepsilon z} + 1]}$ (light-dependent growth rate), where $Y = \dfrac{H(z)}{H_i}$

$r_i(z) = k_R(1.5 g_a(z) + 1)$ (respiration)

$G_i(z) = G_{max}\, \dfrac{P_i(z)}{1 + \Sigma P_i(z)}\, P_i(z)\, C(z)$ (losses due to grazing)

Change of consumer biomass in depth z, $C(z)$:

$$\frac{dC(z)}{dt} = \Sigma G_i(z)\,(1 - 2k_R) - (k_R + k_L)C(z) \qquad (4.28)$$

Variation of light and values of parameters:

$H(z) = H_0\, e^{-\varepsilon z}$
$H_0 = H_{mean} + H_{amp}\, \sin(\omega t)$

where $\omega = 2\pi/360$, $H_{mean} = 1$ and $H_{amp} = 0.8333$

H_i = light intensity for maximal growth rate of species P_i, $H_i = 1.3333$; $H_2 = 0.333$; $H_3 = 0.08333$
G_{max} = 0.05, maximal grazing rate
k_R = 0.05, rate of respiratory losses
k_L = 0.2, rate of external losses
γ = {0, 0.1, 1, 10}, rate of exchange between adjacent layers = D/L^2
ε = 0.33, extinction per depth layer

---

regarding community organization. The different shadings indicate a spatial segregation of different species assemblages. In (b) the absence of a temperature gradient reveals that mixing leads to a homogenization of any spatial pattern. All species experience the same light climate when we assume that they behave like passive particles and that they are transported in a similar fashion by the water movements. In this case, the spatial dimension does not contribute to organization.

In our model, the source for spatio-temporal variability is given by the seasonal changes in insolation together with the exponential decay of the light intensity as a function of depth. In combination with spatial movement we may conceive of a range of complex light climates thus arising. These variations in light intensity mediate a spatio-temporal resource partitioning of the three phytoplankton species, the growth rates of which vary as

(a)   **(b)**

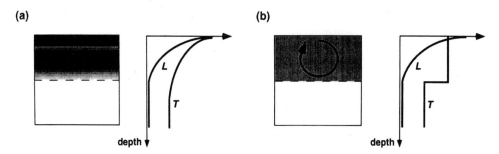

Figure 4.18 Euphotic zone of a lake during (a) stratification and (b) turbulent mixing. During stratification a spatial segregation of different species ensembles may be present as indicated by the different shadings. The profiles for light, L, and temperature, T, depict schematically what one would measure as a function of depth

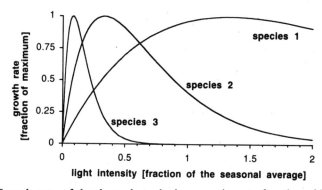

Figure 4.19 Growth rates of the three phytoplankton species as a function of light intensity

a function of light intensity. Figure 4.19 shows the relationships between growth rate and light intensity for the three phytoplankton species. Light intensities are expressed as fractions of the seasonal average of insolation. At first sight we might expect species 1 to be the dominant species because its light function covers just the range of variations displayed by the seasonal changes in insolation. However, the situation is more complex because we have to take into account the full spatio-temporal range of light intensities.

In a first simulation run I prevented spatial movement by choosing the exchange rate γ equal to zero. As in Figure 4.18a the phytoplankton populations thus remain at fixed depths in separate depth layers. In this case the light climate experienced by the algae corresponds to the light intensity that we would measure *in situ*. Figure 4.20a shows the light intensity as a function of time and depth resulting from the seasonal variations in combination with the spatial gradient. The primary production obtained for the three phytoplankton species as a function of time and depth is shown in Figures 4.20b, c and d, respectively. We can clearly discern the pattern induced by the seasonal changes in light availability. The extremes of the light climate are encountered close to the surface and at maximum depth. The medium layers, characterized by a broader range of light intensities,

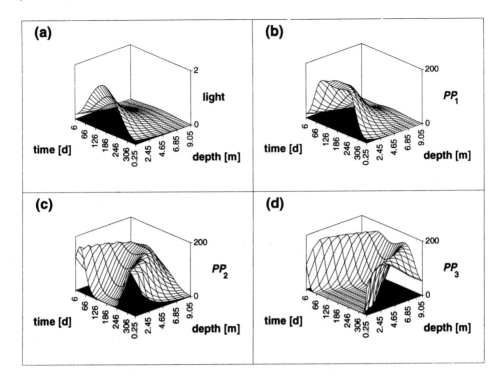

Figure 4.20   (a) Light intensity (b, c, d) primary production PP_i of phytoplankton species i, $i = 1$, 2, 3, as a function of time and depth ($\gamma = 0$)

allow the growth of all three species during different phases of the seasonal cycle. Contrary to our initial expectations, species 1 has, with 23%, the smallest share of the total primary production. Species 2, having intermediate light requirements, holds the highest share of 42%, whereas species 3 has a share of 35%.

To investigate the effects of spatial movement, I performed further simulations with different exchange rates. In any real system the mixing regime is also subject to seasonal variations. The intensity and frequency of mixing events and the presence of stratification depend on a system's morphological characteristics and on the local meteorological conditions. To avoid excessive complexity I did not account for such seasonal changes and assumed the exchange rate to be constant over a seasonal cycle. Figure 4.21 represents the changes in the measures of spatial and/or temporal organization (cf. Eqn 4.14) as a function of the exchange rate γ. For an exchange rate of zero the contribution of the spatial dimension exceeds the one along time. As in the Brusselator model, the combined effect of spatio-temporal organization (ΔI_{ts}) largely exceeds the sum of the organization along the single dimensions ($\Delta I_t + \Delta I_s$). Again this derives from the fact that the spatial distribution of the species' activities is time dependent in contrast to the fixed pattern of our first model of resource partitioning (cf. Figure 4.11). We note further that the spatio-temporal organization (ΔI_{ts}) declines with increasing exchange rate. This has to be attributed to the decline in spatial organization (ΔI_s) caused by the exchange counteracting the spatial pattern imposed by the light gradient (cf. Figure 4.18b). Regarding the temporal organization (ΔI_t), the seasonal variation dominates over the homogenizing effect of

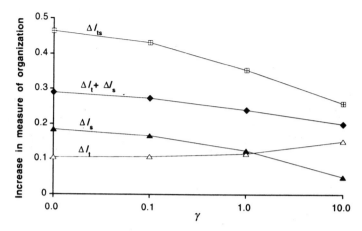

Figure 4.21   Increase in the measures of spatial (ΔI_s), temporal (ΔI_t), and spatio-temporal (ΔI_{ts}) organization as a function of the exchange rate γ

mixing. With increasing exchange rate, the relative contributions of the spatial and temporal dimensions are reversed. We note again that the temporal and the spatial patterns are intricately linked. Both have to be considered simultaneously to understand their functional significance.

To clarify the meaning of the results obtained for the varying contributions of temporal and/or spatial organization, and to show their relevance for different types of ecosystems, let us now discuss the relationship between the spatial and temporal dimensions in a more systematic fashion.

4.8   RELATIONSHIP BETWEEN PATTERNS ALONG THE DIMENSIONS OF TIME AND SPACE

Temporal and spatial patterns and their mutual dependence are important characteristics of ecosystem organization (e.g. Levin *et al.*, 1993). Even when the entire scope of their functional significance remains to be explored, it seems to be evident that spatio-temporal patterns are essential for maintaining species coexistence and biodiversity (comprehensive discussions in Schulze and Mooney, 1993a). In the models introduced earlier, we encountered different ways in which patterns along the dimensions of space and time may be related. The example of spatio-temporal resource partitioning serves as base for discussing in a more systematic fashion the various possibilities that may arise. More explicitly, I am going to use the simple conceptual model of Figure 4.6 comprising two functionally redundant primary producers where spatio-temporal organization results from a segregation of the species' activities along the dimensions of time (t) and/or space (s). Figure 4.22 illustrates a variety of possible spatio-temporal activity patterns. I designed these patterns such that all are equivalent when being averaged and when being resolved over both time and space simultaneously. However, the patterns differ in the relative contributions of the temporal and spatial dimensions and their mutual relationship.

To simplify the representation I assumed that only one species was active at a certain location at a certain time, and further that the spatial activity pattern can be represented along a single dimension. Four intervals are distinguished along both time and space. The fact that a species is growing and utilizing the resource pool is indicated by either a light (species 1) or a dark (species 2) shaded pattern. A blank space signifies the absence of activity of either species. An absence of activity may be due to the fact that no species is present or that a species is present but that it is in a nonactive state. The patterns chosen represent extreme cases regarding their regularity and the complete segregation of the species' activities. Even when we can hardly expect to encounter exactly the same patterns in real systems, we may find at least qualitatively similar behaviours. When making comparisons with real systems, I thus refer to the overall characteristics of a pattern determined by the importance of spatial and temporal patterns and their mutual relationship.

When averaged over space and time, both species have an equal share of the total activity in all the patterns depicted in Figure 4.22. I, the measure of organization in the reference state averaged over space and time, is therefore zero in all cases. Resolved along space and time a complete segregation of the activities is achieved for all patterns. I_{ts}, the measure of spatio-temporal organization, is therefore equal to $\log_2 2 = 1$ in all cases. The patterns (a) to (d) depict the four extreme distributions whereby these limiting states may be realized. This becomes evident in the behaviour of the measures of spatio-temporal organization listed in Box 4.10.

Figure 4.23 helps to clarify this point by showing the spatial and temporal averages of the activity patterns in Figure 4.22. The left column shows the spatial averages as a function of time whereas the right column shows the temporal averages as a function of space. I discuss now the implications for spatio-temporal organization with reference to examples from real systems. In the following the labels (a) to (f) refer to the patterns depicted in Figures 4.22 and 4.23.

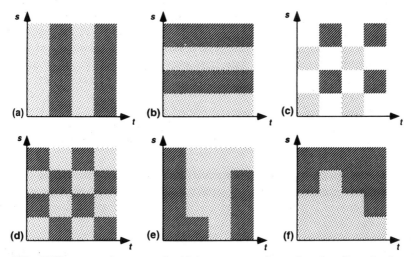

Figure 4.22 Different spatio-temporal activity patterns of two functionally redundant species indicated by the light (species 1) and dark (species 2) shadings. It is assumed that only one species is active at a certain time and location. A blank space denotes the absence of activity of either species. Further explanation in the text

Box 4.10   Measures of organization derived for the patterns
depicted in Figure 4.22

	I	ΔI_t	ΔI_s	ΔI_{ts}	$\Delta I_t + \Delta I_s$
(a)	0	1	0	1	1
(b)	0	0	1	1	1
(c)	0	1	1	1	2
(d)	0	0	0	1	0
(e)	0	0.6	0.1	1	0.7
(f)	0	0.1	0.6	1	0.7

In (a) the species' activities are segregated along time, but not along space. Averaged over time, the activities of both species are evenly distributed over space. Averaged over space, the temporal pattern is retained. Hence, resolution along the spatial dimension alone does not lead to a reduction in redundancy. Such behaviour is expressed in the measures of organization as $\Delta I_{ts} = \Delta I_t = 1$ and $\Delta I_s = 0$.

As an example for such a type of behaviour we may conceive of a seasonal succession of phytoplankton assemblages in a pelagic system under conditions of intense mixing preventing spatial organization. Such seasonal successions have been documented for a variety of temperate lakes (e.g. Reynolds, 1984b; Sommer, 1985).

(b) represents a situation in reverse to the one depicted in (a). Now we have fixed spatial pattern where the species' activities are segregated along space, but not along time. Averaged over space the activities of both species are evenly distributed over the whole time period. Averaged over time the spatial pattern is retained. Hence, resolution along the dimension of time alone does not lead to a reduction in redundancy. Such behaviour is expressed in the measures of organization as $\Delta I_{ts} = \Delta I_s = 1$ and $\Delta I_t = 0$.

In general, we might expect such fixed spatial patterns in terrestrial or benthic rather than in pelagic systems. An example from a terrestrial system displaying a similar type of behaviour is represented in Figure 4.24. Sterling *et al.* (1984) observed the spatial patterns of vegetation in a Mediterranean pasture several years after abandonment of ploughing. Detailed investigations revealed that microtopography, given here by ridges and furrows of ploughing, determined vegetation structure in the early phases of ecological succession and fostered species diversity.

In (c) and (d) the species' activities are segregated along both time and space. In (c) the spatial distribution of activities is time invariant and the temporal pattern is independent of space, whereas in (d) the patterns along the dimensions of space and time are mutually dependent.

In (c), only one species is active at a certain location in space regardless of the time interval under consideration. During a certain time period only one species is active regardless of the spatial location under consideration. Such behaviour is expressed in the measures of organization as $\Delta I_{ts} = \Delta I_t = \Delta I_s = 1 \Rightarrow \Delta I_t + \Delta I_s > \Delta I_{ts}$.

The first model of resource partitioning displayed such a type of behaviour (cf. Figure 4.11). Recall that the model was inspired by observations that plant species differed in their spatio-temporal pattern of nitrogen uptake. McKane *et al.* (1990) distinguished two guilds corresponding to early-active, shallow-rooted and late-active, deeper-rooted

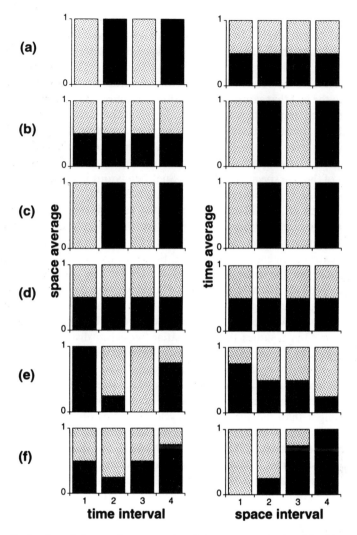

Figure 4.23 Projection of the spatio-temporal activity patterns depicted in Figure 4.22 on the dimensions of time and space, respectively. The light shading refers to the share of species 1, the dark shading refers to the share of species 2

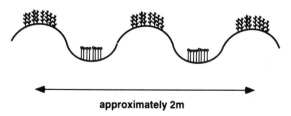

Figure 4.24 Spatial pattern of vegetation observed in a Mediterranean pasture several years after abandonment of ploughing (modified from Sterling *et al.*, 1984)

species (cf. Figure 4.7). Venrick (1990, 1993) analysed 17 years of phytoplankton data from the central North Pacific to show that two stable associations of phytoplankton species can be distinguished: the shallow-living species that are mainly active in summer and the deep-living species that are mainly active in winter. The vertical shift in species composition seems to be related to physiological strategies adapted to the different habitats. Due to the fact that samples from within the same zone were more similar to each other than were deep and shallow samples collected at the same instant of time, Venrick concluded that community organization must be determined by strong biotic interactions.

In (d), averaging over time results in an even distribution of the activities of the two species along space. Correspondingly, averaging over space results in an even distribution of the activities of the two species along time. Such behaviour is expressed in the measures of organization as $\Delta I_t = \Delta I_s = 0$ and $\Delta I_{ts} = 1 \Rightarrow \Delta I_{ts} > \Delta I_t + \Delta I_s$.

A pattern of type (d) could arise from spatial units being subject to a temporal succession in mosaic-like ecosystems. Such is the basic idea for the gap dynamics of forests (Shugart, 1984). Due to the fact that tree species differ in their mineral requirements and their ability to extract minerals from the soil, the same species may not be able to grow immediately again on the same site (e.g. Uhl, 1987). The mosaic cycle concept of ecosystems even predicts that spatial patches of an ecosystem go through repetitive cycles of successional stages (reviews in Remmert, 1991a). The concept is based on the idea of steady state ecosystems where one encounters stable distribution patterns in a dynamic equilibrium state. One assumes that the spatial frequency distribution of successional stages is equivalent to the temporal frequency distribution. The spatial distribution is obtained if a temporal snapshot of the successional states of the various locations is recorded over an area much larger than the single patch size. The temporal distribution is obtained if the time sequence of the successional stages at one location is recorded over a period much longer than the successional cycle. Evidence of patch dynamics has been provided for ecosystems as different as temperate and tropical forests (Remmert, 1991b) and marine benthos (Reise, 1991). In a kelp forest, for example, one observes an alternation of the sequence: barren rock with sea urchins, annual and young perennial algae, canopy of perennial kelp with understory – sea urchins ... Whereas patch dynamics are accepted to be common, the notion of cycling trajectories for mosaic elements is an ideal that will rarely be met. A perfect mosaic cycle should thus be looked upon as an idealized standard against which real systems may be contrasted.

The patterns depicted in (e) and (f) can be derived from the extreme cases depicted in (a) and (b), respectively. In both cases spatial and temporal patterns are mutually dependent. In (e), the temporal dimension makes a higher contribution to organization whereas in (f) we have the reverse case. Correspondingly we obtain for (e) $\Delta I_{ts} > \Delta I_t + \Delta I_s > \Delta I_t > \Delta I_s > 0$ and for (f) $\Delta I_{ts} > \Delta I_t + \Delta I_s > \Delta I_s > \Delta I_t > 0$.

Comparison with Figure 4.21 reveals that we encountered such behaviour in the pelagic model for different mixing regimes. Pattern (e) corresponds to the case where mixing impeded spatial organization ($\gamma = 10$) whereas pattern (f) corresponds to the case where stratification allowed a spatial segregation of species along the light gradient ($\gamma = 0$).

The examples provided show clearly that it is the combined effect of the patterns along time and space that is of functional significance. Averaging over either dimension does not provide an appropriate dynamic picture, in particular when the dimensions are mutually dependent. The case that temporal and spatial patterns are entirely independent is rather unlikely in real systems. It may to some extent be encountered when the physical environment imposes extremely severe limitations on the vegetation that develops on a time-invariant spatial template.

The type of pattern to be observed in a practical situation is also influenced by the observation period and the resolution chosen along the dimensions of time and space, respectively. Pattern (a), for example, corresponds to the first spatial interval of pattern (d). We obtain the same spatio-temporal distribution of activities if the spatial axis in (d) is stretched by a factor of four. Conversely, pattern (b) corresponds to the first time interval of pattern (d). We obtain the same spatio-temporal distribution of activities if the time axis in (d) is stretched by a factor of four. Obviously, the scales of observation chosen might seriously influence the results to be obtained. There exist hardly any rules on how to choose appropriate scales matching in time and space. Understanding the mutual dependence of the temporal and spatial dimensions regarding their effect on ecosystem organization might prove to be helpful to derive guidelines. This problem will be dealt with in detail in Chapter Six.

4.9 A MEASURE EMBRACING QUANTITATIVE AND QUALITATIVE CHARACTERISTICS

The measures of spatio-temporal organization are intensive measures that are independent of a system's size. They depend only on structural properties, but not on the physical magnitude of the energy and/or matter transferred within a network. This is an advantage when we are interested in comparing structural system properties independent of size. When we look at a system's development on a longer term perspective we might, however, be interested in having a measure that expresses both a system's size and its structural organization. As already discussed in the previous chapter, Ulanowicz (1980, 1986) introduced what he called a system's ascendency as a measure combining both intensive and extensive properties. The ascendency is obtained by multiplying the measure of a network's structural organization with the total system throughflow. Ulanowicz limited his attention to static networks averaged over time and space. It should be clear by now that organization is something inherently dynamic. Correspondingly the ascendency measure should also take into account spatio-temporal flow patterns instead of being limited to spatio-temporal averages. Following Ulanowicz's definition, the measures of spatio-temporal organization can be scaled with the total system throughflow to obtain the corresponding ascendency measures: the temporal ascendency, Asc_t, the spatial ascendency, Asc_s, and the spatio-temporal ascendency, Asc_{ts}. The mathematical expressions are summarized in Box 4.11.

The ascendency measures characterize a system as a whole. One may equally well derive the contribution of a transfer between two compartments and/or of an individual compartment to the overall system's ascendency. The contribution of a flow is given by the product of its physical magnitude and a structural "weight" term. The latter depends again on the resolution of the network along space and/or time. Temporal "weights" seem to be far more sensitive indicators than pure physical magnitude to describe a flow's influence on the dynamics of an ecosystem as a whole. A prey species may, for example, constitute only a minor contribution to a predator's diet when averaged over a seasonal cycle. It may, however, be the only food available during critical stages of a keystone predator's life cycle. Despite its little quantitative importance it may thus be essential for

---

Box 4.11   Ascendancy measures

Ulanowicz (1986) gave I, the measure of organization in time- and space-averaged networks, physical dimensions by multiplying it with T, the total system throughput, to generate what he called a system's ascendency, Asc:

$$Asc = TI = T\{H(A/B) - H(A/B) - H(B/A)\} \qquad (4.29)$$

According to Eqn (4.5) the Asc is bounded by:

$$C \geq Asc \geq 0 \qquad (4.30)$$

where

$$C = T\,H(AB) \qquad (4.31)$$

C may be called the development capacity.

In analogy the temporal ascendency, Asc_t, can be defined by scaling I_t defined in Eqn (4.8) with T:

$$Asc_t = T\,I_t \qquad (4.32)$$

According to Eqns (4.9) and (4.31) the Asc_t is bounded by:

$$C \geq Asc_t \geq Asc \geq 0 \qquad (4.33)$$

In analogy, the spatial, Asc_s, can be defined by scaling I_s and the spatio-temporal ascendency, Asc_{ts}, can be defined by scaling I_{ts} with T.

---

the recruitment success of this predator and for community organization as a whole. Such analysis may be applied to bridge levels of organization by studying, for example, the sensitivity of the network structure to changes in the individual flows. A more detailed outline of this approach and examples from simulation models and real energy networks are given in Appendix Two.

The ascendency measures represent the product of a measure of "quantitative" (= throughput) and "qualitative" (= organization) growth. These two factors can be thought of to characterize different phases of ecosystem development that may be viewed to proceed as follows. In agreement with the general perception of ecosystem maturation stated first by Odum (1969), the early phase of an ecosystem's development is dominated by the growth component in terms of increase in physical size. But quantitative growth becomes progressively limited by finite resources. Increases of diversity and specialization within the biological community result in enhanced efficiency of both energy and nutrient utilization. This latter phase is characterized by qualitative in contrast to the early phase of quantitative growth. A famous example for a highly developed system is the tropical rain forest, which is characterized by tight nutrient recycling and a low net production (yield). The measure of spatio-temporal organization describes the characteristics of this latter phase of qualitative development. Diversity and specialization refer to the richness in the spatial and temporal patterns of functional interactions in the ecological network.

Ulanowicz (1980, 1986) claimed that ecosystems optimized their ascendency over evolutionary time scales. Whereas optimization principles in evolutionary strategies are readily accepted at the level of individuals and/or species (e.g. concept of fitness, optimal foraging theory) they find little recognition at the level of an ecosystem. Such a situation derives from the fact that optimization is closely linked to the principle of natural selection which may act on organisms or species but not on ecosystems as a whole. One may argue in general about the existence and/or usefulness of any optimization criterion in ecology, in particular, because optimization requires a type of equilibrium state to be attained. Nevertheless, an optimization hypothesis may be perceived as a model against which an ecosystem's development can be contrasted and compared. Thus it may prove useful for structuring and comparing observations. Along these lines of reasoning, I posit as a working hypothesis that ecosystems tend to acquire a state of the highest spatio-temporal organization and ascendency possible under the limiting constraints of a given situation. This optimization criterion refers now to a dynamic pattern rather than to a specific state. As stated it can never be false, because any major deviation can always be attributed to some specific constraints. Let us look at it as a reference state rather than as an ultimate goal function. It remains to be shown that such a reference state is useful to organize our perception of an ecosystem's function and structure, to derive an understanding for specific species strategies and for the presence of certain spatio-temporal patterns in a given environment.

4.10 SUMMARY AND CONCLUDING CONSIDERATIONS

I sketched a relational perspective emphasizing context and pattern as a counterpart to the prevalent mechanistic perception of natural systems. An important consequence is the need for a simultaneous consideration of the mutual relationship between the component populations and the system as a whole. Spatio-temporal self-organization was identified as an essential characteristic of natural ecosystems. Organization requires some pathways of communication that were identified here with the pathways of energy and matter flows. Thus, the concept of spatio-temporal organization in ecological networks can be given an operational definition and quantitative measures. I used simple models to illustrate meaning and behaviour of the measures introduced. The models focused on two major sources for the generation of spatio-temporal patterns: resource partitioning and pulsed consumption.

We realized that spatio-temporal flow patterns may differ largely in the relative importance of the temporal and spatial dimensions and their mutual relationship. In this respect, the measures of spatio-temporal organization proved to be very useful for characterizing spatio-temporal patterns and their functional significance. The relationship and the interdependence between the spatial and temporal dimensions and their influence on system performance remain to be explored in more detail. We need to find out how spatial and temporal patchiness within a network extending over a range of temporal and spatial scales depend on each other and how they might be related to self-organization.

The concept of spatio-temporal organization has finally been integrated in a broader framework of ecosystem development. Ecosystem development is perceived to proceed by a gradual transition from an early phase of "quantitative" to a latter phase of "qualitative" growth. As a working hypothesis I assume here that this latter phase of ecosystem development is associated with an increase in spatio-temporal organization as quantified by the measures introduced before. To make such a claim more intelligible we need to explore in more detail the effects of increasing spatio-temporal organization on system function. This will be accomplished in the next chapter using a set of multi-species models. Let us recall that the final goal is to obtain a description of the spatio-temporal organization of ecological networks across the hierarchy of spatio-temporal scales and function.

Chapter Five

(*Spatio-*)*Temporal organization in simple communities*

Spatio-temporal organization implies the presence of spatial and temporal variations and is thus only possible in systems operating far away from a stable equilibrium point. This chapter aims now to investigate such systems in more detail in order to elucidate the functional significance of spatio-temporal organization for whole system performance and species coexistence. To do so I will use a variety of multi-species models to illustrate the mutual relationship between the characteristics of species within an ensemble and community organization. In our considerations, we are going to focus on overall regularities that can be discerned irrespective of any mechanistic details of a particular model.

5.1 COMMON CHARACTERISTICS OF THE MODEL SYSTEMS TO BE INVESTIGATED

Figure 5.1 depicts schematically the characteristic flow diagram of a general type of model, to be discussed in this chapter. Comparison with Figure 4.6 reveals that these models can be looked upon as an extension of the simple model of resource partitioning comprising two species only. In analogy to Figure 4.6b, the separate feedback cycles in Figure 5.1 will prove to generate feedback spirals linking the individual units across time and/or space. These units may represent, for example, primary producers or primary producer–herbivore pairs. The resource is equivalent to a nutrient, which is recycled within the system. The systems are open with respect to the exchanges of energy and nutrient with an external environment.

As outlined in the previous chapter, we can distinguish three different levels in an ecological network (cf. Figure 4.2): the microscopic level of the organisms, the mesoscopic level of the compartments equivalent to ensembles of organisms, and the macroscopic level of the network as a whole. I refer to ensembles of organisms rather loosely as species without making explicit reference to taxonomic species. The notions of species and compartment will be used interchangeably depending largely on whether properties of organisms or network properties are referred to.

I will focus here on the mutual relationship between the level of the compartments and the level of the network as a whole. A species is assumed to have certain characteristics representing an aggregated average of the component organisms' functional and dynamic properties. Function refers to links realized within the network of the *inter*compartmental transfers including exchanges with the nutrient pool. Dynamic properties refer to a

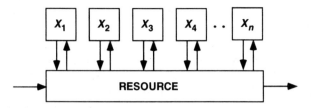

Figure 5.1   Characteristic flow diagram of the type of models to be used in this chapter

species' characteristic growth rate. They may also include preferences for environmental variables in a seasonal environment. The characteristics at the level of the organisms that a population consists of may be accounted for by introducing *intra*specific degrees of freedom such as age structure, life history effects etc. The macroscopic level of the system itself is characterized by the network's overall organization and by global properties as, for example, total biomass, total production, nutrient levels and their variability.

The resource dependence of the growth rates is given by a saturating functional response as shown in Figure 5.2. For low resource concentrations, the relationship between resource level and growth rate is approximately linear. With increasing resource availability the increase in the growth rate saturates, and the rate approaches its maximum.

Wherever appropriate, the models were made dimensionless by expressing all rates in multiples of the maximum growth rates within a species ensemble. Time is then scaled in units of the turnover time of the metabolically most active species. Biomass densities and nutrient concentrations are expressed in multiples of the half saturation constant for resource-limited growth. Hence, unit labels will be missing at most axes with the exception of models with seasonal variability, where I chose the time unit equal to a day to facilitate a comparison with real systems.

At first we limit our attention to temporal organization. The last model will then comprise a spatial dimension as well. The models to be investigated differ in the nature of the interacting units and in the source of the temporal variations that are either imposed

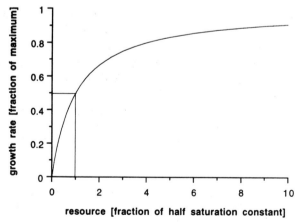

Figure 5.2   Relationship between growth rate and resource availability. The half saturation constant, K, corresponds to the concentration of the resource, where the growth rate achieves half of its maximum. For low resource concentrations ($R/K \ll 1$), one obtains approximately a linear relationship

onto the system exogenously or that are generated endogenously. I refer to variations as being exogenous if their source is a function of time that cannot be influenced by the system's dynamics, e.g. seasonal variability, pulsed nutrient additions. I refer to variations as being endogenous if they are generated within the system by the dynamic interactions of the component species in a homogeneous abiotic environment. In any real ecosystem these sources of the temporal variations can hardly be separated. Exogenously imposed and endogenously generated patterns occur simultaneously, albeit with a changing degree of relative importance.

I am going now to study the behaviour of three sets of models by investigating the relationship between number and dynamic characteristics of the component species and organization and functional properties of the network as a whole. Organization is quantified with the measures introduced in the last chapter. In the first set of models the temporal variations are given by changes in a seasonal environment. The second and third sets of models, where the variations are generated endogenously, illustrate that chaotic population dynamics is compatible with, or even more that it mediates, temporal organization at the community level.

5.2 TEMPORAL ORGANIZATION MEDIATED BY ENVIRONMENTAL VARIABILITY

The patterns and causes of species succession have been the subject of intensive research in terrestrial communities (e.g. Diamond and Case, 1986). Changes in community structure and species dominance patterns may arise as a consequence of the interactions among species. They may equally well be brought about by variations in the abiotic environment. In terrestrial systems, the successional cycles seem to be mainly driven by the endogenous dynamics of species interactions on a physical template (e.g. Remmert, 1991a). However, one can also recognize a strong seasonal component regarding the activities of the species. Recall the spatio-temporal pattern of nitrogen uptake in grasslands that we discussed in Chapter Four (cf. Figure 4.7).

In spite of considerable research efforts it seemed at first impossible to establish similar regularities in the seasonal patterns of phytoplankton communities in aquatic systems. The recent years, however, witnessed an increase in the experimental evidence for the presence of regular successions of phytoplankton assemblages in both marine and lacustrine pelagic systems (e.g. Reynolds, 1984b; Williamson, 1989). Reynolds (1984a, b) analysed such successional patterns in temperate lakes. He distinguished two types of change: autogenic, unidirectional subsequences, regulated by specific responses to critically changing resource-ratio gradients; and allogenic changes, regulated by variability in the physical environment. By ascribing individual species to assemblages, Reynolds demonstrated a high incidence of similarity among annual cycles. Phytoplankton periodicity seems to be the outcome of morphologically, physiologically, and behaviourally mediated responses of phytoplankton to the various dimensions of environmental variability. Environment refers to physical, chemical and biotic variables. The periodic patterns of variability arise from a complex set of interactions among these variables, driven by the seasonality of the physical environment.

Inspired by such observations, I developed a simple conceptual model of an environmentally mediated temporal succession in species' activities. Instead of accounting for any mechanistic details, the model description is reduced to the essentials of a repetitive successional behaviour by characterizing species by what I call their "temporal niche".

5.2.1 Structure of the model with a "temporal niche"

The model describes the seasonal dynamics of an ensemble of primary producers (Pahl-Wostl, 1990). Following the example of phytoplankton periodicity, I assume that the species differ in their period of activity over an annual seasonal cycle. Such an activity period is introduced by assigning to each species a "temporal niche" that is characterized by its shape and its optimum in time. Hence, we may consider the dimension of seasonal time to be the axis of a one-dimensional niche space along which species separate. As shown in Figure 5.3, the temporal niche is described by a Gaussian-shaped growth function along the seasonal time axis. The shape of the temporal niche of a species i is determined by the width, referred to as w_i. The maximum of the growth rate is related to $1/w_i$. The time of maximal growth, referred to as *topt*, determines a species' location in the seasonal cycle. Any seasonal variations in the maximum of the growth rate were neglected.

To give all species the same "fitness", I kept the annual integral over the growth function constant by introducing a trade-off between maximum and width of the temporal niche. Species may hence be assigned to represent "specialists" or "generalists" by choosing the appropriate shapes of the temporal niches. Figure 5.3 depicts an example for three different shapes. An increase in width results in an extension of the activity period with a concomitant decrease of the overall growth rate. The dashed curve represents the case of a constant growth rate corresponding to an infinite width of the temporal niche. A decrease in w results in a shorter period of activity, but during this period higher growth rates are achieved. Such a trade-off is based on the assumption that increasing

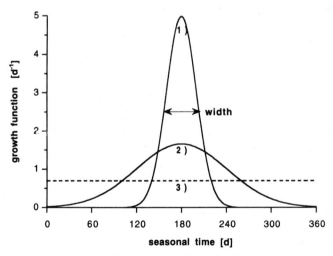

Figure 5.3 Gaussian-shaped growth functions that determine the temporal niche of a species. The widths (2w) were chosen to be: (1) $w = 20$, (2) $w = 60$, (3) $w = \infty$. Reproduced from *Oikos* by permission of Munsgaard

specialization results in an improved exploitation of an increasingly narrower range of environmental patterns.

The growth function r determines only the potential temporal niche of a species. The growth period that can in fact be realized over an annual cycle depends as well on the nutrient availability in the common pool. Thus, the overall performance of an individual species depends on the combined effect of its temporal niche that is determined by the physical environment and the temporal pattern of nutrient availability that is determined by the activity of the species ensemble.

Box 5.1 summarizes (a) the mathematical equations of the simulation model and (b) the symbols and the numerical values of the parameters.

5.2.2 Results of model simulations for different species ensembles

These model simulations aim at illustrating the relationships among the growth characteristics of the species ensemble, the temporal organization of the network, and whole system performance. To investigate such relationships more systematically, I generated different species ensembles by varying the total number of species and the shapes of their temporal niches. Whereas the species ensemble is thus characterized by the number and the temporal niches of the species, the behaviour of the system as a whole is expressed by total system throughput, temporal organization, temporal ascendency, and nutrient efficiency.

At first, let us investigate the performance of a single species as a function of the shape of its temporal niche. Even if temporal organization should be meaningless in this case, it is of interest as a reference state against which results obtained with species ensembles can be contrasted. I varied the shape of the growth function from one having a narrow temporal niche with a high maximum corresponding to a temporal "specialist" ($w < < p$) to one having a constant value corresponding to a temporal "generalist" ($w = \infty$). For practical purposes, I assigned the limiting case of a constant rate to $w = 500$. Due to the fact that the integral of the growth function is kept constant, no species is privileged. Remember that the growth function was designed to exhibit this property by including the trade-off between specialists and generalists. The results of model simulations in Figure 5.4 show that the realized production is highest for the species with a constant growth rate. When a single species is studied in isolation, an increase in the temporal variation of the growth rate (corresponding to a decrease in w) obviously results in a decrease of the production realized. Such behaviour, deriving from self-inhibition inherent in nutrient limitation, is characteristic for any type of saturating functional response (cf. Figure 5.2). Armstrong and McGhee (1980) showed that in a system exhibiting temporal variations of resource and consumer more resource is required to maintain the same consumer biomass than in a system where temporal variations are absent. Thus, pulsed consumption is disadvantageous for a single species in a homogeneous environment when, as in the present model, saturation comes into play. Limiting our attention to single species (or single predator–prey pairs) only, we might conclude that the nutrient pool is used most efficiently in a time-invariant steady state.

The situation changes once the number of species is increased. Then we may expect some type of temporal organization to be required for several species to coexist. Experimental support for such a statement was, for example, provided by Sommer (1984). He showed laboratory experiments that pulsed nutrient addition mediated the

Box 5.1   Description of the simulation model

(a) Differential equations

Changes X_i, in the ith species' biomass, are calculated as:

$$\frac{dX_i}{dt} = \left\{ r_i(t_p)s(N) - k_R - k_E \right\} X_i \tag{5.1}$$

where

$$r_i(t_p) = f_i \frac{1}{w_i} e^{-exp}, \ \exp = \frac{t_i^2}{2w_i^2} \tag{5.2}$$

$$t_i = \begin{cases} |t_p - topt_i| & \text{if } |t_p - topt_i| \le 0.5p \\ |t_p - topt_i| - p & \text{if } |t_p - topt_i| > 0.5p \end{cases}$$

and $\ 0 \le t_p \le p\ $ refers to the time during a seasonal cycle.
In simulation runs starting at $t=0$, t_p is calculated according to:

$$t_p = t \ (\text{mod } p) \tag{5.3}$$

The integral of r_i over one cycle is normalized via the parameter f_i to remain constant independent of w_i.

$$\int_0^p r_i dt = \text{const. for all } i \tag{5.4}$$

$$s(N) = \frac{N}{N+1} \tag{5.5}$$

Changes in N, the nutrient concentration of the pool, are calculated as:

$$\frac{dN}{dt} = \sum_{i=1}^n \left\{ k_R + f_R k_E - r_i(t_p)s(N)X_i \right\} + N_{in} - k_{EN}N \tag{5.6}$$

To maintain the nutrient balance in steady state over a seasonal cycle requires that the sum of all exports from the system equals the total input:

$$k_{EN}\int_0^p N dt + k_E(1 - f_R)\sum_{i=1}^n \int_0^p X_i dt = p\,N_{in} \tag{5.7}$$

Nutrient efficiency is then defined as:

$$NutEff = \frac{k_E(1 - f_R)\sum_{i=1}^n \int_0^p X_i dt}{pN_{in}} \tag{5.8}$$

Following Hurlbert (1978) and May (1974), the temporal niche overlap among several species is defined as the convolution integral of their growth functions:

$$OvL_i = \frac{\sum_{j=1}^{n \ne i} \int_0^p r_i r_j dt}{\int_0^p r_i dt} \tag{5.9}$$

Box 5.1   *(continued overleaf)*

Box 5.1   *(continued)*

(b) List of symbols and parameters used in the model

The biomass and the nutrient concentrations are expressed in multiples of the half saturation constant for nutrient-limited growth – Eqn (5.5).

n	number of species
t_p	time during the seasonal cycle
p	length of the period of a seasonal cycle = 360 [d]
r_i	Gaussian-shaped growth function for species i [d^{-1}]
$topt_i$	optimum in time of growth function r [d]
w_i	width of growth function r [d]
d	distance between the time optima of two neighbouring species = p/n [d]
k_E	rate of export of a species = 0.05 [d^{-1}]
k_R	rate of respiration of a species = 0.05 [d^{-1}]
f_R	recycling ratio = 0.75
k_{EN}	rate of nutrient loss from the pool = 0.0125 [d^{-1}]
N_{in}	external nutrient input = 0.1 [d^{-1}]

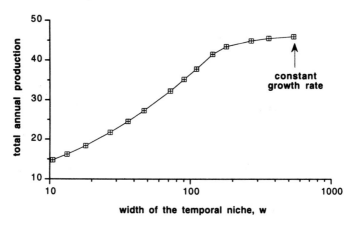

Figure 5.4   Simulation results from a system with one single species. The total annual production is represented as a function of the width of the temporal niche, w. For practical purposes, the limiting case of a constant growth rate, corresponding to $w = \infty$, was assigned to $w = 500$

coexistence of a variety of phytoplankton species that would not coexist under steady state conditions. The temporal variations in nutrient levels prevented competitive exclusion becoming effective. Species that dominated the time-invariant steady state were only of minor importance in the pulsed regime where those species dominated that were able to cope with a variable environment and to engage in temporal resource partitioning.

In the present model, the characteristics of a species' response to temporal variations in the environment are already inherent in the shape of the temporal niche. Following the customs established in competition theory, the degree of competitive pressure on a species can be quantified by the temporal niche overlap of this species with the other species present (cf. Eqn (5.9) in Box 5.1). Reduction of temporal overlap results in a reduction of competition for the nutrient as well as in a reduction of functional redundancy. Since a reduction of functional redundancy corresponds to an increase in temporal organization, we expect temporal organization to be positively correlated with species coexistence.

This expectation is confirmed by results from a set of simulation runs with ensembles comprising an increasing number of species. Within such an ensemble, all species are characterized by the same shape of the temporal niche but differ with respect to the location of the growth optimum along the seasonal time axis. As shown in Figure 5.5, I distributed the species within an ensemble evenly over the annual period, p, by setting the distance between the maxima of the growth functions for two neighbouring species equal to p/n for all species.

It is evident that for an increasing number of species the widths of the temporal niches must decrease to prevent an increase in niche overlap. The overlap of the growth functions between two neighbouring species depends on w/d, the ratio of the temporal niche width, w, to the distance between the optima, d. The two ensembles depicted in Figure 5.5 are characterized by the same ratio $w/d = 0.2$ where the overlap approaches zero. Given species ensembles with three and six species we might thus expect the maximum of temporal organization to be achieved when the width of the temporal niche is around 24 and 12, respectively. In ensembles with a large number of species the overlap shows a similar behaviour in that it approaches zero for a w/d ratio of about 0.2 (see also Pahl-Wostl, 1990).

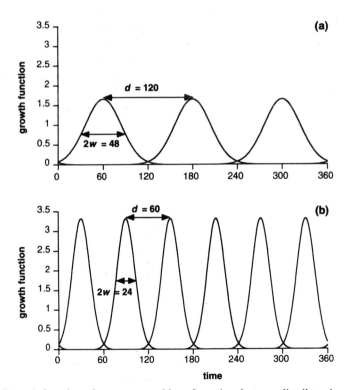

Figure 5.5 Growth functions for two ensembles of species that are distributed evenly over the seasonal cycle. (a) $n = 3$, $d = 120$, $w = 24$; (b) $n = 6$, $d = 60$, $w = 12$

This expectation is confirmed by results from model simulations comprising the same number of species but differing in the shape of the temporal niche. If the ratio w/d exceeds 0.2, the species experience niche overlap. If w/d is smaller than 0.2, the growth period is not used efficiently since there are gaps in the annual cycle where no species is active. Figure 5.6 shows the temporal ascendency, Asc_t obtained for ensembles with three, five and six species. Recall that the Asc_t is calculated as the product of the total system throughput and the measure of temporal organization. It reflects thus the joint behaviour of the system's total activity and its temporal organization.

The maximum in ascendency corresponds to the point where both the total activity and temporal organization are at their maximum (cf. Pahl-Wostl, 1990). At this point the increase in self-inhibition (cf. Figure 5.4) is balanced by the reduction in the temporal niche overlap with other species. A further narrowing of the temporal niche results in an increase of self-inhibition without any corresponding gain from decreasing niche overlap. It may be helpful to recall here the picture of feedback spirals along the time axis (cf. Figure 4.6): the situation of maximum ascendency corresponds to the case where the next species appears just in time to make full profit of the nutrient becoming available from the preceding species. Just in time means, not too early thus being impeded by the growth of the other species, and not too late thus leaving the nutrient accumulating in the pool.

We can infer from Figure 5.6 that given a certain number of species there exists an optimal range of the widths of the species' temporal niches. If the niches are broadened, corresponding to an increase in w/d, the concomitant increase in overlap causes a decrease in both throughput and organization, and correspondingly in Asc_t. The arrow indicates the point where the six-species system was observed to collapse into a three-species system (cf. Pahl-Wostl, 1990). This means that in simulation runs with initially six species (1–2–3–4–5–6) only an ensemble of three species (1–3–5 or 2–4–6) survives. In a five-species system the distances between the species are not favourable for a three-species system to become established and the Asc_t drops below the one obtained for the three-species system. Such behaviour can again be visualized by the picture of feedback

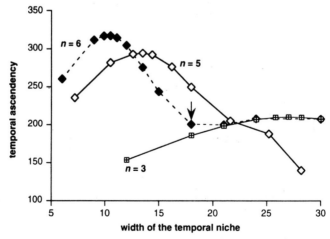

Figure 5.6   Temporal ascendency, Asc_t, obtained in model simulations with $n = 3, 5$ and 6 species. The arrow indicates the point where the six-species system collapses into a three-species system

spirals. With increasing width of the temporal niches, the turn of a spiral along the time axis increases. The nutrient pulses that render possible the growth of the next species become progressively delayed. Obviously, the persistence of a given number of species is critically dependent on the shape and the distribution of the temporal niches and thus on the temporal organization of a species ensemble.

An increase in temporal organization leads to an increase in whole system performance. More specifically, the efficiency of utilizing the available nutrient seems to hinge on the system's ability to organize in time. To express this quantitatively, I defined a measure for nutrient efficiency equivalent to the ratio of the amount of nutrient stored in biomass to the total amount stored in the system:

$$NutEff = \frac{\text{nutrient in biomass}}{\text{nutrient in [biomass + pool]}} \tag{5.10}$$

An increase in nutrient efficiency thus defined is concomitant with a decrease of the amount of nutrient lost from the system, since in our model the majority of the external losses occurs via the pool. When we compare *NutEff* in model simulations with species ensembles having the same number of species but differing in the shapes of the temporal niches, the maximum of *NutEff* coincides with the maximum in temporal ascendency (Pahl-Wostl, 1990). Figure 5.7 shows that these maxima of *NutEff* increase with increasing temporal organization mediated by an increasing species diversity.

NutEff measures realized relative to maximum possible growth. The maximum of 1 would be obtained if the nutrient concentration of the pool was zero and all the available nutrient was fixed in the biomass. Figure 5.7 shows that the total community activity comes closer to its theoretical maximum with increasing temporal organization that is mediated by an increasing number of species. The envelope over the activities of all species increasingly approaches its upper limit that is given by a nutrient efficiency of one. Viewed from the perspective of the system as a whole, the efficiency of nutrient use, the overall productivity, and the biomass increase with increasing number of species. The nutrient

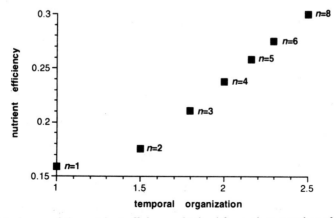

Figure 5.7 Maximum of the nutrient efficiency obtained for a given number of species, *n*, as a function of the measure of temporal organization

input per species and the population biomass, however, decrease with increasing species number, because the increase in the overall efficiency does not compensate for the fact that more species have to share the same limited amount of nutrient. These contrasting trends are shown in Figure 5.8. What appears as a disadvantage to the single species turns out to be an advantage for the system as a whole. It would here be rather difficult to argue that a species optimizes its reproductive success by specializing into a narrower niche. Such specialization must rather be seen as a cooperative development at the community level.

To give further support for this statement, I performed a number of what may be called "competition experiments". These consist of monitoring the performance of species in model simulations with large species ensembles where the individual species differed in shape and location of their temporal niches.

In a first simulation, I distributed an ensemble of 40 species along the axis of the temporal niche width. The widths for the single species were assigned incrementing w stepwise by 10% from $w = 10.5$. At an upper limit of $w = 500$, the growth rate was assigned a constant value. Thus, the species ensemble covered the whole range from temporal generalists to temporal specialists. I assumed that all species had their maximum growth rate at the same time of the seasonal cycle. As in Figure 5.3, the species differed only in the shape but not in the location of the temporal niches. In such an ensemble temporal organization is not possible. Starting simulations with the same initial biomass for all species, I quantified the long-term success of a species by its share of the total biomass over successive annual periods in model simulations. Figure 5.9a shows the results obtained for the first and for the seventh period. During the first seasonal cycle species that are characterized by a narrow temporal niche with a high maximum in growth rate are favoured. In the long-term, species with a constant growth rate come to dominate, while the others go extinct. To understand such findings we have to recall the results from simulations with a model version comprising but a single species (cf. Figure 5.4). Whenever temporal organization is not possible, a time varying growth rate becomes a disadvantage.

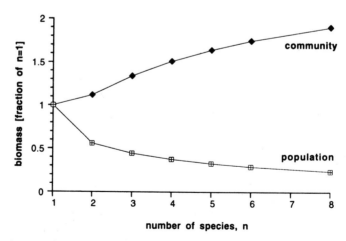

Figure 5.8 Community and population biomasses as a function of the total number of species expressed as a fraction of the corresponding results obtained in simulations with a single species ($n = 1$)

To enable temporal organization I allowed species to vary as well in the seasonal timing of the optimal growth rate. The widths of the niches were assigned as in the previous example. For each width, 80 species were distributed at random over the seasonal cycle with respect to the optimum of their temporal niche, *topt*. The total ensemble hence comprises 40×80 species distributed in a two-dimensional, *w–topt*, parameter space. Figure 5.9b shows the results obtained for the first and for the seventh period in simulations that started with equal biomass for all species. To facilitate comparison with Figure 5.9a, I summed the biomass within all the 40 ensembles of species having the same

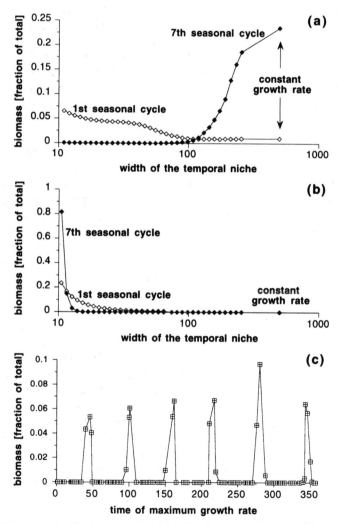

Figure 5.9 Biomass distributions obtained in simulations with ensembles of species being distributed along the *w*-axis with equal biomasses for all species at $t = 0$. *w* was incremented stepwise by 10%, starting from $w = 10.5$. (a) All species had the same optimum of the growth curve, *topt* = 180. (b and c) Species were distributed in a two-dimensional parameter space of *w* and *topt*. (b) shows the biomass summed over all different *topt* for a given *w*. (c) shows the biomass as a function of *topt* for the smallest width $w = 10.5$

niche width. Contrasting with the previous simulation we note that now that temporal organization is possible, the situation is reversed and species with a narrow niche come to dominate. That temporal organization has actually taken place can be inferred from Figure 5.9c showing the relative biomass of the species ensemble with the smallest width ($w = 10.5$) as a function of the optimum in time. The irregularities derive from the species being randomly distributed over the seasonal cycle. We should note that a width of 10.5 corresponds to the width where the optimum of the temporal ascendency, Asc_t, was obtained in simulations with six species (cf. Figure 5.6). The dominance of six species groups implies that the system organized itself into a configuration where the Asc_t is maximal. We may thus conclude that whenever temporal organization is possible, the system comes to be dominated by an ensemble of species engaging in cooperative interactions.

These examples illustrate also the problems associated with too restricted a perspective relying on monocausal explanations. Obviously, we could argue that competition is the sole driving force for community organization leading to a minimization of temporal niche overlap. Such an argument neglects that the persistence of a single species depends on the presence of the other species in the ensemble, it depends on the presence of its competitors. It is the activity profile of the community as a whole that generates the temporal gaps to be utilized by the single species. Community organization can only be understood when we become aware that competitive interactions at the level of the single species and "cooperation" at the level of the whole community work in concert.

The results show thus that the behaviour of a group of species engaging in temporal organization can be understood only by taking into account the system as a *whole*, not by observing a single species in isolation. Similar conclusions can be drawn from investigating processes of invasion. Model simulations revealed that new species may invade successfully into a community without causing any extinctions if they are able to exploit the temporal gaps provided by the resident species. After a successful invasion, the whole community adopts another configuration with a new distribution of the temporal share of the resources among the coexisting species. Figure 5.10 shows the simulation results of an illustrative example for a new species invading into a group of three coexisting species. Assuming the width of the invading species' temporal niche to remain constant ($w = 30$), I monitored the success of invasion for different locations of its optimal growth period along the seasonal axis. In Figure 5.10a, the curve represents the envelope of the temporal niches of the species ensemble to be invaded. The squares represent the invading species' share of the total biomass as a function of its seasonal optimum. The success of invasion depends critically on a favourable timing. Favourable means that the new species enters when there is a temporal gap in the activity profile of the resident species. As can be inferred from Figure 5.10b, this is reflected in an increase in the measure of temporal organization, I_t, as result of an invasion. The change in total biomass and a species' success are positively correlated to the change effected in I_t. Successful invasion does not lead to the extinction of a resident species, only to a rearrangement of the relative biomass shares.

The results obtained from model simulations are supported by empirical results reported by Williamson (1989). He analysed the nature of temporal variations in a plankton community based on an extensive collection of surface plankton in the Irish Sea. First, he could show that the repetition of an annual cycle – albeit slightly different each year – led to the selection of a set of species. Second, he showed that the invasion of an

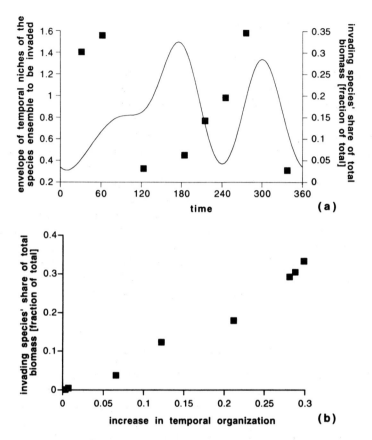

Figure 5.10   Results of the invasion of one species into a group of three species with parameters: *topt* 90/180/300 and *w* 50/30/30. The curve in (a) represents the integral over the growth functions of the species ensemble to be invaded. The scattered points represent an invading species' share of total biomass as a function of (a) the time optimum of the invading species, (b) the increase of temporal organization effected upon invasion

exotic diatom species led to a temporal reorganization of the whole community. None of the resident species became extinct due to the successful establishment of the invader. However, principal component analysis revealed significant temporal rearrangements of the activity periods. Williamson concluded that this indicated the importance of weak and indirect effects for the community structure as a whole, even when he had to admit that one was yet far away from a detailed mechanistic understanding of the population dynamics. I should like to add on this that instead of emphasizing detailed mechanisms it might be more fruitful to focus on the general patterns of ecosystem organization.

5.2.3   Some first conclusions

This simple conceptual model was explicitly designed to focus on temporal organization in a seasonal environment. We observed temporal organization and the efficiency of nutrient utilization to increase with increasing species diversity. The measures of temporal organization and of temporal ascendency proved to be sensitive tools to characterize whole system performance.

As the number of species is increased, specialization (= narrowing of the temporal niche) confers an advantage. Increasing diversity mediates a species ensemble's approach to the growth limits that were given by nutrient availability. Specialization and the concomitant temporal organization can be looked upon as a cooperative process. We can talk here about "cooperation" without invoking any notion of deliberate altruistic behaviour. Cooperation arises simply from the effects of positive feedback spirals in a dynamic network where an ensemble of species shares a limiting resource. The survival of a single species depends on the presence of the species ensemble. Correspondingly, it would also make no sense to derive a species' fitness in isolation.

Results from model simulations revealed that the success of an invading species is contingent on a favourable timing with respect to the activity profile of the resident species. Whether a species arrives at a certain location at a favourable time may be largely determined by chance events. However, an improved understanding for the essentials of spatio-temporal organization as a mutual dynamic relationship between the species level and the level of the ecological network as a whole should lead to an improved understanding of species' evolutionary strategies in the context of a dynamic picture of ecosystem organization.

5.2.4 The seasonal dynamics of a phytoplankton community

Admittedly, the model discussed is remote from the complex dynamics of a natural system where, for example, we cannot expect to find such extreme temporal regularities. However, an idealized and simplified reference state helps us to clarify and study more systematically essential characteristics of system behaviour. The main elements of the behaviour just described are not contingent on the presence of an idealized temporal niche. This is illustrated by results obtained from a model where the seasonal succession of phytoplankton species is described more realistically. Instead of assigning a fixed temporal niche to a species, I defined it indirectly by assuming that a species' growth rate depends on characteristics in the physical environment that vary over a seasonal cycle. More specifically, a species' growth rate depends on temperature (T) and daylength (L), as expressed by the functions $r_i(T)$ and $s_i(L)$, the characteristic shapes of which are depicted in Figure 5.11. The mathematical expressions are summarized in Box 5.2.

The dependence on a limiting resource was introduced indirectly by keeping constant the total growth of all species (cf. Eqn 5.11). At a certain time, a species' share of this limited environmental capacity is determined by the ratio of its potential activity, equal to $r_i\{T(t)\} \times s_i\{L(t)\} \times B_i(t)$, to the sum of all potential activities of the whole species ensemble.

In model simulations I distributed species ensembles at random in a parameter space of the optima in temperature ($-2 < Topt_i < 22$), and daylength ($4 < Lopt_i < 20$). The initial biomasses were drawn from a uniform random distribution. Figures 5.12a and b represent the results from two representative simulation runs with ensembles comprising each 600 species. The simulations differed only in the initial distribution of the species ensembles in the parameter space. Species that were still present after several years of model simulations are represented by large squares the areas of which reflect the corresponding species abundance. The dots represent those species that either became extinct during

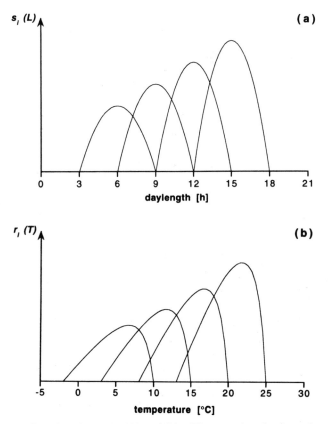

Figure 5.11   Shape of the functions (a) $s_i(L)$ and (b) $r_i(T)$ expressing the dependence of the growth rate of a species X_i on daylength (L) and temperature (T), respectively. The functions were obtained for different values of the optima

simulation runs or achieved a biomass more than eight orders of magnitude lower than the dominant species. The driving force for community organization is given by the seasonal variation of temperature ($0 < T(t) < 20$) and daylength ($6 < L(t) < 18$) shown in Figure 5.12c. The time lag of the temperature maximum relative to the maximum in daylength is caused by the large heat capacity of a water body retarding its response to a change in insolation. This time lag leads to an elliptic shape of the seasonal variation when represented in a T–L-plane as in Figures 5.12a and b.

The surviving species ensemble consists of those species that were able to organize in time by densely occupying the temporal niches provided by the environmental variations. A comparison of Figures 5.12a and b shows that the overall shape of the surviving species distribution is consistent and seems therefore to be largely determined by the seasonal variation. If the parameter combination of a species comes to lie on the ellipse of the seasonal variation, this species experiences the optimal conditions in temperature and daylength simultaneously. For all other species the time of optimal daylength and the time of optimal temperature do not coincide. The time gap increases with increasing distance from the ellipse. Hence, we might at first expect the biomass of a species to decrease in proportion to the distance of its parameter combinations from the ellipse. However, the

Box 5.2   Equations of the simulation model

$$\frac{dX_i}{dt} = \left(\frac{r_i(T)s_i(L)}{R} - m\right)X_i \tag{5.11}$$

where

$$R = \sum_{i=0}^{n} r_i(T)s_i(L)X_i$$

$$r_i(T) = \begin{cases} e^{T\ \text{max}/20}\dfrac{(T - Tmin_i)(Tmax_i - T)}{(Tp_i - T)} & \text{if } Tmin_i \le T \le Tmax_i \\ 0 & \text{if } T < Tmin_i \text{ and } T > Tmax_i \end{cases} \tag{5.11a}$$

where

$$Tmin_i = Tmax_i - 12, \; Tp_i = Tmax_i + 2$$

The optimum can be derived to be: $Topt_i \approx Tmax_i - 2.47$

$$s_i(L) = \begin{cases} (9 - (L - Lopt_i)^2)\left(0.5 + \dfrac{Lopt_i}{6}\right) & \text{if } Lmin_i \le L \le Lmax_i \\ 0 & \text{if } L < Lmin_i \text{ and } L > Lmax_i \end{cases} \tag{5.11b}$$

where

$$Lmin_i = Lopt_i - 3, \qquad Lmax_i = Lopt_i + 3$$

The seasonal dynamics of temperature and daylength follow:

$$T(t) = T_{\text{mean}}(1 + \sin(\omega t - \phi_T)) \tag{5.12a}$$

$$L(t) = L_{\text{mean}}(1 + 0.5\sin(\omega t - \phi_L)) \tag{5.12b}$$

where

$$L_{\text{mean}} = 12\ [\text{h}], \; T_{\text{mean}} = [10°\text{C}], \; \omega = 2\pi/360, \; \phi_L = 0.5\pi, \; \phi_T = 0.67\pi$$

situation is slightly more complicated. Due to the phaseshift between daylength and temperature, two different daylengths exist for a given temperature and vice versa. In Figure 5.12c this is indicated by the dashed lines for the temperatures, 10°C and 17.5°C. For a given temperature, both the temporal distance between the occurrences over the seasonal cycle and the difference between the corresponding daylengths decrease the closer the temperature is to the seasonal maximum (or minimum). Except for the region of the seasonal maxima (minima), two matching pairs of T and L are too distant apart to be exploited by a single species (e.g. $L_1(10°)$ and $L_2(10°)$ and a species with $Topt = 10°$). Hence, those species whose $Lopt$–$Topt$ parameter combinations come to lie in the region of the turning points of the seasonal ellipse are privileged. This explains the position of the two dominant groups in Figures 5.12a and b.

Whereas the gross pattern is determined by the nature of the seasonal variation, the realized distribution over the whole seasonal cycle is determined by the temporal organization of the species ensemble. The ensemble organizes into guilds of coexisting

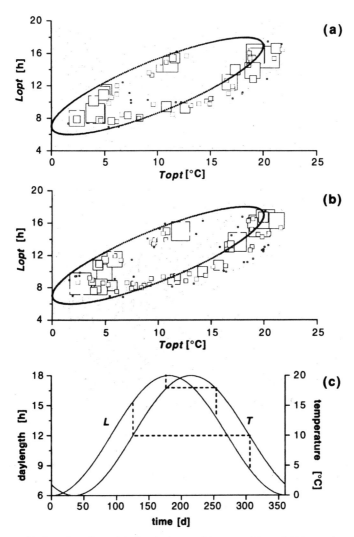

Figure 5.12   (a, b) Results of model simulations with ensembles of 600 species distributed at random in the two-dimensional parameter space of *Topt* and *Lopt*. The two simulation runs differed only in the initial distribution of the phytoplankton species. The areas of the open squares reflect the annual biomass of a species observed during simulation runs with a seasonal variation of *T* and *L* as indicated by the ellipse. Species represented by the dots became either extinct or achieved an abundance at least eight orders of magnitudes lower than the dominant. (c) Seasonal variation of daylength (*L*) and temperature (*T*). Due to the phase shift between *T* and *L* we observe two different daylengths for a given temperature and vice versa. Further explanations in the text

species which are also reflected in the shape of the biomass distribution function. Figure 5.13a shows the relative biomass of a species as a function of its rank in biomass. The biomasses are close to being log-normally distributed. As shown in Figure 5.13b, the distribution is skewed to the right indicating the organization of the species into subgroups.

Log-normal curves have frequently been observed for the relationship between species abundance and rank order. McGowan and Walker (1985) and Venrick (1990), for

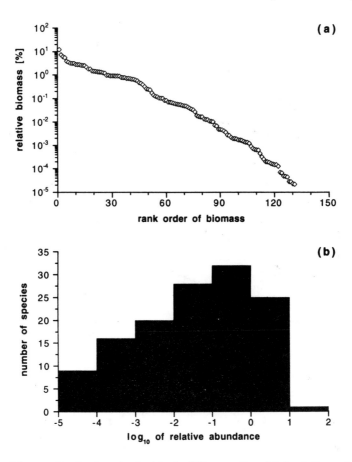

Figure 5.13   Biomass distribution obtained in the fifth year of model simulations where the initial biomass was evenly distributed over an ensemble of 600 species distributed at random in the parameter space of *Lopt* and *Topt*. (a) Relative biomass as a function of the rank order. (b) Number of species in different categories of relative biomass on a logarithmic scale

example, obtained log-normal distributions for the annual averaged biomasses of phytoplankton and zooplankton communities in oceanic ecosystems. Sugihara (1980) suggested accounting for such observations in terms of a hierarchical community structure represented by a sequentially divided niche space. Sugihara's model is based on a static description that does not consider any temporal dynamics. In the model presented here, the niche space of the seasonal variation is shaped by the two environmental variables of temperature and light. The abundance pattern arises from the species' dynamics in a temporally varying niche space. The seasonal period is broken up into progressively smaller intervals by species that engage in temporal partitioning of a limited environmental capacity. Such behaviour suggests that we may reduce a multi-dimensional static niche space of resources to a one-dimensional niche axis that accounts for the dynamic characteristics. I will follow up this idea in more detail in Chapter Six.

Let us now explore in more detail the way in which the temporal niches can also be generated by the complex dynamics of species interactions. In the two previous models the temporal variations in the activities of the species were mediated by seasonal variations in the environment. In the models to be discussed in the following sections, temporal variations are generated endogenously by the dynamics of species interactions in a constant environment.

5.3 TEMPORAL ORGANIZATION MEDIATED BY SPECIES-INTRINSIC VARIABLES

5.3.1 Intrinsic variables describe the internal state of a population

In models at population level like, for example, the Lotka–Volterra models, species are represented by homogeneous aggregations of organisms. Such practice is based on the assumption that the dynamics of a population can accurately be described by referring to one averaged value for biomass or abundance. We may account for the fact that a population represents heterogenous aggregations of organisms by giving it an "internal degree of freedom", as average size or age, density of species-specific parasites, life history memories etc. The addition of such intracompartmental states accounts for the hitherto neglected basic level in the triadic structure of an ecological network, the level of the organisms (cf. Figure 4.2 and comments in the introduction to this chapter). Each population represents itself an interaction network of organisms that may exhibit changes due to internal dynamics (e.g. social structure, age) and/or changes in the environment. The exercise here will show us that accounting for such internal dynamics by the introduction of a dynamic intracompartmental state renders the idea of stable time-invariant equilibrium points prevailing in ecosystems rather unlikely.

A dynamic intracompartmental state means that a population is now described by two dynamic variables (X_i, Y_i), where X_i is the biomass and Y_i is an additional dynamic variable describing an internal state. The internal variables may hence be looked upon as a type of dynamic feedback. To contrast such a model with those discussed in the previous section I sketched the interaction diagrams of both in Figure 5.14.

The populations are again coupled via the resource pool. The dynamic variables describing the internal states act on the resource pool indirectly by affecting the populations and their exchanges with the pool. Changes of the internal state may depend on population biomass (X_i) as well as the change in biomass (dX_i/dt), because the internal state responds to a change in biomass rather than to an absolute value. The age structure of a population at a given biomass, for example, may differ largely depending on how fast the population has grown. These internal variables store thus a certain memory and may show time lags in the response. This gives them a meaning and effect different from simple density-dependent feedback effects that are directly related to a change in biomass (e.g. logistic model).

5.3.2 Structure of the model

Let us assume as a tangible example that the species-intrinsic variables refer to host-specific parasites. Due to the negligible biomass of a parasite, there is no need to account explicitly for energy and matter transfers from host to parasite. Instead, a parasite's effect can be reduced to an additional mortality term of its host. The derivation of such a term and the resulting system of equations are summarized in Box 5.3.

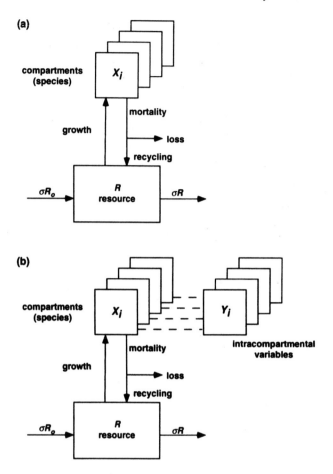

Figure 5.14   Comparison of two types of network models where an ensemble of functionally equivalent species shares a common resource pool. (a) Conventional model as discussed in the previous section; (b) model with intracompartmental variables

We can thus distinguish two types of loss processes: The loss rate, μ_i, is a constant accounting for all parasite-independent loss processes such as respiration or natural mortality. The parasite-induced mortality rate, Y_i, is a dynamic variable that increases with increasing host biomass when parasites thrive and that decreases with decreasing host biomass when parasites perish. Effectively the species-intrinsic variables, Y_i, represent dynamically varying mortality rates, where aside from parasite infestation other interpretations are possible – it expresses a dynamic density-dependent negative feedback. Therefore I will now simply refer to it as dynamic mortality. The introduction of such dynamic mortality terms leads to the onset of instability in species ensembles that would otherwise be characterized by a time-invariant steady state.

Box 5.3   Equations for the host–parasite model

The host–parasite interactions are described by a Holling-type II functional response:

$$\frac{dZ}{dt} = \beta' Z \frac{X}{1 + \alpha X} - \gamma Z$$

$$\frac{dX}{dt} = \left(h(R) - \mu - \frac{Z}{1 + \alpha X} \right) X$$

where Z represents the normalized parasite biomass. Y may be chosen to reflect the parasite-induced rate of mortality on the host X:

$$Y = \frac{Z}{1 + \alpha X}$$

The time derivative for Y yields:

$$\frac{dY}{dt} = \left(\frac{\beta' \; X - \alpha dX/dt}{1 + \alpha X} - \gamma \right) Y$$

and after some modifications and simplifications one obtains:

$$\frac{dY}{dt} = ((\beta - \alpha Y)X - \gamma)Y$$

where $\beta \approx \beta' - \alpha(h - \mu)$ and α is neglected in the denominator. The parameter β determines parasite growth, whereas γ determines the critical host density for parasite growth to occur.
The system of equations for an ensemble of n host–parasite pairs yields:

$$\frac{dX_i}{dt} = (h(R) - \mu_i - Y_i)X_i \quad i = 1,2 \ldots, n \tag{5.13}$$

$$\frac{dY_i}{dt} = ((\beta + \alpha Y_i)X_i - \gamma)Y_i \tag{5.14}$$

$$\frac{dR}{dt} = \sigma(R_o - R) - \sum_i (h(R) - \eta(\mu_i + Y_i))X_i \tag{5.15}$$

where h(R) was chosen as: $h(R) = \dfrac{R}{1 + R}$.

As formulated the model is dimensionless. Biomass densities and resource concentrations are expressed in multiples of the half saturation constant for resource-dependent growth. Time is scaled by setting the maximum growth rate for all host species equal to 1.

5.3.3   Intrinsic variables foster coexistence and temporal organization

I performed model simulations for ensembles with an increasing number of species. As outlined in Box 5.4a, the species within an ensemble differed with respect to their constant loss rates, μ_i, and were arranged in the sequence: $0 < \mu_1 < \mu_2 < \ldots < \mu_n$. At first sight we may expect this sequence to be equivalent to a ranking order due to the disadvantage conveyed by an increasing loss rate. However, results from model simulations reveal that an increase in the loss rate is compensated by a decrease in the time average of the dynamic

Box 5.4   Parameter values and results from model simulations

(a) Numerical values of the parameters

$$R_o = 2$$
$$\sigma = 0.1$$
$$\eta = 0.9$$
$$\alpha = 0.1$$
$$\beta = 1$$
$$\gamma = 1$$
$$\mu_1 = 0.2$$
$$\mu_2 = \mu_1 + \Delta\mu$$
$$\mu_3 = \mu_2 + \Delta\mu$$
$$...$$
$$\mu_n = \mu_{n-1} + \Delta\mu$$

Initial conditions: $R = 0.5$, all $X_i = 1$, all $Y_i = 0.1$, $i = 1,2 ..., n$.
   Simulations were performed for $n = 2, 3, 4$, $\Delta\mu = 0.05$, and for $n = 20$, $\Delta\mu = 0.01$.

(b) Biomasses and dynamic mortality terms ($n = 3$)

i	μ_i	X_i	Y_i
1	0.2	0.97	0.23
2	0.25	0.97	0.18
3	0.3	0.99	0.12

The values for X_i and Y_i were obtained by averaging in simulation runs over a period of 200 time steps.

mortality. Hence, an ensemble of species may coexist in a dynamic situation where the biomass densities vary in time.

Unlike the model with the Guassian-shaped temporal niches, this model obtains not only a periodic but a chaotic alternation of population abundances. An illustrative example is given in Figure 5.15a for a simulation run with three species. The time period that was rescaled to start at $t=0$ actually shows a later phase of a simulation run to prevent any distortions that may be caused by a memory of the initial conditions chosen. We observe a succession of the three species where the temporal order shifts irregularly. The instability of the equilibrium state is caused by the dynamic nonlinear nature of the dynamic mortality terms. Figure 5.15b shows a typical oscillation of population biomass and the feedback in the dynamic mortality term. As listed in Box 5.4b, the advantage of a lower loss rate is compensated by an increase in the dynamic mortality leading thus to an even distribution of the average biomass over all species.

In such a type of models the internal degress of freedom mediate the onset of oscillations and species coexistence. To monitor the associated changes at the system level, I calculated the measure of temporal organization, I_t, for successive time periods starting at $t=0$. The results obtained for a total number of $n=2$ and 4 species are represented in Figure 5.16 as a function of time. I_t increases with progressive development of the final temporal pattern to reach a plateau that indicates a dynamic steady state. An increase in temporal resource partitioning and in temporal organization is observed for an increasing number of species. Whereas the pattern was still periodic in the case of two species,

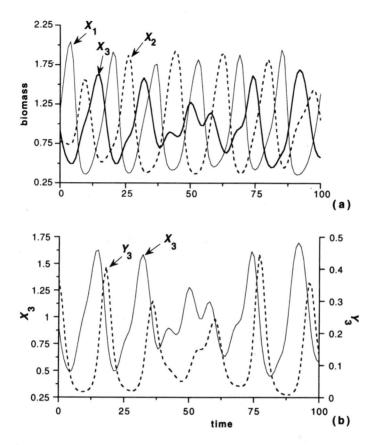

Figure 5.15 (a) Population biomasses as a function of time. The species differ in their constant loss rates: $\mu_1 = 0.2$, $\mu_2 = 0.25$, $\mu_3 = 0.3$. (b) Oscillations of population biomass, X, and the dynamic mortality term, Y, obtained for the third species

irregular chaotic fluctuations were observed in the case of four species. Obviously, the temporal organization was determined to be higher for the chaotic than for the periodic pattern. Some readers might feel somewhat uncomfortable because intuitively one may argue that a chaotic pattern is incompatible with any type of organization. However, the chaotic nature of the oscillations prevents the entrainment into a rigid temporal pattern. Any small resource peak may thus be effectively utilized resulting in a smoothed envelope over the activities of all species. That is what I refer to here as temporal organization, which is entirely compatible with a chaotic population dynamics.

To summarize, the example presented provides evidence for temporal organization and the coexistence of similar species mediated by the interplay of both intra- and interspecific interactions. Temporal organization refers here to the density of the succession of oscillations in species' activities. Despite the impression of coherence and regularity imposed by the notion of organization, temporal organization is thus observed equally well in systems exhibiting irregular or even chaotic fluctuations. In the long term a temporal pattern of activity prevails, where the envelope over the activities of all species is smoothed, where every available resource peak is utilized.

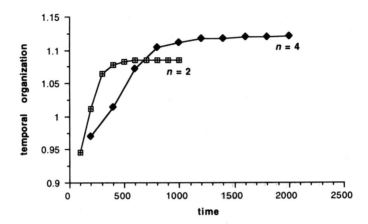

Figure 5.16   Measure of temporal organization, I_t, calculated for successive time intervals in model simulations with models of host–parasite pairs. The length of one time interval was chosen to equal 100 for $n = 2$ and 200 for $n = 4$

5.4   SPATIO-TEMPORAL ORGANIZATION MEDIATED BY A MULTIPLICITY OF TIME SCALES

5.4.1   Body weight and time scales

In this model the "units" sharing a common nutrient pool are identical to predator–prey pairs that differ in body weight and thus in the time scale they operate on. Body weight is a major determinant for both an organism's physiological and ecological characteristics. Allometric relationships provide the foundations for quantifying the implications of weight on rates of, for example, energy use, productivity and nutrient cycling in ecosystems (e.g. Calder, 1984; Peters, 1983). They were originally derived from observations that organisms' physiological rates, R, vary as a power function of body weight:

$$R = aW^{-\varepsilon} \qquad (5.16)$$

where R has dimensions $[T^{-1}]$ for mass-specific rates, W is body weight $[M]$, ε is the allometric exponent and is dimensionless, and a is the rate coefficient and has dimensions $[T^{-1}M^{\varepsilon}]$. ε proved to be remarkably constant at a value of around 0.25 to 0.3 for a large range of organisms of different weights and taxa. Eqn (5.16) implies that growth processes and metabolic costs of maintenance per unit body mass decrease with increasing body weight. The regular scaling of a variety of organismal properties with body weight is not based on empirical observations only. One can derive explanations based on physiological considerations as, for example, the ratio of surface to volume, or proportional growth processes at the level of the organism (Calder, 1984; Peters, 1983).

Such relationships are of particular practical interest for the description of aquatic food webs where there is a net movement of matter and energy upward along a gradient of increasing body weight from small to large organisms (e.g. Platt and Denman, 1978; Sheldon *et al.*, 1972). Hence, body weight may be used as a central variable to describe the energy transfer and the biomass distribution in such a food web. Instead of specifying taxonomic species, organisms are characterized by their body weight, and by their being primary producers or consumers. Consumers are distinguished by their own body weight and by the weight range of the prey they feed on. Experimental data from a pelagic food web support the adequacy of such assignments (Gaedke, 1992a). The organisms within a weight class proved to be more similar with respect to their feeding habits than organisms belonging to a given genus but being spread over a wide range of body weights.

These regularities inspired the development of energy transfer models that are based on body weight as central dynamic variable. Allometric relationships provide a basis for a systematic approach to modelling multi-species systems. The number of parameters can be reduced by choosing a basic parameter set for process rates and by using body weight as an independent criterion to estimate the dynamic characteristics. Models of this type may be entirely abstract, expressing the biomass density as a continuous function of body weight and using weight-specific transfer probabilities to specify the likelihood for a biomass transfer along the body weight axis (e.g. Silvert and Platt, 1978, 1980). The transfer probabilities may be derived from processes of predation, growth, cell division etc. In recent years, models have been developed that include more ecological details. Moloney and Field (1991) and Moloney et al. (1991), for example, developed a model for marine pelagic ecosystems by distinguishing functional groups in different weight categories.

5.4.2 Structure of the model

This conceptual model does not aim to depict mechanistic details. It was designed to investigate the influence of dynamic diversity on the dynamics of a species ensemble and on a system's spatio-temporal organization. The dynamic diversity increases with the number of different time scales present in a species ensemble and thus with the number of predator(P)–prey(B) pairs differing in body weight. Figure 5.17a shows the network of nutrient flows. The size of the arrows representing internal exchanges indicates that most of the nutrient is recycled within the system. This network resembles the one depicted in Figure 5.14b referring to models with intracompartmental variables. However, in contrast to intracompartmental variables that act only on the dynamics of the species they refer to, the predators are now an integral part of the energy and matter network, thus making the feedback structure more complex. A predator–prey pair comprises two pathways of recycling: a fast, short one deriving from direct losses from the prey species to the nutrient pool, and a slower, longer one deriving from the recycling via predators.

As depicted in Figure 5.17b, the predator–prey pairs are distributed along the body weight axis which is partitioned into weight classes equally spaced on a logarithmic scale. The average body weight in the kth weight class is expressed as a fraction of W_0, the weight in class 0:

$$W_k = 2^k W_0 \qquad (5.17)$$

The weight ratio between neighbouring classes, W_{k+1}/W_k, equals two. The choice of a logarithmic instead of a linear scale becomes more intelligible when we consider that the weight of a predator's prey is determined by the predator–prey weight ratio rather than by the predator–prey weight difference. A ratio is constant on a logarithmic scale. Similar considerations can be made with respect to the time scale associated with body weight. We

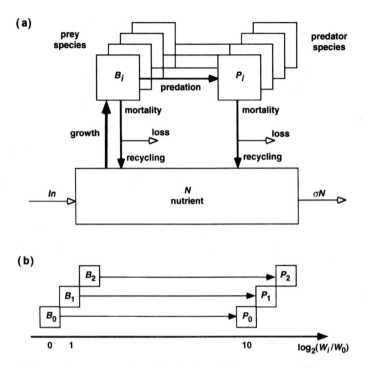

Figure 5.17 (a) Network representation of the basic model components. The arrows representing internal exchanges are set in bold type to emphasize the quantitative dominance of recycling over the external exchanges. The exchanges with the environment are denoted by open arrow heads. (b) Arrangement of predator–prey pairs along the body weight axis. Weight is expressed in fractions of W_0, the average weight in class 0. The weight class difference between a predator and its prey was chosen to be equal to 10 corresponding to a weight ratio of 1000

could conceive of the body weight axis representing the axis of a one-dimensional niche space along which species occupy niches according to their dynamic characteristics. Temporal organization implies that species organize themselves along this axis. The widths of the niche that a species occupies can be assumed to increase in proportion to the species-typical time scale – providing thus again an argument in favour of a logarithmic rather than a linear scale. A more detailed discussion of this issue is postponed until we have developed a better understanding for temporal organization. Now, we simply take advantage of the fact that a difference in body weight implies also a difference in the characteristic process rates and time scales. Hence, a hierarchy of time scales can be generated by distributing a number of predator–prey pairs along the body weight axis. This way the whole feedback network becomes progressively more complex and dynamically diverse.

I assumed the weight ratio between a predator and its prey to be constant and equal to 1000, corresponding to a weight class difference, of $q = 10$. The ith pair in our model comprises then the prey, B_i, with body weight, w_i, and its predator, P_i, with body weight, W_{i+q}. As a further simplification, I assumed a predator to feed on one weight class only. Variations in the predator–prey weight ratio and in the width of the prey windows may have important consequences for the dynamics of a system as will be discussed in Chapter

Six. However, the simplified model chosen is better suited to focus on the qualitative characteristics of system behaviour mediated by a multiplicity of times scales.

The metabolic rates of a species in weight class k are derived from the basic rates of the smallest species in weight class 0 by combining Eqns (5.16) and (5.17). This yields for the respiration rate, ρ_k, and the growth rate, γ_k:

$$\gamma_k = \Psi_k \gamma_0 \text{ and } \rho_k = \Psi_k \rho_0 \tag{5.18}$$

where the weight ratio W_k/W_o was replaced by $\Psi_k = 2^{-\varepsilon k}$, the allometric factor of the weight class k. Eqn (5.18) implies that the metabolic rates decrease with a factor $2^{-\varepsilon}$ from one class to the next. We can now define a characteristic time scale as the inverse of the rate of a certain process. Correspondingly the characteristic time scales increase with a factor 2^{ε} from one class to the next. The mathematical equations and the major symbols and parameters are summarized in Box 5.5.

Again, the model variables are dimensionless. All concentrations are expressed in multiples of the half saturation constant of nutrient-dependent prey growth. Time is expressed in multiples of γ_0^{-1}, the turnover time of a species in weight class 0. The basic version of the model consists of only one predator–prey pair, the prey species B_0 and its predator P_0. Extended versions comprise ensembles of pairs where a pair with index d differs from the basic pair by being shifted up the body weight axis for d weight classes.

5.4.3 Effects of increasing dynamic diversity

To investigate the influence of dynamic diversity, I performed simulations for an increasing number of pairs and monitored the resulting changes in system behaviour. Figure 5.17b shows that a new pair is added by shifting it up the weight axis by one class relative to the previous pair. With an increasing number of species, an ensemble thus comprises pairs with progressively higher body weights operating on progressively slower time scales.

I adjusted the parameters so as to obtain a stable equilibrium point for the model version comprising a single predator–prey pair only. The stability of the equilibrium point is not affected by most parameter changes or by the pair's location along the body weight axis. However, as soon as a second pair is added, the point equilibrium becomes unstable, and we observe periodic oscillations as shown in Figure 5.18a. The onset of instability caused by the presence of a second pair derives from the two pairs sharing the nutrient pool in turns. A single pair approaches the equilibrium point in an oscillatory fashion. The presence of a second pair prevents the first pair settling into the equilibrium point and persistent oscillations are observed. I have discussed such behaviour in more detail for a similar model including a detritus and a nutrient pool (Pahl-Wostl, 1993c).

The addition of further pairs leads to a progressive breakdown of the periodic oscillations and finally to the onset of irregular behaviour. This is revealed by Figures 5.18a and b representing the temporal variations in prey biomass obtained for equilibriums with two and four pairs, respectively. If we compare Figure 5.18b to Figure 5.18a, the loss in the regularity in the biomass fluctuations is the most conspicuous change. At first sight, such findings do not convey the impression of any temporal organization whatsoever. However, we may start to discern some pattern when we look at the metabolic activities depicted in Figure 5.18c. The biomass itself does not adequately reflect a population's energy and nutrient requirements which depend rather on the

---

Box 5.5   Equations and parameters of the simulation model

(a) Mathematical equations

$$\frac{dB_i}{dt} = \Psi_i\{f(N)B_i - \rho B_i - \Omega h(B_i)P_i\} \tag{5.19}$$

$$\frac{dP_i}{dt} = \Psi_{i+q}\{h(B_i)P_i - (\rho + \lambda P_i)P_i\} \tag{5.20}$$

$$\frac{dN}{dt} = In - \sigma N - \sum_{i=0}^{n} \Psi_i f(N)B_i + rec \sum_{i=0}^{n} \Psi_i \rho(B_i + \Omega P_i) \tag{5.21}$$

$$\text{where } f(N) = \frac{N}{N+1}, \ h(B_i) = \frac{B_i}{B_i + K}, \ i = 0, 1, \ldots, n\colon n \leq q-1.$$

The quadratic loss terms of the predators simulate density-dependent limitations of growth. They exert a stabilizing influence onto the system by preventing violent predator–prey oscillations. These may occur due to the absence of predation by higher trophic levels that are not included in the present version of the model. The stability of the model including one pair is only affected by omitting the predator's quadratic loss term and/or by decreasing the half saturation constant of the predator's functional response (unstable for $K \leq 1$). A critical nutrient concentration, N_{cr}, can be determined from Eqn (5.19) as the threshold where in the absence of predation the net growth of primary producers becomes zero:

$$\frac{N}{N+1}B - \rho B = 0 \Rightarrow N_{cr} = \frac{\rho}{1-\rho} \equiv 0.7 \tag{5.22}$$

(b) Symbols and parameters

n total number of predator–prey pairs in a simulation run
$<>$ time average
i index of a predator–prey pair with B_i being in weight class i and P_i being in weight class $i + q$

Model parameters and numerical values used in the model simulations presented here are as follows:

ρ	$= 0.41$	rate of respiration
σ	$= 0.10$	rate of loss from the nutrient pool
rec	$= 0.90$	degree of recycling
K	$= 2.55$	half saturation constant for predation
In	$= 1.00$	external input of nutrient
λ	$= 0.05$	quadratic loss term of predator
ε	$= 0.25$	allometric exponent
Ψ_k	$= 2^{-\varepsilon k}$	allometric factor for the kth weight class
q	$= 10$	predator–prey weight class difference
Ω	$= 2^{-\varepsilon q}$	allometric factor for a predator relative to its prey

Allometric factors are denoted with capital Greek letters, rates are referred to with small Greek letters.

---

biomass turnover than on standing stocks. The biomass turnover can be expressed by the metabolic activity that is obtained by multiplying a population's biomass, B_i, with the allometric factor of its weight class, Ψ_i (cf. Eqn 5.18). We account thus for the fact that the metabolic requirement per unit mass decreases with increasing body weight. A species in a higher body weight class needs less resources to maintain a given biomass than species in

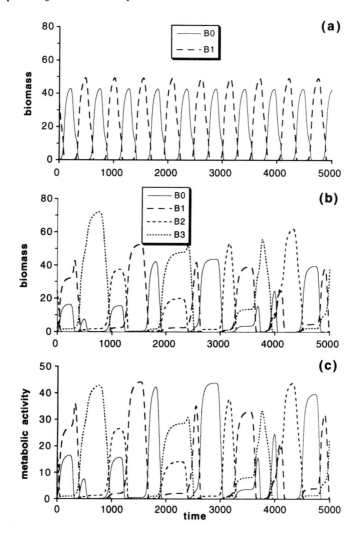

Figure 5.18   Temporal variations of the prey biomasses obtained in simulation runs for (a) 2, and (b) 4 pairs. (c) Metabolic activity derived from (b) by weighting B_i the biomass of a species with Ψ_i, its allometric factor. The periods depicted were chosen during latter phases of the simulation runs when the influence of the initial conditions had vanished. The time axis was arbitrarily rescaled to zero

the weight classes below. Contrasting with the fluctuations in biomass, we note that the maximum amplitudes of the metabolic activities are of equal magnitude. The maximum seems to be constrained by an upper threshold. This threshold is given by the critical nutrient level in the pool where in the absence of predation the net growth of a prey population equals zero (cf. Eqn 5.22 in Box 5.5). In our model this threshold is time invariant and is the same for all species. In a real system we would expect any such threshold to be variable due to environmental variability in combination with species-specific physiological characteristics. Our simple model does nevertheless allow us to make some interesting observations with respect to the relationship between the

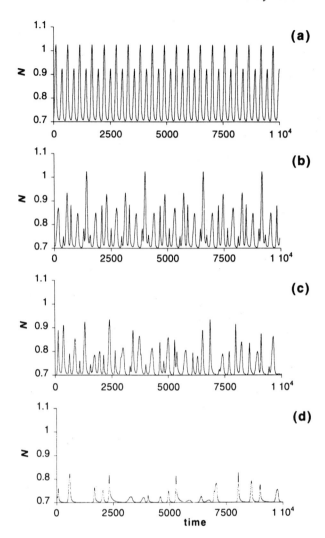

Figure 5.19   Temporal variations of nutrient concentrations obtained in model simulations for (a) 2, (b) 3, (c) 4, and (d) 10 pairs. The periods depicted were chosen during latter phases of the simulation runs when the influence of the initial conditions had vanished. The time axes were arbitrarily rescaled to zero

dynamic diversity of a community and its potential to efficiently use the nutrient available by maintaining a state close to the limiting threshold. To reveal this we have to investigate in more detail the behaviour of global system properties.

Figure 5.19 shows the temporal variations of the nutrient pool obtained for ensembles of 2, 3, 4 and 10 pairs. Again, we note the breakdown in the regularity of the oscillations which are periodic for ensembles with 2 and 3 pairs (Figures 5.19a and b), whereas the addition of further pairs results in an irregular behaviour. This first visual impression can be substantiated by an autocorrelation analysis. The autocorrelation coefficient measures the correlation of a time series as a function of the lag time. Practically one takes a time

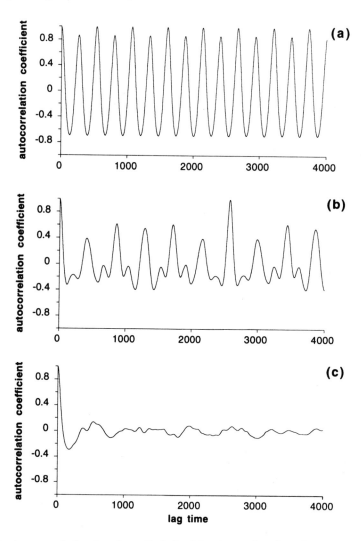

Figure 5.20 Autocorrelation functions, C, derived for time series of nutrient concentrations that were obtained for ensembles with (a) two (b) three and (c) four pairs

series and compares the values measured at a time t with those measured at a time $t + t_k$. A coefficient of one for a certain lag time t_k indicates a repetitive pattern with a period of t_k. A pure sine wave with an oscillation period, p, yields autocorrelation coefficients of one for all lag times $t_k = a \times p$, and coefficients of minus one for lag times $t_k = (a + 0.5) \times p$, where $a = 0, 1, 2, 3 \ldots$ Figure 5.20 represents the autocorrelation functions obtained for the time series of nutrient concentrations depicted in Figures 5.19a to c. The autocorrelation analysis confirms that the temporal variations obtained for ensembles with two and three species are periodic with periods of 534 and 2578 time steps, respectively. We observe a breakdown of the long-term correlation indicative for chaotic behaviour when the number of pairs exceeds three. (I could also show that neighbouring trajectories diverged

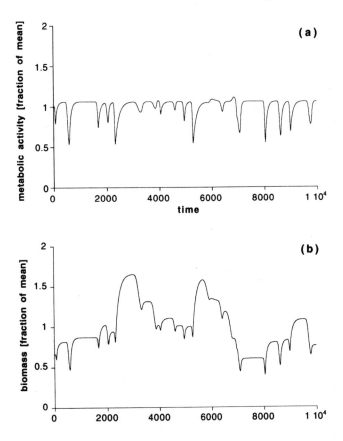

Figure 5.21 Temporal variations of the total (a) metabolic activity that was derived from the (b) biomass of all primary producers. Both are expressed as a fraction of the temporal average. The results were obtained in simulation runs with 10 pairs during periods where the influence of the initial conditions had vanished. The time axis was arbitrarily rescaled to zero

exponentially for $n > 3$, thus giving further evidence for the chaotic nature of the dynamics – cf. Figure 2.4 of the Lorenz model in Chapter Two.)

The autocorrelation function indicates a system's "memory". In the case of chaotic dynamics the length of the memory extends approximately over the period until the correlation coefficient decays to zero. Over this time period we can still trace back the influence of previous states on the system dynamics. It is not possible to predict the temporal development of individual variables, such as the population biomass, for times exceeding this period. From Figure 5.20c we can infer that the memory of a model with four pairs lasts approximately 100 time steps. As a consequence the prediction horizon is limited to this period.

These results support our first impression of an increase in irregularity with an increasing number of pairs. However, a closer look at Figure 5.19 reveals as well that both the amplitudes and the frequency of the fluctuations in the nutrient pool decay when the number of pairs increases. We can further discern the lower threshold beyond which the nutrient concentration cannot drop (cf. Eqn 5.22 in Box 5.5). This threshold sets an upper limit to the total activity of all prey species. The closer a system remains to this limit the higher is its efficiency of utilizing the available nutrient.

With an increasing number of pairs, our model system approaches more and more this threshold. This is illustrated in Figure 5.21a showing the temporal variations of the total activity of all primary producers obtained with an ensemble of 10 pairs. The total activity that is derived by summing over the metabolic activities of the individual populations remains rather constant close to the upper activity threshold determined by the critical nutrient concentration. In the context of Figure 5.18 we noted already that the variations in biomass do not reflect the true dynamics and the nutrient uptake due to the different body-weight-dependent metabolic rates. This becomes also evident when we compare the temporal variations observed for the metabolic activity (Figure 5.21a), with the ones observed for the total biomass (Figure 5.21b). The wide range of fluctuations observed for the biomass does not reveal the regularities detected in the activity pattern.

We may thus conclude that despite the initial impression of a progressively irregular behaviour, closer inspection reveals that the system's temporal organization increases with an increasing number of pairs. The increase in temporal organization becomes evident in the smoothing of the envelope over the activities of all species. What manifests itself as an increase in the irregularity at the level of the single species proves to be an increase in the regularity at the level of the system as a whole. These qualitative trends can be expressed in more quantitative terms when we look at some statistical properties of system behaviour.

I should mention beforehand that numerous simulation runs provided evidence that given a certain parameter set, the model converged to the same stationary state whatever the initial conditions. For periodic oscillations the same amplitudes and periods of the oscillations were obtained. For chaotic behaviour, statistical properties such as long-term averages, probability distributions, power spectra and autocorrelation coefficients were observed to be independent of the initial conditions chosen.

We inferred from Figure 5.19 that both the mean and the variance of the nutrient concentration in the pool decreased with an increasing number of pairs. Figure 5.22 shows that this proved generally to be the case. Figure 5.22a reveals that the decrease in the mean is positively correlated with the recycling ratio. Figure 5.22b shows that the decrease in variability, quantified by the ratio of the standard deviation to the mean of the nutrient concentration, is mediated by an increase in temporal organization, quantified by ΔI_t. We can infer from the increase in the recycling ratio that the increase in temporal organization results in a progressively higher retainment of the nutrient in the system. These observations support the experimental results obtained by Van Voris *et al.* (1980). They observed in microcosm experiments that the calcium loss upon cadmium treatment decreased with an increase in the frequency richness of the microbial community. This frequency richness was determined by analysing the dominant frequencies in the power spectrum of respiratory activity. Frequency rich communities proved to be more resistant towards perturbations. Van Voris concluded thus that an ecosystem's capacity to maintain a functional state depended on the functional complexity and dynamic richness of its component populations. Even when the populations in our model are not yet functionally complex, we can conclude that the richness in time scales is an important component of ecosystem diversity. However, we have to be aware that it is not simply the number of different time scales that is important, but as well the evenness of their distribution.

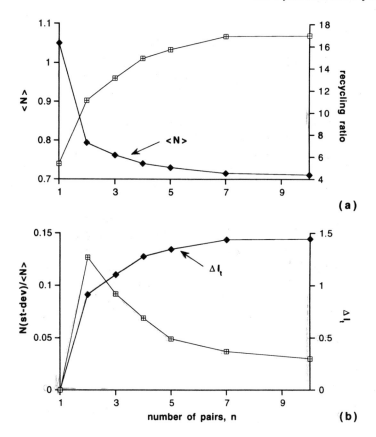

Figure 5.22 Global system parameters as a function of the number of pairs. (a) $<N>$, time average of the nutrient concentration in the pool and the recycling ratio that is expressed as the ratio of nutrient recycled within the system to the external nutrient input. (b) Ratio of the standard deviation to $<N>$ and temporal organization quantified by ΔI_t

5.4.4 Some considerations on the relevance of the results obtained

Temporal organization reflects the dynamic diversity of a system. In our model dynamic diversity derives from the fact that the pairs are spread over a range of weight classes and correspondingly over a range of time scales. The dynamic diversity increases with the number of differently sized pairs that actually contribute to the total system activity. In model simulations with an increasing number of species I observed a regular distribution of the biomasses among the coexisting pairs. Depending on the parameter values chosen, the biomass was either evenly distributed or showed a regular increase with increasing weight class. Due to the lower turnover, the slowly growing species reach a peak biomass significantly higher than the one of the fast growing species (cf. Figure 5.18b). However, whereas the amplitudes of the prey's biomass fluctuations increased with increasing body weight, the frequencies of the oscillations were observed to decrease. These opposing effects resulted finally in the regular biomass distributions observed (cf. Pahl-Wostl, 1993c).

This must not necessarily be the case. To illustrate the importance of dynamic diversity for system behaviour and species coexistence, Figure 5.23 shows the results of a model simulation obtained for three pairs in weight classes 0, 4 and 9. In this case with a large gap between the classes, the large pair dominates the system and the small pairs are entrained. Entrainment and dominance are prevented in ensembles with a larger number of pairs that are more closely spaced along the weight axis: e.g. 0–1–2–3–4–5–6–7–8–9 or 0–2–4–7–9.

We may pose the question whether the regularity of the distributions arises indeed from effects of temporal organization or whether it reflects merely a statistical distribution of activities. To answer this question, I performed simulations with large ensembles of pairs which were distributed evenly along the axis of turnover time on both a linear and a logarithmic scale. The turnover time is the inverse of a species' growth rate. A distribution along the axis of a characteristic time scale is more appropriate than body weight because in our model the main effect of body weight on temporal organization lies just in generating a range in time scales. If the effects observed were of a mere statistical nature, we would expect an even distribution over the whole range of pairs irrespective of their being distributed along a linear or a logarithmic scale. Figure 5.24 shows (a) the number of pairs and (b) the total biomass in categories of turnover time on a linear scale. A linear scale means that category 1 comprises all pairs the turnover time of which lies in the range between 1 and 1.99999, category 2 comprises all pairs the turnover time of which lies in the range between 2 and 2.99999 etc. We note that the number of pairs in each class is approximately the same whereas the total biomass per class shows a clear decrease with increasing turnover time. This is a strong indication that the distributions observed arise from temporal organization of the pairs along the axis of turnover time on a logarithmic rather than a linear scale.

These results may be of interest regarding regularities observed in the biomass–weight distributions in pelagic ecosystems (We may talk of pelagic ecosystems in marine systems and/or large lakes where the pelagial, the open water body, is sufficiently large to be regarded as an autonomous entity with respect to the littoral, the region along the shore.) What has intrigued researchers was the evenness of the biomass distributions across the body weight gradient and the absence of major gaps (Ahrens and Peters, 1991, Sheldon *et*

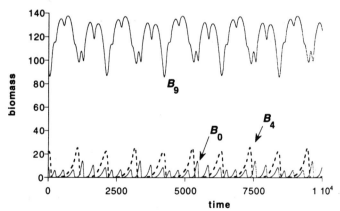

Figure 5.23   Temporal variations of the biomasses of the prey species obtained in a simulation with three pairs in weight classes 0, 4 and 9

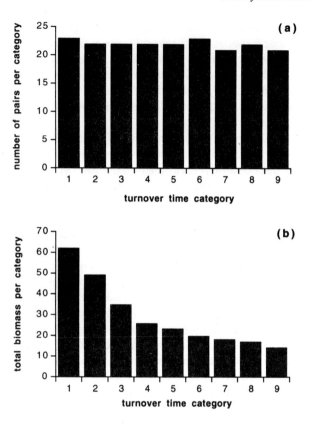

Figure 5.24   Results from simulations where an ensemble of 200 pairs was distributed evenly along the axis of the turnover time on a linear scale. A turnover time category *i* comprises the range of turnover times from *i* to $i + 0.99999$

al., 1972, Sprules and Munawar, 1986). Explanations were sought in physical forces imposing constraints on the community (Legendre and Demers, 1984; Levasseur *et al.*, 1984; Therriault and Platt, 1981). The distribution of biomass was mainly explained by steady state considerations of energy balance (e.g. Platt and Denman, 1978). However, the systems under investigation are hardly ever in steady state. Gaedke (1992b) observed in Lake Constance that the annually averaged biomass of the plankton community was evenly distributed despite pronounced seasonal variations. Such observations are not compatible with steady state considerations. The results obtained from our model simulations suggest that the regularities observed arise from dynamic processes of self-organization in systems operating far from a stable equilibrium point. We have further to be aware that the pairs in the model do not represent taxonomic species but dynamic classes. In any real system each class may comprise a number of taxonomic species. Dominance patterns that characterize most species ensembles may then arise from organization within functional dynamic groups.

Finally I should like to comment on some possible shortcomings of these model simulations. The suspicion may be raised that the model results obtained depend on the special conditions chosen. To refute such objections, I performed a variety of model simulations where I relaxed some of the restrictive assumptions made. One may, for example, take issue with the current version of the model where the same set of basic rates for both autotrophs and heterotrophs was used to reduce the overall number of parameters. Indeed Moloney and Field (1989) found that the estimated rate coefficients of both ingestion and respiration were about an order of magnitude higher for heterotrophs than for autotrophs, whereas the allometric exponent proved to be constant. The conclusions derived from model simulations do not hinge on any of these conditions. In a similar model comprising an additional inactive nutrient pool I showed that relaxing the assumptions of a constant exponent and constant rate coefficients does not change the qualitative characteristics of system behaviour. However, the overall dynamics may be speeded up considerably (Pahl-Wostl, 1993c). Quantitative characteristics, such as frequencies and amplitudes of the fluctuations, were quite sensitive with respect to special assumptions in the model. Changes in the allometric rules or the presence of an inactive nutrient pool were observed to damp the oscillations and to lead to their extension over longer time periods (Pahl-Wostl, 1993c). However, the qualitative nature of the phenomena observed proved to be robust and insensitive to parameter changes. Based on such observations I suggest that inherent in the sensitivity of quantitative characteristics resides a regulatory potential influencing details of the system dynamics, whereas global organizational properties are retained.

The results from the model simulations show that chaos and order may be closely entwined. I even argue that order at the macroscopic level, expressed most clearly in temporal organization, is mediated by the chaotic nature of the interactions at the level of the individual species. This does not imply that any type of chaotic dynamics may always lead to improved system performance. First results from a one-dimensional spatial model show that the effects on system performance hinge critically on a system's potential to organize in space and time.

5.4.5 Introducing a spatial dimension

In the models discussed in this chapter, any spatial dimension has so far been neglected. Doing so is justified when we focus on a local patch where the environment is homogeneous, and the species are distributed evenly over space. To be honest, modellers have to admit that space is neglected in most models for mathematical and conceptual simplicity rather than for its being justified. The introduction of space enhances the complexity of both the models and the interpretation of the results by orders of magnitude. One can still argue that a similar simplification is made in considering species as homogeneous ensembles of organisms neglecting individual variability. In analogy, the effects of spatial patterns could be assumed to neutralize each other on average. In such a situation we would not have to account for the spatial structures and could limit ourselves to dealing with spatial averages only. However, numerous model simulations have provided evidence for this assumption to be hardly ever warranted. Tilman (1994), for example, emphasized the importance of spatial interactions for understanding the coexistence of plants and the maintenance of biodiversity. He cites numerous data to support the results that he obtained with a model combining his earlier ideas on resource competition and between-species trade-offs with dispersal and patch turnover. Holmes *et al.* (1994) summarize results showing that simple organism movement can produce striking large-scale patterns in homogeneous environments.

To investigate the interplay between spatial movement and temporal dynamics and their combined influence on spatio-temporal organization I performed a number of model simulations. Results from a spatial version of the allometric model provide evidence that movement and pattern in space may come to play an important role. I derived the spatial model from the predator–prey model of the previous paragraph by assuming that the species disperse at random along one spatial dimension. In the context of the Brusselator we encountered earlier such a type of random movement that can be described by diffusion models (cf. section 4.6.2). The rate of dispersal is determined by the diffusion coefficient D, with units distance^2/time $\equiv L^2/T$. This implies that the distance covered increases with the square root of time. If a dispersing organism needs for a distance ΔL a time of ΔT units, it needs for a distance $n \times \Delta L$ a time of $n^2 \times \Delta T$ units. Obviously, diffusion is not a very effective means of locomotion to proceed in one direction. For a directed movement with constant velocity, the spatial extension increases linearly with time. When we wish now to investigate the influence of diffusion on model behaviour it is convenient to change the diffusion coefficient. A decrease of D by an order of magnitude implies that the time needed to cover a certain distance increases by an order of magnitude and that the distance covered in a certain time decreases by a factor of $10^{0.5}$.

In a one-dimensional spatial model we focus on a defined interval in space that adjoins at its two ends with the surrounding environment. In the present model, I have chosen the conditions at these two boundaries so as to allow no exchange with the environment, thus preventing the migration of species. To help the imagination, we could think of the model as representing the dimension of depth in a lake where the losses due to sedimentation are neglected. Strictly speaking, we would have to account for spatial gradients in environmental variables such as temperature or light (cf. section 4.7). We may also think of the model as representing an extended lake where we account for the movement along one horizontal dimension only and neglect depth. In a first simplifying approach, I applied the same diffusion coefficient to all species as well as to the nutrient in the pool. This assumption seems to be justified for a system of phytoplankton–herbivore pairs when the influence of the physical mixing processes dominate the active movement of the organisms. However, even when I consider it helpful to have a picture in mind, I have to emphasize again that the model was developed as a conceptual model. It should be looked upon as illustrative of the changes on system behaviour effected by the introduction of a spatial dimension rather than as an accurate model of a real situation.

Combining movement with the allometric equations for resource utilization and multi-species interactions, we obtain the set of reaction–diffusion equations listed in Box 5.6. In such a model the population biomasses and the nutrient are a function of both time (t) and space (x): $B_i(t,x)$, $P_i(t,x)$, $N(t,x)$. The temporal dynamics are governed by the same equations as in the spatially homogeneous model. But now the interactions are local, they are a function of space. Superimposed onto the local dynamics of the interactions are the effects of movement within a neighbourhood the size of which is determined by the speed of dispersal relative to the rate of the biological processes under consideration. Practically, the spatial dimension is partitioned into discrete intervals. The diffusion coeffient determines the exchange rate between neighbouring patches. If the exchange rate is very slow, the patches are essentially independent and the influence of the neighbourhood on local interactions vanishes. If the exchange rate is very fast, the whole spatial dimension becomes homogeneous. What is fast and slow is again determined by the rates of the biological processes under consideration.

Box 5.6   Equations of the spatial model

The equations are obtained by adding diffusion terms to Eqns (5.19), (5.20) and (5.21):

$$\frac{\partial B_i}{\partial t} = \Psi_i \{ f(N)B_i - \rho B_i - \Omega h(B_i)P_i \} + D\frac{\partial^2 B_i}{\partial x^2} \tag{5.23}$$

$$\frac{\partial P_i}{\partial t} = \Psi_{i+q} \{ h(B_i)P_i - (\rho + \lambda P_i)P_i \} + D\frac{\partial^2 P_i}{\partial x^2} \tag{5.24}$$

$$\frac{\partial N}{\partial t} = In - \sigma N - \sum_{i=0}^{n} \Psi_i f(N)B_i + rec \sum_{i=0}^{n} \Psi_i \rho(B_i + \Omega P_i) + D\frac{\partial^2 N}{\partial x^2} \tag{5.25}$$

The reflecting boundary conditions chosen yield at $x = 0$ and $x = 1$:

$$\frac{\partial B_i}{\partial x} = \frac{\partial P_i}{\partial x} = \frac{\partial N}{\partial x} = 0 \text{ for all } t \text{ and for } i = 0,1 \ldots, n \tag{5.26}$$

The spatial dimension was scaled by setting the length of the interval of interest equal to 1. The equations were solved numerically using 100 spatial grid sites. Hence, the exchange rate between adjacent sites equals $D*10^4$. The initial values of all state variables were assigned at random.

If the final goal of the simulations had been the exact numerical solution of the partial differential equations, it would have been necessary to use a progressively finer discretization scheme along space with a decreasing diffusion coefficient. However, for the purposes of demonstrating the effect of spatial movement on spatio-temporal organization, the coarse resolution is satisfactory.

Let us now discuss as representative examples the results obtained from simulations with ensembles of two and five pairs, respectively. To investigate the influence of spatial movement more systematically, I varied the diffusion coefficient for a given set of biological time scales which are determined by the body weight of the component species. We can distinguish two extremes where the spatial dimension has no influence on the network's functional organization. One extreme is given by a diffusion coefficient of zero corresponding to the absence of spatial movement. In this case the system dynamics is identical to a superposition of independent, isolated patches. The other extreme is given when the spatial movement is sufficiently fast that we can consider the whole spatial dimension one single homogeneous patch.

We move now between these extremes by discussing the results of model simulations obtained for different diffusion coefficients ranging from $D = 10^{-3}$ to $D = 10^{-11}$ and finally $D = 0$. The simulations showed that for $D \geq 10^{-3}$ any spatial pattern is immediately homogenized leading to a uniform spatial dimension. A decrease of the diffusion coefficient effects first correlated spatial movements and then a progressive breakdown of the spatial correlations. The transition to uncorrelated spatial patterns is illustrated with results from a model comprising five pairs. Figure 5.25 shows temporal snapshots of spatial profiles of the nutrient concentration in the pool obtained for different diffusion coefficients. For $D = 10^{-5}$ we still observe correlations extending over the whole spatial dimension. With a further decrease of D, the local dynamics of the biological processes become much faster than the movement across space. As a consequence of the local spatial fluctuations become progressively independent and irregular.

The transition in the spatial pattern is even more pronounced for the model version

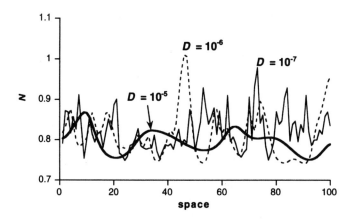

Figure 5.25   Temporal snapshots of spatial profiles of N, the nutrient concentration in the pool, obtained with ensembles of five pairs for different diffusion coefficients, D

comprising two pairs only. Figure 5.26 represents the nutrient concentration in the pool as a function of time and space obtained in model simulations with two pairs. Recall that periodic oscillations were observed for the model without spatial structure (cf. Figures 5.18a and 5.19a). For $D = 10^{-5}$ the periodic temporal oscillations become spatially synchronized and oscillate in phase (Figure 5.26a). For $D = 10^{-6}$, however, we observe a breakdown of the spatial correlation (Figure 5.26b).

The change in system behaviour can be traced back to the onset of diffusion-mediated chaotic behaviour in time. This can be inferred from the autocorrelation functions in Figure 5.27a that were determined for time series of N obtained at a single spatial location. The dashed curve, obtained for $D = 10^{-5}$, is characteristic for periodic temporal oscillations (cf. Figure 5.20a), whereas the solid curve, obtained for $D = 10^{-6}$, reveals the chaotic nature of the local temporal oscillations. In the spatially homogeneous model, the onset of chaotic oscillations, mediated by an increasing number of species, was associated with an increase in temporal organization. However, in the case of the diffusion-mediated chaos here, the onset of irregular behaviour seems to be associated with a deterioration of the system's potential to organize in space and time. This statement is corroborated by the shape of the distribution functions of the nutrient in the pool. Figures 5.27b and c show two distribution functions that were derived from time series of the nutrient concentration at fixed spatial locations. To obtain these distributions, I partitioned the whole range of the concentrations observed (0.7–1.1) into eight discrete intervals, and determined the number of observations falling into each interval. For $D = 10^{-5}$ the distribution is highly skewed. N remains close to the threshold N_{cr} ($= 0.7$) most of the time, indicating thus an effective nutrient utilization. The shift towards a Gaussian type of distribution for $D = 10^{-6}$ reflects the decline in spatio-temporal organization.

Figure 5.28 summarizes the influence of the spatial movement by representing the behaviour of the measures of spatio-temporal organization, and of the spatio-temporal averages of the biomasses and the nutrient concentration as a function of D. Figures 5.28a and b show results obtained for two pairs, whereas Figure 5.28c shows results obtained for five pairs.

For a diffusion coefficient of 10^{-5} we observed spatial synchronization in the model

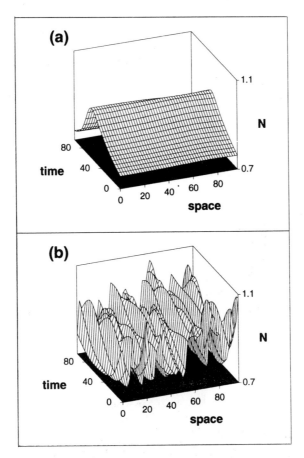

Figure 5.26 Nutrient concentration in the pool as a function of time and space obtained in model simulations with two pairs for different diffusion coeffients. (a) $D = 10^{-5}$, (b) $D = 10^{-6}$

with two pairs. A diffusion coefficient of zero corresponds to an absence of spatial movement and hence to a superposition of 100 independent patches. Due to the fact that the initial conditions for each patch were chosen at random the phases of the temporal oscillations are randomly distributed. In between these extremes of either spatial synchronization or complete independence we observe a transition in system performance with decreasing D. Figure 5.28a reveals that the transition from periodic to chaotic behaviour $(D = 10^{-5} \Rightarrow D = 10^{-6})$ is associated with a pronounced drop in total biomass and a significant increase in the nutrient concentration of the pool. We already inferred from the shape of the nutrient distribution depicted in Figure 5.27 that efficient spatio-temporal organization is impeded. This observation is substantiated by the results obtained for the measures of organization represented in Figure 5.28b. The transition to chaos results in a drop in ΔI_{ts} quantifying the network's spatio-temporal organization. A further decrease in D results in a recovery of system organization to finally exhibit in the limit $(D < 10^{-11})$ the behaviour obtained for $D = 0$. In Chapter Four, I explained that the measure of spatio-temporal organization does not distinguish between random and correlated patterns. Hence, the random superposition of 100 patches exhibiting periodic

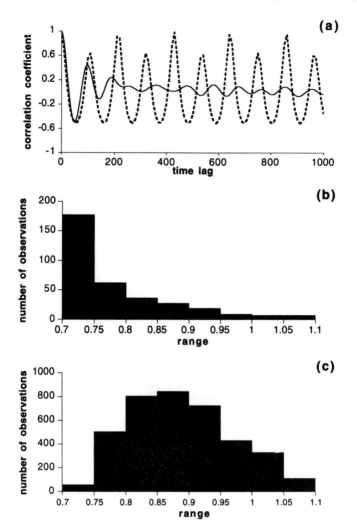

Figure 5.27 (a) Autocorrelation functions derived for the time series of N at a single spatial location for: $D = 10^{-5}$ – dashed curve, and $D = 10^{-6}$ – full curve. (b and c) Distribution functions of the values observed for N for (b) $D = 10^{-5}$ and (c) $D = 10^{-6}$

oscillations gives the same spatio-temporal organization as the synchronized oscillation extending over the whole spatial dimension. Effectively, the same ensemble of spatio-temporal network configurations is generated in both cases. ΔI_t, the measure of temporal organization in the time-resolved network averaged over space, reflects the random distribution of the temporal pattern across space. ΔI_t is high for the synchronized motion, but remains close to zero for a random superposition of independent patches.

Similar results were obtained in model simulations with five pairs. Figure 5.28c shows the spatio-temporal average of the nutrient concentration in the pool and the measure of spatio-temporal organization ΔI_{ts} as a function of D. A major difference compared with the two-species model lies in the fact that the transition is more gradual and that it

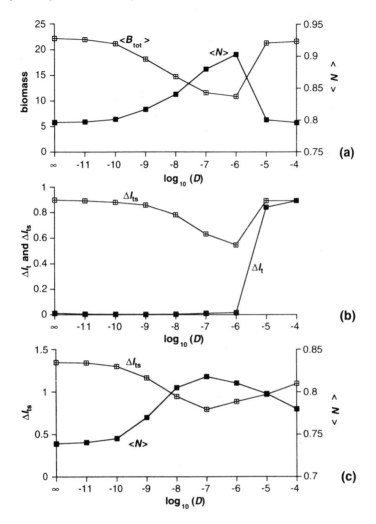

Figure 5.28 The behaviour of variables describing whole system behaviour as a function of D. The results were obtained in simulations with (a) and (b) two pairs and (c) five pairs. $<N>$, and $<B_{tot}>$ refer to the average over space and time of the nutrient concentration and the total biomass, respectively. Temporal and spatio-temporal organization are quantified by ΔI_t and ΔI_{ts}

proceeds from a temporally irregular but spatially correlated behaviour, where organization is possible, to a both temporally and spatially irregular behaviour, where organization is impeded.

We may thus conclude that under the special conditions chosen, any spatial pattern proved to be detrimental to the system's potential to organize in time and space. The organization was at maximum when either dispersal was fast enough for the whole spatial dimension to behave as a single homogeneous patch, or when dispersal was slow enough that the spatial dimension could be looked upon as being partitioned into an ensemble of independent patches. We have, however, to be aware that the conditions chosen comprise only a tiny fraction of possible combinations of temporal and spatial characteristics. In particular, we might expect that species operating at different time scales have different spatial patch extensions. By choosing one single scale of movement along space we force the hierarchy in time scales into a rather rigid scheme. Further, space cannot be reduced to one single dimension.

Can we nevertheless draw some conclusions for the real world from these preliminary observations? The results obtained indicate that the potential to engage into effective spatio-temporal organization depends on the relationship of the temporal and spatial scales of a system. The fact that spatial effects depend on the ratio between rates of interaction and rates of spatial movement and that effects may be observed for certain spatial and temporal scales only is hardly new. Kareiva (1990) gave a comprehensive review on population dynamics in spatially complex environments ending with a strong plea for experimentalists to pay more attention to spatial patterns. The approach chosen here offers a tool to investigate more systematically the relationship between the spatial and temporal scales of the component populations and the spatio-temporal organization of the system as a whole. Instead of emphasizing mechanistic details, I focus here on patterns of distribution, patterns of spatio-temporal diversity in a system. I conjecture that given a certain spatio-temporal pattern in the environment, an ecosystem adopts a configuration with respect to the temporal and spatial scales of its component populations that enables spatio-temporal organization to become effective. Here might be a clue to understanding functional patterns in space and time and their mutual relationship. This is quite an intriguing idea even when we cannot as yet derive any general conclusions from these first preliminary results. The characteristics of the relationship between space and time remain to be investigated and conditions for effective spatio-temporal organization remain to be derived. In the next chapter I will present a general approach for both experimental and model investigations to foster progress in this direction.

5.5 SUMMARY AND CONCLUDING COMMENTS

In this chapter it was my intention to convey further understanding for the nature of spatio-temporal organization in ecological networks, to illustrate the interplay between system-level constraints and species interactions, and to reveal which factors are important in this context. To do so I designed a set of structurally similar models with different sources for spatio-temporal variability. The following properties are common to all the models studied:

- An ensemble of functionally equivalent units (primary producers, predator–prey pairs), X_i, shares a common resource, R.
- The dependencies between the variables follow the relationships:

$$\frac{\delta X_i}{\delta R}<0 \text{ and } \frac{\delta R}{\delta X_i}>0 \Rightarrow \frac{\delta X_j}{\delta X_i}<0$$

hence

$$\frac{dX_i}{dt}>0 \Rightarrow \text{negative feedback of } X_i \text{ on } X_j$$

$$\frac{\mathrm{d}X_i}{\mathrm{d}t} < 0 \Rightarrow \text{positive feedback of } X_i \text{ on } X_j$$

In words, the above relations imply that the units exert a negative influence on the global resource pool by depleting the resource. Conversely, the resource exerts a positive influence on the units by fostering growth. Hence, the obvious conclusion would be that the sharing of the common resource results in indirect competition among the units. However, in a dynamic situation a unit exerts a negative influence on others during phases of net growth only. In phases of decline a unit exerts a positive influence on others by rendering resource available for growth. (This applies to any finite resource be it a nutrient that is recycled or physical space that becomes available to be colonized again.) In Figure 4.6 I visualized such dynamic interactions by the picture of feedback spirals linking units along the dimension of time (this could equally be space or both). When the units are equivalent to primary producers, the situation of spatio-temporal resource partitioning is easy to understand. In the case of predator–prey pairs, the situation becomes more involved since each pair is organized itself in terms of pulsed consumption. But essentially the same overall pattern of spatio-temporal organization applies.

We used the models to study systematically changes in temporal organization and resource utilization brought about by changes in the dynamic diversity of the species ensemble. In the first model such a behaviour was realized by distributing primary producers along a temporal seasonal gradient. In the last model predator–prey pairs were distributed along a gradient of body weight corresponding to a gradient of dynamic characteristics. Regardless of the source of temporal variability, we observed an increase in temporal organization to effect an improvement in whole system performance. The latter was expressed in the efficiency of resource utilization. An increase in dynamic diversity effected a smoothing of the envelope over the activities of all species and a decrease in losses from the system. We may thus speak of a trade-off between local and global variability. Whereas the variability of the individual populations increased, there was a decrease in the variability of global system properties such as total primary production or the nutrient concentration in the pool.

Such observations support the hypothesis that ecosystems must be viewed from a dualistic perspective. Processes at the species level permanently change macroscopic properties of the system, which then imposes new constraints onto the species themselves. A species receives its identity on how it is embedded in a network context. The latter is not simply given by a static linkage pattern. It is just the temporal and spatial variations that render possible species coexistence and spatio-temporal organization. The ability to organize in time and space hinges on the presence of a diversity of functionally similar but dynamically distinct units. More emphasis should thus be given to diversity regarding dynamic characteristics both in model development and in the investigation of natural systems.

Further investigations along the lines of reasoning pursued here should also be useful in discussions on another contested topic in ecology, the relationship between species abundance and body weight (size). There seem to exist two conflicting lines of empirical evidence (e.g. Currie, 1993; Lawton, 1989, 1990). Figure 5.29 depicts schematically the major features of the two different patterns obtained. In (a), if data are collated for populations from a wide variety of published studies across many ecosystems, average population densities tend to be inversely correlated with body weight. In contrast, data from local assemblages resemble more the triangular shape depicted in (b). The seeming contradiction may be resolved by taking into account the body weight range covered by the different approaches. The ranges are denoted by the double-headed arrows. A linear relationship as depicted in (a) seems to hold for global species ensembles encompassing body weights that are spread over 10 to 20 orders of magnitude. The triangular distributions sketched in (b) characterize local assemblages of taxonomically related organisms, e.g. beetles from the canopies of rain forest trees, where the body weights extend over one or two orders of magnitude only. Currie (1993) showed that the two distributions are compatible if one takes into account sampling bias and the difference in range. Whether such gross patterns are of any ecological relevance and how they should be interpreted has been a matter of considerable dispute (e.g. Currie, 1993; Gaston *et al.*, 1993; Lawton, 1989). There are possible biases and pitfalls in collecting and reporting data. Problems are also associated with linear regressions in log–log plots. The variables may scatter over several orders of magnitude and still yield a significant correlation close to one. However, despite certain reservations, the idea of general regularities is too intriguing to be easily dismissed.

I argue here that the two distributions depicted in Figure 5.29 may reflect different levels of ecological organization. Imagine the species "sausage" in Figure 5.29a to be cut into thin slices as denoted by the rectangle. Within such restricted groups we could conceive of species ensembles organizing themselves within guilds of similar dynamics and function. The overall pattern is then nothing else but the superposition of such overlapping ensembles. Such patterns can be made more intelligible in the light of spatio-temporal self-organization. Distributions that can be considered to be similar to those depicted in Figure 5.29 were obtained in the model simulations presented in this chapter. We should recall that the biomass was distributed evenly among the pairs in the allometric predator–prey model corresponding to an inverse correlation between numerical abundance (=density) and body weight as depicted in Figure 5.29a. Considering that a

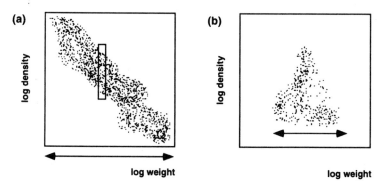

Figure 5.29   Generalized relationships between body weight and population density obtained for (a) global, and (b) local species assemblages. The "wealth" of data points represents an idealized situation. Distributions as depicted in (a) are generally obtained for global species ensembles encompassing body weights that are spread over 10 to 20 orders of magnitude. The triangular distributions sketched in (b) seem to characterize local assemblages of taxonomically related organisms where the body weights extend over one or two orders of magnitude only. The rectangle in (a) indicates that the distribution in (b) may be looked upon as a blown-up slice of the larger distribution

pair represents a species ensemble within a weight class rather than a single population this can be looked upon as example for the temporal organization of species ensembles along the body weight axis. In the seasonal phytoplankton model the species distribution was nearly log-normal (cf. Figure 5.13). This can be looked upon as an example of the organization of species within an ensemble. In the allometric model we focused on the self-organization of predator–prey modules operating on a range of different time scales. In the phytoplankton model the emphasis was on the organization of a large ensemble of dynamically similar species along a seasonal gradient. Even when the models are not very representative for the organization in large ecological networks we may tentatively draw some analogies to the distributions depicted in Figure 5.29. We may conceive of a modular structure, the functional pattern of which varies in time and space. To support these first vague statements a more coherent theoretical framework is required to gather further empirical and theoretical knowledge that might foster progress in elucidating the spatio-temporal hierarchical structure of ecosystems.

Chapter Six

On how to delineate the structure of ecological networks

When I introduced the measures for a network's spatio-temporal organization, I simply assumed the network structure comprising compartments and flows as being given. However, in natural ecosystems the structure of the networks is by no means self-evident. In the meantime, we have developed an understanding of the nature of spatio-temporal organization, its requirements, and its relevance for system function, but we should now take a step back and ask what is it that organizes itself? How do we have to conceptualize an ecosystem in terms of a network to reveal patterns of self-organization? We face the problem that what we detect depends to some extent on the description chosen. Thus we have become aware of potential bias. Before dealing with the questions posed it might be useful to have a closer look at the current practice and debate in ecological research to comprehend the intricate problems facing us when we try to investigate ecological networks.

6.1 ANALYSES OF PATTERNS IN ECOLOGICAL NETWORKS – POTENTIALS AND PITFALLS

We may distinguish two major approaches for investigating patterns of trophic interactions in ecosystems based on the importance attributed to quantitative aspects. The strength of an interaction between two network compartments may be quantified by the amount of energy and/or matter exchanged. A network may also depict the topological linkage pattern where a link denotes simply the presence of a trophic interaction, whereas its magnitude is neglected. The theoretical edifice relying on the topology of webs is generally referred to as food web theory (comprehensive introduction in Cohen *et al.*, 1990). Structural attributes such as the number of links per species or average chain length seem to exhibit regularities in a whole collection of webs from different communities (Cohen *et al.*, 1990; Lawton and Warren, 1988; Pimm *et al.*, 1991). The surge of interest in food web structure and organization has largely been motivated by the work of theoretical ecologists concerned with the relation between complexity and stability. We have already discussed the major issues in the complexity–stability debate in Chapter Two. Recall that May (1973) defined an upper bound for an ecosystem's connectivity beyond which stability breaks down in model systems. May's argument can be extended to the prediction that food web structures observed in nature ought to correspond to the structures that enhance the likelihood of stability in models. Subsequent theoretical considerations added further attributes of stable systems as, for example, the absence of omnivory or the prevalence of short chains (Cohen *et al.*, 1990;

Pimm, 1991). Indeed, real food webs seem to conform to the predictions that structural attributes which decrease stability in models are rare in nature.

In recent years, however, a hot debate has started about the validity of the theoretical results and the underlying assumptions. Researchers have become increasingly aware of the problems arising from the lack of methodological standards on how to describe a food web (Paine, 1988; Peters, 1988; Polis, 1991; Schoener, 1989b). These problems cast doubt on the relevance of the general food web patterns that were derived from a catalogue of published webs. Instead of repeating the arguments in detail I confine myself to pointing out major points at issue. These are:

- The arbitrariness in the choice of the spatial and temporal scales both with regard to resolution and with regard to the boundaries of the observations.
- The neglect of spatial and temporal variations in species interactions.
- The highly heterogenous level of aggregation across the webs. Most taxa are highly aggregated, but some are identified to taxonomic species.

Contradictory conclusions have been drawn from different investigations addressing some of the problems stated above. Interestingly enough proponents of the food web approach seem to find evidence supporting at least partially the results from earlier investigations (e.g. Pimm and Kitching, 1988; Pimm *et al*, 1991; Schoenly and Cohen, 1991). The critics emphasize the discrepancy between results from more detailed studies with those derived from catalogued webs, and tend to reject most theoretical results as artifacts of oversimplified descriptions of real communities (e.g. Martinez, 1991; Paine, 1988; Peters, 1988; Polis, 1991; Winemiller, 1990). Facing these conflicting lines of evidence it is quite difficult to come to any conclusions. To illustrate the nature of the points at issue it is worth mentioning the results from a fews studies.

Box 6.1 displays an illustrative example for the problems associated with aggregation in a network of a tidal marsh creek adjacent to Crystal River (Ulanowicz, 1986). It is quite evident that the researchers were particularly interested in fish. Such an inhomogeneous aggregation may not really be disturbing as long as one focuses on a single system and its changes over space and time. Artifacts of aggregation may, however, become a source for pitfalls when one attempts to derive general properties of food webs and/or ecosystems (e.g. Pahl-Wostl, 1992b).

Sugihara *et al*. (1989) were the first to address the effects of trophic aggregation systematically. They concluded that most of the food web statistics were scaling in the sense that they remained constant over a wide range of data resolution. Martinez (1991)

Box 6.1   Compartments of the network of Crystal River (Ulanowicz, 1986)

microphytes	bay anchovy	moharra
macrophytes	needlefish	silver jenny
zooplankton	sheepshead killifish	sheepshead
benthic invertebrates	goldspotted killifish	pinfish
blacktip shark	gulf killifish	mullet
tingray	longnosed killifish	gulf flounder
triped anchovy	silverside	detritus

however, provided evidence to the contrary by studying the effects of progressively increasing aggregation in a lacustrine food web. In his opinion the absence of significant effects of aggregation on some food web patterns detected by Sugihara *et al.* originated from their applying an incomplete analysis and a not very thorough scheme of aggregation.

Other studies emphasized the effect of temporal variability. Schoenly and Cohen (1991), for example, compared cumulative webs (averaged over a certain time period) with time-specific webs of the same communities. They showed food web structure to vary substantially over time due to a successional as well as a seasonal turnover of species. According to their results, most quantitative food web patterns derived for the time-specific webs did not deviate significantly from those catalogued for prior food web collections. Particular details like counting detritus as a single basic trophic species or the inclusion of opportunistic species were identified to cause the major deviations observed. Hence, they concluded that there was no reason to assume fundamental flaws in current theoretical concepts.

A rather contrary position was advocated by Polis (1991). In his study of a desert food web he found hardly any correspondence with the highly aggregated webs studied previously. Consequently he concluded that most of the theoretical derivations were simply wrong, resulting from an oversimplified representation of real communities, which he perceived as extraordinarily more complex than those webs catalogued by theorists. The catalogued webs typically comprise 10 to 50 (mainly trophic) species whereas in the study of the desert food web several hundreds of taxonomic species were resolved with several thousands (!) remaining still aggregated. Such numbers raise the question whether any attempt to resolve a food web into taxonomic detail would in fact be reasonable.

This discussion gives us an idea about the difficulties facing us when we search for general patterns in ecological networks. In defence of the food web theorists it must be mentioned that it is notoriously difficult to find data on food webs as a whole. Most data sets in the catalogued webs were not assembled for the specific purpose of a food web study and thus reflect the biases of investigators interested in particular taxonomic groups. Further, if we are interested in general properties, such as the relationship between complexity and stability, and if we intend to infer dynamic consequences from structural patterns, we need to pay more attention to spatial and temporal scales. The notions of both stability and complexity hinge entirely on how the system of interest is defined, in terms of both the boundaries placed around a system and the variables the attention is focused on.

The original definition of an ecosystem by Tansley (1935) is essentially devoid of any considerations of temporal and spatial scales. Such an attitude seems to prevail as well in network or food web analysis. Cousins (1990) made an attempt to overcome this deficiency by suggesting a new food web entity that he called the ecotrophic module. It is delimited in space and time by the scales determined by the social group of a food web's largest predator. Such a unit that gives a guideline for delineating the overall spatial and temporal range of observation is important. However, in order to elucidate network patterns, we need to resolve the internal structure of an ecosystem. We should like to answer questions such as: What is the relationship between a system's adaptability and the spatio-temporal organization and dynamic diversity of the ecological network? To what extent does the nature of the physical environment determine the internal organization of an ecological network? To address questions of this type in a meaningful way, requires a coherent concept of the identity of the network compartments and the nature of spatio-temporal interaction patterns. Such a concept is of paramount importance for expressing theoretical considerations in a precise way as required for comparisons with empirical observations, and for the construction of adequate models.

6.2   A NETWORK CONCEPT EMBRACING TIME, SPACE AND FUNCTION

Any network concept is based on some theoretical foundations. Describing ecological networks in terms of taxonomic species can be looked upon as a bottom-up approach since it is assumed that the network structure is determined by the properties of its basic constituents that are identified with taxonomic species. On the contrary, describing a network in terms of trophic levels is rather a top-down approach. The compartments are defined with respect to their trophic function in the network context irrespective of the properties of the component organisms. A possibility to start bridging these divergent descriptions is the introduction of functional groups which categorize species according to features such as body plan, behaviour or life history strategies. Hence, we encounter a wide array of possibilities of how to describe an ecological network. Depending on the description chosen, a network's structure and its organization in space and time may present themselves rather differently.

The concept to be introduced here is an attempt to provide a more coherent framework for establishing ecological networks. It has been developed with the goal of accounting for the dynamic and functional diversity of a network's component populations and the spatio-temporal variations in the network structure. Recall that the dynamic nature of ecosystems is perceived as a continuous interplay between the populations generating their network environment and the organization at the level of the network as a whole imposing constraints on the populations. Based on our understanding of spatio-temporal organization the essential characteristics of network compartments can thus be formulated as: the typical temporal and spatial scales on which a unit operates and the relational context in the network. Due to the fact that the network is subject to continuous change any such unit can only have a finite lifetime and a finite spatial extension.

The main elements of the network concept thus arising are sketched in Figure 6.1. Part (a) depicts the level of the system as a whole viewed as a dynamic network where energy and material are processed and that has finite extensions in time and space. Both energy and nutrients enter the system locally at the level of the primary producers indicated by the dark shaded interface to the environment. Most of the solar energy fixed by primary producers is degraded into heat via respiration by all living organisms. The degradation of energy is associated with a recycling of nutrients. Part (b) depicts the level of a single compartment representing an ensemble of organisms. Function is determined by the location along the trophic gradient and by the embedding in the relational context of the trophic web. The dynamic characteristics refer to the typical spatio-temporal scales characterizing the interactions of an ensemble of organisms with its environment. The dynamic characteristics are determined by the component organisms, e.g. by body size. Recall that an ecosystem's structure may be viewed as comprising the three successive levels of the organisms, compartments and the system as a whole (Figure 4.2). A compartment is now defined by constraints from the level of the organisms (dynamics) and the level of the ecosystem as a whole (functional embedding). Spatio-temporal

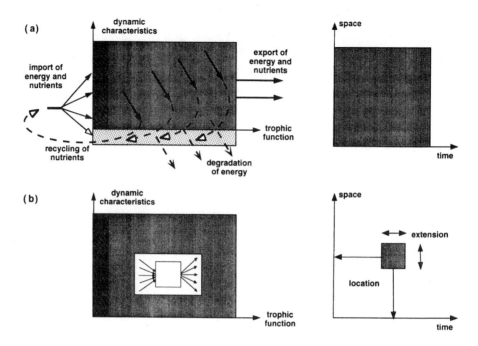

Figure 6.1   The main elements of the network concept. (a) The system as a whole is viewed as a dynamic network where energy and nutrients are processed and that has finite extensions in time and space. Both energy and nutrients enter the system locally at the level of the primary producers indicated by the dark shaded interface to the environment. The energy leaves the system distributed over many compartments. Most of the solar energy fixed by primary producers serves for maintenance and is degraded into heat via respiration by all living organisms. The degradation of energy is associated with a recycling of nutrients. (b) A network compartment is characterized by it embedding in a network context and by its dynamic characteristics referring to the typical temporal and spatial scales of a compartment's activities. A compartment is assumed to have a physical identity characterized by the extensions and locations along the dimensions of time and space

organization means that the internal network structure organizes itself at any moment in space and time to fill the envelope determined by these constraints in combination with the boundaries imposed by the physical environment. The rigid box in Figure 6.1 has therefore to be looked upon as something dynamic changing its shape to interact with its environment. Within this dynamic structure an ensemble of functionally equivalent organisms is a tangible unit that has a physical identity characterized by the extensions and locations along the dimensions of time and space.

In its simplest form such a description could be rather trivial, reducing to a food chain in terms of trophic levels as sketched in Figure 6.2. The dynamic characteristics are expressed by body weight, since this can be looked upon as the essential property determining the spatio-temporal scales of an organism's interactions with its environment. Figure 6.2 corresponds to the traditional assignment of a pelagic food chain where the energy transfer is directed along a gradient of increasing body weight. The input- and output-environments of the compartments are reduced to the transfers between trophic

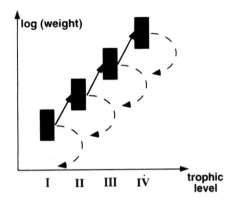

Figure 6.2 The simplest possible description of an ecological network where the dynamic characteristics are expressed by body weight and function is identified with trophic level. Such a description corresponds to the familiar picture of a food chain in a pelagic system where energy is transferred along a gradient of increasing body weight. The dashed arrows indicate the recycling of nutrients

levels the dynamic characteristics of which are determined by average body weights.

Such is quite a coarse and rigid description leaving few possibilities to express the dynamic nature of ecosystems. To describe spatio-temporal organization we need to resolve the coarse structure in more detail and take into account that any specific network configuration has only a finite extension in time and space. Based on these considerations I introduce *trophic dynamic modules* (*TDM*) as the basic units of ecological networks. The name implies their being defined on both dynamic and trophic function and their forming the base of a modular structure of an ecological network. A trophic dynamic module comprises all organisms that have the same dynamic characteristics and the same function in the trophic web and that coexist over the same finite period in time and space. A description in terms of TDMs must be seen as an observational grid decomposing an ecological network into a spatio-temporal mosaic of interaction units. To give an example I return again to the simple picture of the pelagic food chain. A TDM could for example comprise all phytoplankton within a certain size range coexisting over a certain period in spring at a certain location.

The notions of a certain time and a certain location are not arbitrary. To investigate the spatio-temporal organization of the whole network requires a coherent resolution across the wide range of spatial and temporal scales typically encountered in an ecosystem. A coherent resolution can be obtained by introducing relative spatio-temporal coordinates that are measured in the intrinsic scales of the organisms involved. Relative units such as generation time or habitat range are more appropriate than absolute units such as days or metres for describing and comparing the dynamics of organisms and their environment. The typical scales in time and space determine what I refer to here to as the "*dynamic niche*". The trophic dynamic modules of a food web will then be defined in a hierarchical (= covering a wide range of spatial and temporal scales) spatio-temporal framework according to their dynamic niche and according to their "*functional niche*" determined by the characteristics of their input- and output-environment.

6.2.1 The dynamic niche

Empirical and theoretical investigations have revealed the importance of variability in space and time for species coexistence and community organization (e.g. DeAngelis and Waterhouse, 1987; Diamond and Case, 1986). This view was also supported by the results from model simulations discussed in the previous chapters. Based on such findings, I make the assertion that a major characteristic of organisms concerning their role in community organization is how they perceive and/or generate patterns in time and space. The temporal and spatial scales of variability are of importance rather than the nature of what varies (e.g. biomass, resource concentration, environmental factor). The total activity of a system is partitioned among the organisms. The share of an ensemble of organisms is determined by what I refer to as their dynamic niche. Hence, the traditional concept of the ecological niche that is based on a static description within a multi-dimensional resource space, is replaced by a dynamic perspective where the niche space is reduced to the nature of a system's variability in time and space.

Regarding its dynamic properties, I assume that a population of organisms can be characterized by a typical time scale which refers to a population's response time. The dynamic niche of a population can thus be viewed as being located along a one-dimensional niche space of a typical response time. The axis of this niche space is more appropriately described on a logarithmic scale where the ratios of equally spaced time intervals remain constant. Organisms perceive time relative to their intrinsic scale rather than on an absolute linear scale. This means that the shape of such a dynamic niche varies in proportion to the typical response time it refers to. Such behaviour can also be made intelligible with a simple model of logistic growth:

$$\frac{dB}{dt} = \lambda B \left(1 - \frac{B}{K}\right) \tag{6.1}$$

B is population biomass, λ is the growth rate and K is the carrying capacity of the environment. The process of growth can be characterized by a typical time scale $\tau = \lambda^{-1}$ that is called the response time. The biomass as a function of time, $B(t)$, can be derived from Eqn (6.1) to yield:

$$B(t) = \frac{K}{1 - (1 - K/B_0)e^{-\lambda t}} \tag{6.2}$$

where B_0 is population biomass at the time $t = 0$. We note that the biomass tends towards the carrying capacity K for times that are much larger than the response time τ (for $t > > \tau \Rightarrow e^{-\lambda t} = e^{-t/\tau} \to 0$). How fast the biomass changes is determined by the rate λ in the exponent. Figure 6.3 illustrates the difference in the dynamics of two populations with different growth rates equal to 1 and 0.1, respectively. Figure 6.3a shows the response of population biomasses to a doubling of the carrying capacity at $t = 0$. As expected, the change of population 2 with the smaller growth rate and hence the larger response time is much slower. Figures 6.3b and c show the change in biomass on relative time scales given in units of the reponse times τ_1 and τ_2. We note that the populations respond to changes in their environment in proportion to their time scale. The dynamic characteristics of a population thus scale with the logarithm of the time scale. It is a typical property of linear models that we obtain exactly the same shape for the temporal variation when they are depicted in units of the response time. We would not expect such a perfect coincidence in

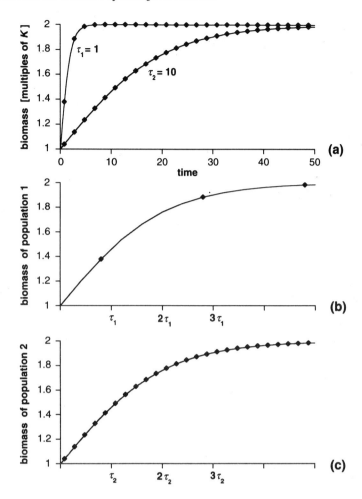

Figure 6.3   The response to a doubling in the carrying capacity, K, obtained with the logistic model for two populations with different growth rates: $\lambda_1 = 1$, $\lambda_2 = 0.1$, and correspondingly different response times $\tau_i = \lambda_i^{-1}$. The changes in population biomass are shown (a) on an absolute time scale, and (b) and (c) on relative time scales given in units of the response times. The diamonds always indicate intervals of two time units on an absolute time scale

complex systems where the response is determined by the environment and nonlinear feedback. However, such a simple linear model is a convenient tool to derive a comparative classification of the dynamic characteristics of groups of organisms – a classification that extends as well to complex feedbacks and nonlinear interactions.

We must now derive an operational measure for the typical time scales of a population of organisms regarding their interactions in the ecological network. Biomass turnover time seems to be an appropriate choice. It determines ecologically relevant interactions such as starvation when food is absent or response to slow changes in the environment that may lead to the replacement of one species by another. Changes that occur on time scales much faster than the turnover time are reflected in physiological and/or behavioural responses at the level of the individual organisms. Hence, I consider the biomass turnover time an appropriate measure to characterize the dynamics of ecologically relevant interactions among populations (see also Calder, 1984). The biomass turnover time can be determined as:

$$\tau = \frac{B}{P} = \frac{NW}{P} \tag{6.3}$$

where B is population biomass, P is net production, N is the number of individuals and W is body weight.

In Chapter Five we made use of the fact that the metabolic requirement of an organism depends on body weight. For a wide range of taxa, the biomass turnover time, τ, can be expressed as a function of body weight, W, according to allometric relationships of the type (Calder, 1984; Peters, 1983; Reiss, 1989):

$$\tau \sim W^\varepsilon \tag{6.4}$$

where the experimentally derived values for the allometric exponent ε varied from 0.25 to 0.33. Body weight may thus be used as a classification criterion to substitute for the biomass turnover time.

As the notion of spatio-temporal organization already implies, an ecological network's organization is not only determined by the temporal dynamics but also by the patterns in space. A classification of organisms based on turnover time includes as well a classification with respect to resolution in space. In our discussions of models including a spatial dimension, we noted that the time scale and the spatial scale of a phenomenon are not independent. The spatial distance covered increases with the time scale under consideration. It is hence reasonable to assume that the typical spatial range increases with increasing biomass turnover time of a population. When we consider the link between time scales and body size this relationship becomes even more obvious. The relationship between body weight and body size, L, follows approximately:

$$W \sim L^3 \Rightarrow \tau \sim L^{3\varepsilon} \tag{6.5}$$

The exponent 3ε may range from about 0.75 to 1. The response time changes in regular fashion with body size that itself determines the resolution of the spatial dimension.

Small organisms perceive, use and partition environments on much finer spatial scales than large organisms. Morse *et al.* (1985) observed an inverse proportionality on a logarithmic scale between the numerical abundance of insects and body size. They explained their observations by the increase in available home range with decreasing body size. In their case home range was identical to plant surface, which exhibited a fractal structure. Swihart *et al.* (1988) observed for mammals that the amount of time required to traverse the home range was related to W^ε. Thus, even though small mammals traverse their home ranges more rapidly in a chronological sense, during a lifetime, which is related to W^ε as well, small and large mammals appear to use their home ranges with equal intensity.

Hence, in a coarse approximation the dynamic niche as defined above includes implicitly the scaling along the dimension(s) of space even when details of body shape and/or the characteristics of the spatial perception of the organisms are not taken into account. Body weight is not central to the description of ecological networks presented here. It constitutes, however, a convenient and experimentally accessible substitute for the dynamic characteristics of organisms.

6.2.2 The functional niche

As indicated in Figure 6.1, the function of an ensemble of organisms is defined by its embedding in an ecological network. In Chapter Four we discussed that such an embedding can be characterized by a compartment's input- and output-environments. The input-environment determines a population's location along the trophic gradient. Here it is useful to make a first coarse distinction between autotrophic and heterotrophic organisms.

As shown in Figure 6.4a, the embedding of heterotrophic organisms in a network can be characterized by the location and range of their input- and output-environments along the axis of turnover time. If we identify the axis of dynamic characteristics with the body weight axis, heterotrophic organisms would be characterized by the body weight range of their prey and by their susceptibility to predation by differently sized organisms.

The location along the trophic gradient is given by a compartment's trophic position. Contrasting with trophic levels the trophic position is not confined to integer values. The trophic position of a consumer is equivalent to the average number of trophic transfers an energy unit has passed through from being stored by primary production until serving as a food item to this consumer. In a network where the magnitudes of energy flows are known it is possible to assign to each compartment its trophic position based on the amount of energy received via pathways of different lengths (Ulanowicz, 1986). Figure 6.5 illustrates such an assignment for the example of two simple networks. The trophic positions that are derived from the fractional contribution of all pathways to a compartment's total input

(a)

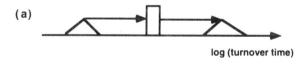

log (turnover time)

(b)

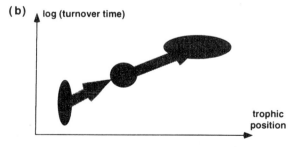

log (turnover time)

trophic position

Figure 6.4 (a) A population of organisms can be characterized by its own location and by the location of its input- and output-environments along the turnover time axis. (b) Taking the network context into account a population and its input- and output-environments can be characterized in the two-dimensional framework of dynamic characteristics and trophic position. The situation sketched is most likely to be encountered for a population defined by taxonomic criteria where the input- and output-environments of the component organisms may extend over a large range of both turnover time and trophic position

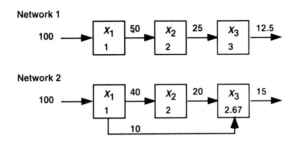

Figure 6.5   Simple networks to illustrate the derivation of a compartment's trophic position. Only net energy transfers are shown. Losses via respiration are omitted. The trophic position is denoted below a compartment's label, X_i

are denoted below the biomass labels, X_i. Network 1 represents an ideal food chain with well-defined trophic levels whereas network 2 is branched. This difference in structure is reflected in the trophic positions of the compartments. Network 1 agrees entirely with the trophic level concept. In network 2, compartment 3 has a trophic position of 2.67 because it receives one-third of its input via a pathway comprising one transfer and two-thirds via a pathway comprising two transfers. At a later stage we are going to discuss in more detail that the distinction of trophic positions along a continuous trophic gradient is important for our understanding of ecosystem dynamics.

As shown in Figure 6.4b, the input- and output-environments of a population of organisms defined by taxonomic criteria may extend over a large range regarding both turnover time and trophic position. The range covered is especially large for species where the body weight of adult organisms is several orders of magnitude higher than the one of juveniles (e.g. many fish). This derives from the fact that body weight determines both turnover time and trophic position. In order to derive a coherent description of an ecosystem's spatio-temporal organization it may thus be more useful to aggregate organisms according to dynamic and functional rather than taxonomic criteria.

One may object that the primary producers do not receive the attention they deserve when the assignment of the functional niche is based on trophic function only. Primary producers are distinguished by their output-environments only, whereas they all serve the same function with respect to their input-environment by fixing solar energy. We thus neglect that primary producers are the major interface to the abiotic environment. Autotrophic organisms are particularly exposed to changes in the environmental conditions such as insolation, or draught. In terrestrial systems the vegetation is a main structuring element of the landscape. This is not meant to imply that heterotrophic organisms are not affected by changes in their physical environment. For example, growth of larval fish and hence reproductive success is known to be largely influenced by vagaries in temperature and transport processes in the water body (e.g. Bollens *et al.*, 1992). However, regarding a classification into functional groups such impacts are more important for primary producers, the energy and nutrient uptake of which depends entirely on the abiotic environment. We have now to ask how we can refine the functional distinction of primary producers without becoming entangled in a jumble of specific classification criteria. This is the more important since the distinction of the primary producers influences the functional assignment in the whole web due to their being at the base of each input-environment of all populations higher up the trophic gradient.

It seems sensible to account explicitly for spatial patterns by making distinctions when we have different types of habitat with largely varying physical conditions. It would, for example, not be reasonable to ignore that the vegetation close to a stream is rather particular. Correspondingly, it seems to be warranted in some marine environments to make the distinction between shallow and deep associations of phytoplankton (Venrick, 1990, 1993). However, what are appropriate criteria for a distinction of functional types within a plant community in a given habitat?

For planktonic algae a distinction according to body size should already result in a reasonable distinction with respect to a large range of physiological differences and ecological strategies (e.g. Reynolds, 1984a, b). Phytoplankton can thus be easily integrated into a general description of the dynamic and functional characteristics of the whole network. In aquatic and especially in pelagic systems there is a gradual transition from autotrophic to heterotrophic organisms regarding body size and the perception of the environment. All planktonic organisms are predominantly passively dispersed. This similarity in the habitat characteristics imposes similar constraints on organisms irrespective of their being autotrophs of heterotrophs.

The distinction between fauna and flora becomes progressively more pronounced when plants start to grow on solid templates as in benthic and even more so in terrestrial systems. In such cases, plants deserve a special treatment. Terrestrial plants may be distributed over a large range of dynamic characteristics both at the level of the species and the level of the individual. As for certain animal species, one reason is given by the different dynamics of various life stages – saplings differ in their dynamic behaviour from adult plants. An individual plant itself may be regarded as an ensemble of semi-autonomous modules regarding the dynamic characteristics and the location in space. A plant's modular organization confers a considerable independence on functional parts, on both an ecological and an evolutionary time scale (Küppers, 1989; Schmid, 1990; Vuorisalo and Tuomi, 1986; Weiner and Thomas, 1992; White, 1979). This statement is supported by the phenotype plasticity in particular traits and by the fact that different parts of a plant channel into different food chains such as granivory, grazing, nectarivory, detritivory (e.g. Odum and Biever, 1984). Plants are preyed upon partially whereas animals are preyed upon as a whole. There are notable exceptions like parasitism, but with regard to the more common predator–prey relationships, for animals no equivalent exists to seeds, leaves and wood all serving as independent food sources of varying resource quality. The various plant parts differ also in their spatial location with the most conspicuous difference occurring between below- and above-ground parts. Körner (1993) gave evidence for the usefulness of distinguishing spatial functional groups in plant communities. In experiments investigating the effects of CO_2-enrichment on artificial ecosystems, Körner and Arnone (1992) observed a strong response for the rhizosphere whereas hardly any effect was detectable in the canopy. Considering that these functional groups serve also rather different functional properties in the whole ecological network, a spatial distinction in the vertical dimension seems warranted.

Hence, I suggest distinguishing terrestrial plants growing on solid templates into functional groups by their composition with respect to tissues differing in turnover time and by their morphological characteristics in the vertical dimension. The examples listed in Box 6.2 show that a distinction according to tissue composition goes in parallel with generation time because the longer lived a species is, the more resources are allocated to infrastructure.

Box 6.2   Leaf weight ratios in major groups of plant life forms (data from Körner, 1993)

Plant life form	Leaf weight ratio (leaf mass as % of total plant mass)
Trees	2–3
Shrubs	5–20
Wild perennials	10–40
Cultivated biannuals	40–65
Annuals	50–80

Regarding the vertical dimension, a functional distinction is obviously given by size. Thus it may partially be accounted for in the distinction according to tissue composition due to the latter's correlation with lifetime. Figure 6.6a shows a further distinction with respect to the morphologies of root systems. The two different morphologies that may be distinguished for desert shrubs result in a major difference in the perception of space. Empirical results suggest that such differences are essential for community organization (Cody, 1986). Recall also the example of the old-field plant community discussed in Chapter Four, where differences in root morphologies mediated the spatio-temporal segregation of nitrogen uptake (cf. Figure 4.7) that was essential for the distinction of plants into different functional guilds.

Figure 6.6b depicts an example for a possible final description of a functional group of trees. The functional distinction is based on tissue composition and hence generation time. Functional modules are distinguished for different tissues according to their biomass turnover time and their location in the vertical. Whereas leaves have turnover times of the order of days to weeks, heterotrophic tissues have turnover times of the order of years.

Recently, Steneck and Dethier (1994) introduced a functional group approach for benthic marine algae that follows similar lines of reasoning. They distinguished functional groups according to characteristic rates of mass-specific productivity, thallus longevity and canopy height. Species ensembles aggregated according to such categories were observed to display a similar behaviour with respect to both their habitat requirements and their being exposed to specific groups of herbivores. Such findings support the assumption that a scheme as outlined above comprises the essential characteristics of primary producers regarding their function in an ecological network.

6.3   DEVELOPING THE NETWORK CONCEPT STEP BY STEP

Let us assume that we had complete knowledge about the spatio-temporal dynamics of the trophic interactions among all organisms in an ecosystem. The conceptual scheme of delineating a network based on trophic dynamic modules is then introduced step by step. First, we classify organisms according to their biomass turnover time by grouping them into dynamic classes. Next we give these ensembles an identity in time and space by distinguishing dynamic modules. A dynamic module comprises all organisms belonging to the same dynamic class that coexist over the same interval in time and in space. Finally, we distinguish the organisms within a dynamic module according to their functional niche. A trophic dynamic module comprises then all organisms that coexist over the same interval in time and in space and that occupy the same functional niche which means that

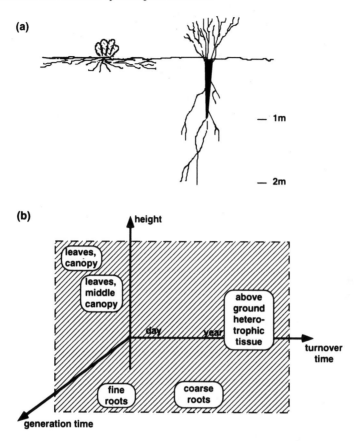

Figure 6.6 (a) Plants may be grouped into functional types by life forms affecting how the plants experience space. The example contrasts two root systems morphologies observed for desert shrubs (redrawn from Cody, 1986). (b) Description of a functional group of trees distinguished by the generation time. The biomass is distributed among several modules as a function of the biomass turnover time, and the location in the vertical dimension

they engage in the same type of interactions in the network. The procedure outlined seems to be a logical sequence of how to proceed in pelagic systems where the dynamic classes can be identified with weight classes. Modifications may be required to find the most appropriate scheme for terrestrial systems – in particular when we intend to describe complex patterns of vegetation. Box 6.3 gives the reader some guidance by summarizing the relevant new expressions and symbols. Let us now discuss the successive steps in more detail.

6.3.1 Classification according to dynamic properties – dynamic classes

Based on the previous considerations, I assume that the dynamic characteristics of an ensemble of organisms, are adequately reflected by the average biomass turnover time, τ. A classification scheme based on dynamic properties can be derived by partitioning the τ-axis into discrete dynamic classes as shown in Figure 6.7. The classes are equally spaced

---

Box 6.3   Major definitions and symbols used in the network concept

Dynamic class, C_i, comprises all organisms the turnover time, τ, of which lies in a defined range which is chosen such that it remains constant on a logarithmic scale (cf. Eqn 16.6). More specifically the width of a class equals one unit on a logarithmic scale to the base K_τ.

Functional niche, C_{ia}, is defined as a characteristic linkage pattern of an ensemble of organisms specifying the input- and output-environments. The input-environment corresponds to a "prey" and the output-environment corresponds to a "predator window" along the dynamic axis. The primary producers with their specified input-environment serve as reference for a recursive assignment. Terrestrial plants may further be characterized by functional niches derived from their modular structure that determines how the plants experience time and space, respectively.

Dynamic module, DM_{ikr}, comprises all organisms belonging to the dynamic class C_i that coexist over the same intervals in time τ_{ik} and in space σ_{ir}.

Trophic dynamic module, TDM_{ikra}, comprises all organisms of the dynamic module DM_{ikr} that have the same functional niche C_{ia}.

Indices

i	index of a dynamic class, $1 \le i \le n$
k	index of time intervals in units of τ_i
r	index of spatial dimension in units of σ_i
a	index of functional niche

Parameters

τ_i	characteristic time scale equivalent to the biomass turnover time of the dynamic class C_i
σ_i	typical spatial extent of a dynamic module in class C_i
K_τ	scaling factor for dynamic classes
T, S	total observation periods of the system as a whole along time and space, respectively

---

on a logarithmic scale. On a linear scale, the τ-range of a class C_{i+1} increases relative to the range of the preceding class C_i with a constant factor that is referred to as K_τ.

$$\frac{\tau'_{i+1}}{\tau'_i} = K_\tau = \frac{\tau_{i+1}}{\tau_i} \quad i = 1 \dots n \tag{6.6}$$

where n refers to the total number of classes. τ'_i and τ'_{i+1} represent the lower and upper boundaries of a dynamic class i with an average turnover time $\log\tau_i = \log\tau'_i + 0.5$ ($\log\tau'_{i+1} - \log\tau'_i$). The derivation of the average implies that the biomass is assumed to be evenly distributed along the τ-axis on a logarithmic scale. The characteristics of the dynamic classes shown in Figure 6.7 ($K_\tau = 2$) yield:

Class	τ'_i	τ'_{i+1}	$\tau_i = (\tau'_i \tau'_{i+1})^{0.5}$
1	$2^0\Delta\tau$	$2^1\Delta\tau$	$2^{0.5}\Delta\tau$
2	$2^1\Delta\tau$	$2^2\Delta\tau$	$2^{1.5}\Delta\tau$
3	$2^2\Delta\tau$	$2^3\Delta\tau$	$2^{2.5}\Delta\tau$

$\Delta\tau$ denotes the lower threshold for the range of turnover times encountered for the whole ensemble of organisms. The range of a dynamic class was chosen arbitrarily as a multiple of $\Delta\tau$.

The dynamic characteristics may also be expressed by body weight (size), and the dynamic classes may be replaced by weight (size) classes. If the dynamic classes in Figure 6.7 were identified with weight classes, body weight would increase approximately by a factor of 10 with a doubling of lifetime (based on the allometric relationship in Eqn (6.4) a factor 2 in time corresponds to a factor $2^{1/\varepsilon}$ in weight). Here, a word of caution may be warranted regarding the choice of the range of a weight class. The base of the logarithm chosen determines the degree of resolution. For a coarse resolution (e.g. base 10) the weight distribution of the organisms within a class may be rather uneven. Cyr and Pace (1993) showed, for example, that the extrapolation of allometric relationships from individuals to communities may be flawed when the body weight distribution in the latter is bimodal. Therefore, for any practical application it may be advisable to choose a fine resolution (base 2 logarithms for weight classes).

6.3.2 Identity of a network unit in time and space – dynamic modules

The composition of an ensemble of organisms belonging to a dynamic class and the functional niches in a network change over time and space. Species with different feeding characteristics may replace each other. The membership of organisms to a dynamic class may change due to the growth of juveniles leading to an increase in weight and biomass turnover time. Populations may migrate to other places. Based on such considerations, I assume that an ensemble of organisms belonging to the same dynamic and the same functional niche has a typical minimal lifetime and a bounded spatial extension over which its identity is retained. These typical extensions in time and space depend on the dynamic properties of the organisms rather than on their function in the network. Regarding the spatio-temporal dynamics a water flea resembles most algae more than it resembles a whale, even when the latter might be closer with respect to trophic function. Therefore, I introduce the notion of a dynamic module (DM) to denote an ensemble of organisms within a dynamic class that are present over a specific time period and a specific spatial area (volume). If dynamic classes are identified with size classes, a DM of a pelagic food web comprises, for example, all equally sized organisms irrespective of their belonging to phytoplankton or zooplankton species, which means irrespective of their trophic function.

The typical lifetime of a DM is identified with the average biomass turnover time, τ,

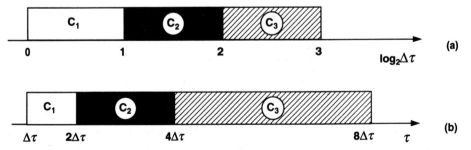

Figure 6.7 (a) Partitioning of the axis of the biomass turnover time τ into classes equally spaced on a logarithmic scale. In this example the base of the logarithm was chosen to be two. (b) On a linear scale, the τ-range of a class doubles from one class to the next

introduced before. The biomass turnover time may be looked upon as some buffering capacity of a population representing an appropriate measure for the time until changes become visible at the population level and thus at the level of the interaction structure in the ecological network. Such a distinction is not meant to imply that all functional niches really change with every biomass turnover time. A functional niche may persist much longer than its component organisms. Assuming that an ecosystem is observed over a total period T, we can distinguish T/τ_i different DMs for a dynamic class i. This implies a decrease in the number of DMs with increasing dynamic class as illustrated in Figure 6.8 depicting a temporal sequence of DMs as obtained for four successive dynamic classes. The first index denotes the dynamic class, the second index denotes the time interval measured in the intrinsic lifetimes of each dynamic class. The time step on the absolute time axis was arbitrarily chosen to be equivalent to τ_1, the typical lifetime of the first dynamic class.

We have now introduced for each dynamic class a relative time axis on which time is measured in a population's own intrinsic units. Correspondingly, we have to introduce relative spatial axes on which space is partitioned in a population's own intrinsic units. To do so it is reasonable to assume that the typical spatial extent of a dynamic module is determined by the average dispersal range of its component organisms. In particular, given the lifetime, τ_i, the appropriate spatial extent, σ_i, is identified with the average radius of activity to be expected for that period. A dynamic class i comprises thus a temporal succession (with a temporal spacing of intevals of τ_i) of DMs on a spatial grid (with patches of diameter σ_i). An example of a DM is schematically depicted in Figure 6.9 where I assumed for simplicity the patches to be rectangular with extension σ_i in each horizontal dimension. Whether different spatial dimensions are equivalent and whether a spatial extent is patchy or continuous depends on the particular properties of the physical and biotic environment and the interactions in the ecological network.

To delineate in practice the spatial extents associated with a certain lifetime is not an easy and straightforward task (Pahl-Wostl, 1993b). To begin with it may be helpful to distinguish among mobile, immobile and sessile organisms. Mobile and immobile refer to the relevance of active movement on the time scale of the biomass turnover time. Many phytoplankton species, for example, are not sessile but essentially immobile, being passively displaced by physical transport processes. Benthic periphyton and terrestrial

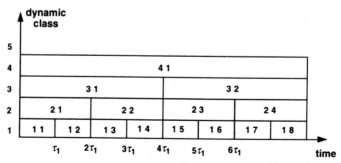

Figure 6.8 Temporal sequence of dynamic modules for the classification scheme of dynamic classes depicted in Figure 6.7. The indices ik refer to the dynamic class (i) and the time interval (k). The latter are measured in units of a DM_i's typical lifetime, τ_i. Each DM_{ik} then comprises all those organisms belonging to a dynamic class i that are present during the kth time interval of length τ_i (Reproduced by permission from Pahl-Wostl, 1993b)

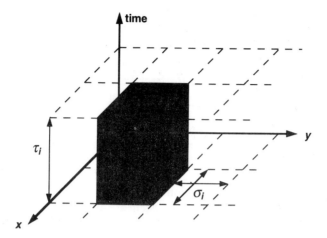

Figure 6.9   Schematic representation of the spatio-temporal succession of dynamic modules, DMs, a dynamic class i consists of. The shaded volume refers to the scales of resolution for a single DM that extends over a time interval τ_i and a patch of the extension σ_i in both x and y. A dynamic module is specified as DM_{ikr} where i refers to the dynamic class, k is the time interval τ_{ik} and r refers to the spatial coordinates σ_{ir}. Both the dimension of time and the dimension(s) of space are assumed to continue further beyond the small section shown in the figure (Reproduced by permission from Pahl-Wostl, 1993b)

plants are both immobile and sessile. Despite such disparities in movement characteristics I derived a relationship expressing the typical spatial scale as a power function of the typical time scale (Pahl-Wostl, 1993b):

$$\sigma_i \propto \tau_i^\gamma \tag{6.7}$$

where the exponent γ may range from about 0.7 for passively dispersed organisms in a pelagic environment to about 2 for actively moving organisms with a defined home range. I obtained the latter by using allometric relationships between home range and body size (Swihart *et al.*, 1988). The steepness of the relationship suggests that actively moving organisms perceive their environment on a fractal rather than on a linear scale. If an organism experienced its spatial environment in a linear fashion as suggested by Euclidian geometry, we would expect the spatial range, S, to be linearly related to the body size, L. However, the structure of the natural environment is fractal rather than linear. In analogy to the famous example of a coastline that becomes longer, the smaller the measuring unit chosen (Mandelbrot, 1983), a habitat may become the larger, the smaller the size of the organism using it. Correspondingly, we expect the spatial range of a population of organisms to be related to body size according to $S \sim L^\alpha$. α is one for a linear scaling; it is larger than one for a fractal scaling. The corresponding relationship between body weight and spatial patch area can be derived as $A \sim W^{2\alpha/3}$ with an exponent larger than 0.67 indicating a fractal scaling.

  Expressing the relation between spatial and temporal scales at this level of generality may run counter to the intuition of many biologists who may have already felt uncomfortable with the application of allometric relationships to derive time scales. Nevertheless, I consider such general relations an essential requirement to reveal overall trends and characteristics that may otherwise be lost in the wealth of detailed knowledge.

An illustrative example of this happening was reported by Wiens *et al.* (1993) in their studies on the movement of beetles in landscapes. Beetle movement proved to be more complex than correlated random walk. Features of movement pathways were clearly affected by the structure of the landscape, and these effects were more pronounced for some species than for others. However, the fractal dimension of the movement pathway did not vary among species or with habitat heterogeneity. Thus, despite obvious differences in movement features when measured on an absolute scale, the pathways exhibited similar characteristics when scaled relative to the movement rates of the individual organisms. Limiting the attention to details can blur the perception of overall patterns that may reveal underlying regularities. General scaling relationships, as derived above, should assist to improve our understanding of scaling patterns in nature and to set up hierachical models for their description.

In summary, the organisms within an ecosystem are aggregated into discrete units, the dynamic modules that have an identity in time and space. The criteria of aggregation are derived from the dynamic characteristics of the organisms. The dynamic modules were defined to be identical in their typical lifetimes and spatial extensions when measured in their intrinsic scales. Thus, a coherent resolution is achieved across the wide range of spatio-temporal scales in an ecosystem. It remains to define the trophic dynamic modules by taking into account the context within the ecological network.

6.3.3 Taking account of function – trophic dynamic modules

An ensemble of organisms in a dynamic module that share the same functional niche is called a trophic dynamic module, TDM. Hence, the number of dynamic modules, DMs, is the lower bound on the total number of trophic dynamic modules, TDMs, that can be distinguished. In the case of the lower limit all organisms within a DM belong to the same functional niche, and the DMs can be identified with the TDMs.

To illustrate the meaning of TDMs, I constructed a simple hypothetical network that may be envisaged to represent part of a pelagic food web. For the sake of a clearer graphical representation, the considerations are restricted to the dimension of time; the dimension(s) of space can be treated accordingly. I focus on the energy network assuming that the functional distinctions made there apply as well to the nutrient network that includes recycling.

Figure 6.10a shows all possible links of the network configuration comprising the first three dynamic classes. The patterns correspond to functional classification that are explained at the bottom of the figure. For simplicity, organisms in class three are assumed to be identical with respect to their output-environment by being preyed upon by the same consumer in a higher class. Let us assume that the network exhibited the temporal variations shown in Figure 6.10b. The time intervals are given in units of τ_2, the biomass turnover time of TDMs in the second dynamic class. The network configuration is assumed to be time invariant over τ_{21} and τ_{22}. In τ_{23} a consumer appears in class 3 that feeds on the organisms in both classes 1 and 2. As a result, we observe in τ_{24} the disappearance of the consumer in class 2 and the emergence of nonedible algae.

These temporal variations result in the description in terms of TDMs represented in Figure 6.11a. As in the previous examples (cf. Figure 6.7) the typical lifetime of a DM doubles from one class to the next: $\tau_{i+1}/\tau_1 = 2$. The area of a pattern indicates a TDM's fraction of the total biomass of a DM. The difference in size is not meant to imply that

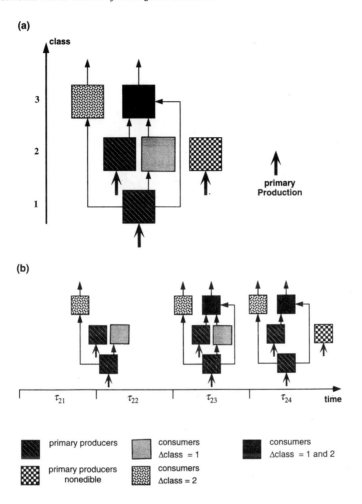

Figure 6.10   (a) Hypothetical network of a pelagic ecosystem including all possible links between functional groups in the first three dynamic classes. (b) It is assumed that three different network configurations can be distinguished as a function of time. Time is given in units of τ_2, the turnover time of the second class. Δ class refers to the difference in dynamic class(es) between a predator and its prey (Reproduced by permission from Pahl-Wostl, 1993b)

TDMs of the same class differ in lifetimes. The first number of the index refers to the time interval, the second number refers to the different functional niches numbered consecutively for each dynamic class. Figure 6.11b shows the functional niches that can be distinguished based on the specificities in both input- and output-environments. We note that the primary producers in class 1 occupy two functional niches due to the changes in their output-environment. A functional niche is equivalent to a network pattern and has at least a lifetime equal to the biomass turnover time τ of the dynamic class the niche is characteristic for. The total time period over which a functional niche persists depends on the overall dynamics of the interactions within the ecological network. We expect a high diversity with respect to functional niches in networks with a highly dynamic network structure.

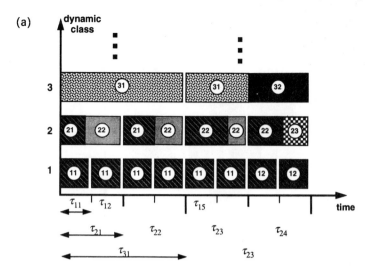

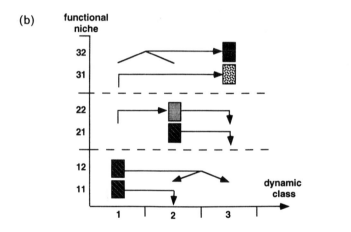

Figure 6.11   (a) Description of the network depicted in Figure 6.10 in terms of TDMs. The time intervals τ_{ik} refer to the typical lifetimes of the functional modules in the ith dynamic class. The shadings correspond to the different functional types explained in Figure 6.10 (Reproduced by permission from Pahl-Wostl, 1993b). (b) The different functional niches that can be distinguished based on the time sequence depicted in Figure 6.10b by taking into account the specificities of both input- and output-environments

To simplify the representation, I assumed trophic transfers to be possible between adjacent classes. If the dynamic classes were identified with weight classes, body weight would increase approximately by a factor of 10 with a doubling in lifetime. Empirical investigations from a pelagic food web showed that the average predator–prey weight ratio is around 1000 (Gaedke, 1992a) which means that in reality the situation would look more complex than in our simplified example.

The treament of the spatial dimension(s) can proceed in the same fashion as described for the dimension of time. Due to the fact that planktonic organisms are predominantly dispersed by physical transport processes, the typical spatial extension can be assumed to increase regularly with lifetime (review by Okubo, 1980; Powell, 1989). A typical patch size does not mean that we expect spatially segregated habitat ranges as might be the case in terrestrial environments. It indicates the spatial scale over which we may detect changes in the functional characteristics of populations and/or in their distribution. For an algal population it gives us, for example, an estimate of the smallest size we might expect for an algal bloom. Since the typical spatial extensions depend on the physical environment they may change over time. The same applies to the characteristics of the vertical dimension. Due to the strong vertical gradients in the habitat characteristics such as temperature and light, it seems warranted to resolve the vertical dimension in more detail. In section 4.7 we observed for a simple phytoplankton model that the presence of spatially distinct functional assemblages may depend largely on the characteristics of the mixing processes. The pronounced temporal variabilities of the spatial characteristics of the environment are a major difference to terrestrial systems where the spatial patterns are fixed.

To summarize, we proceeded in our stepwise procedure in three stages corresponding to different levels of resolution of a network's structure:

1. Dynamics characteristics $\Rightarrow$ dynamic classes
2. Locations in time and space $\Rightarrow$ dynamic modules
3. Functional characteristics $\Rightarrow$ trophic dynamic modules

Before discussing how to evaluate the information that may thus be obtained, I will show that the concept of TDMs allows us to develop a coherent picture of spatio-temporal organization extending over a range of spatio-temporal scales and trophic functions in an ecosystem.

6.4   ORGANIZATION ACROSS A CONTINUUM OF TEMPORAL AND SPATIAL SCALES

To develop an idea of how an ecological network's feedback structure may be organized across a wide range of temporal and spatial scales let us start with the simple food chain depicted in Figure 6.2. This food chain can be given an identity in time and space by describing it in terms of its component TDMs. To simplify the representation, I focus again on discussing the resolution along the dimension of time in more detail. Figure 6.12 shows the food chain with each trophic level being resolved into a temporal succession of its component TDMs. A nutrient pool is included to emphasize the importance of feedback effects. The dimensions of a feedback spiral increase with increasing number of transfers up the food chain regarding both the range of weight classes covered and the time gap between successive turns. This is visualized by the arrows tracing the path of two feedback spirals of increasing temporal extension originating both in the same direct transfer.

As indicated by the dashed lines we may start to discern a hierarchical modular structure where each modular building block can be thought of as representing a turn of a feedback spiral. Effects of positive feedback link the modules across time (and space) with increasing delays the larger the number of transfers along the food chain and thus the body weight gradient (cf. Pahl-Wostl, 1993a).

To emphasize the nested pattern across space and time, I chose another representation in Figure 6.13. The Roman numbers refer to the modules discerned in Figure 6.12. The spatial and temporal scales increase from level to level as illustrated by the change in scale of the coordinate axes. The arrows represent the feedback resulting from nutrient recycling. The decrease in arrow size indicates the decrease in flow intensity with increasing trophic level due to the lower metabolic requirements associated with a higher body weight. Hence, the flow intensity in a feedback spiral is inversely related to its extension along time and space as depicted in Figure 6.14a. We further note the discrete nature of the feedback pattern arising from the chosen food chain structure with defined trophic levels characterized by defined body weights.

Such a discrete feedback pattern contradicts the claim that ecosystems should be characterized by a high dynamic diversity mediating spatio-temporal organization. In the last chapter we discussed a variety of simulation models to develop a more coherent picture of what such organization implies. However, the range of time scales covered by these models was rather limited. A first start to investigate the behaviour of spatio-temporal organization across a range of scales was made with the model comprising ensembles of predator–prey pairs that were distributed along the body weight axis. We observed species coexistence and system performance to be correlated with an even distribution of the pairs across a range of body weight and hence of temporal scales. These observations can be extended to the conjecture that ecological networks exhibit spatio-temporal self-organization across a continuum of spatio-temporal scales. Hence, we should also expect to find a continuous distribution of the feedback effects rather than discontinuous changes, unless specific patterns are imposed by discontinuous environmental gradients. A more continuous distribution can in a first approximation be derived from the simple food chain chosen when we imagine an ensemble of food chains that differ by being shifted along the body weight axis. The predator–prey model discussed in the last chapter

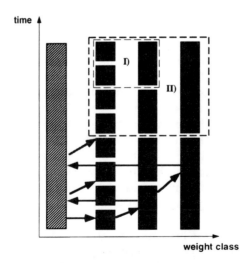

Figure 6.12   Resolution of the food chain depicted in Figure 6.2 into TDMs that are defined by body weight classes. The hatched box denotes the nutrient pool. The arrows trace the path of two feedback spirals of increasing temporal extension. As indicated by the dashed lines, we may discern a hierarchical modular structure. See text for further explanations

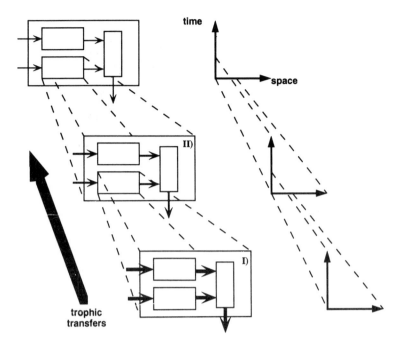

Figure 6.13 Nested modular structure of the food chain depicted in Figure 6.12. The Roman numbers refer to the modules distinguished there. The spatial and temporal scales increase from level to level as illustrated by the change in scale of the coordinate axes. Feedback spirals are indicated by the arrows, the size of which varies to indicate the decrease in flow intensity with increasing spatial and temporal extensions

displayed such a configuration where each chain consisted of two members only (cf. Figure 5.17). Correspondingly, we obtain a continuous distribution of the feedback spirals across spatio-temporal scales as shown in Figure 6.14b.

The overall shape of the distribution of the feedback effects does not hinge on the simple food chain chosen to illustrate its derivation. To obtain such a regularly shaped distribution requires a continuous distribution of biomass and a similar functional diversity across the gradient of biomass turnover times. Similarity implies that the functional niches that can be distinguished for a dynamic class generate the same statistical distribution of spatio-temporal feedback patterns irrespective of the dynamic class under consideration. In such a case spatio-temporal organization is characterized by a continuum of feedback spirals across the dimensions of space and time. Considering that each feedback spiral may be looked upon as storing a certain memory in space and time, a spatio-temporal hierarchical distribution of feedback effects may endow ecosystems with their remarkable capability of simultaneous internal regulation and adaptation.

At first sight one may feel inclined to conclude that due to the decrease in flow intensity long pathways have only a small impact on overall system dynamics. If we look, however, at network dynamics in terms of trophic dynamic modules, the dynamic impact of a feedback pathway has to be assessed by the whole amount of matter transferred over the lifetime of the longest-lived component rather than by the flow intensity only. This can be done by using the product of the flow intensity defined per absolute unit of time and space and the time and/or space interval over which the memory of a cycle extends. This

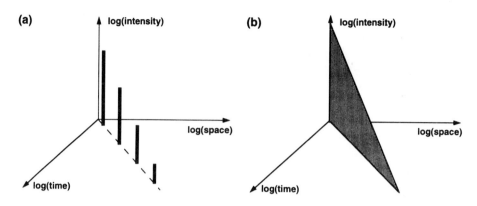

Figure 6.14   Intensity of the feedback spirals in a hierarchical modular food chain as a function of the scales along space and time. The intensity refers to the amount of matter transferred per absolute unit in time and space. (a) Food chain with defined trophic levels characterized by a defined body weight. (b) Ensemble of food chains where each trophic level comprises a range of body weights resulting in a continuum of scales in time and space

memory is determined by the temporal and spatial extensions of the slowest TDM in this pathway. The decrease in flow intensity with increasing dynamic class is thus compensated by the increase in the typical spatio-temporal extensions. Essentially we could conceive of a spatio- temporal and functional organization that results in an even distribution of these dynamic impacts across the TDMs.

We find support for a continuum in spatio-temporal scales by the distribution of biomass as a function of body weight in pelagic systems. Due to the fact that the energy flow is directed along a gradient of increasing body weight, biomass weight spectra are a convenient tool for characterizing pelagic communities. To obtain such spectra organisms are aggregated into logarithmic weight classes and the total biomass is determined for each weight class. To render biomass spectra more amenable to quantitative analysis and suitable for cross-system comparisons requires some type of standardization. Figure 6.15 shows an example for a normalized biomass spectrum derived for the plankton community of Lake Constance, a large lake in Central Europe (Gaedke, 1992b). Such normalized spectra are obtained by dividing B_i, the biomass in the ith weight class by ΔW_i, the weight range of the ith weight class. The resulting biomass density corresponds approximately to N_i, the numerical abundance of individuals in a weight class. Two attributes of these spectra are of major interest: the slope and the evenness of the distribution. Both can be quantitatively assessed by a linear regression. A slope of -1 is obtained when the same biomass is found in the different weight classes ($N \propto W^{-1} \Rightarrow N_i \times W_i = B_i = \mathrm{const.}$). A decrease in the coefficient of determination r^2 indicates an increase in the deviation of the data points from a straight line ($r^2 = 1$ is obtained for an ideal even distribution). We note that the biomass distribution in Lake Constance conforms closely to the requirements of uniformity.

Sheldon *et al.* (1972) were the first to analyse particle-size distributions in open oceans. They observed a uniform distribution of particles over logarithmic size classes. Since then further investigations confirmed the initial observations (e.g. Ahrens and Peters, 1991; Gaedke, 1992b; Rodriguez and Mullin, 1986). It seems that large lakes and marine systems are characterized by even distributions with an absence of major gaps in the

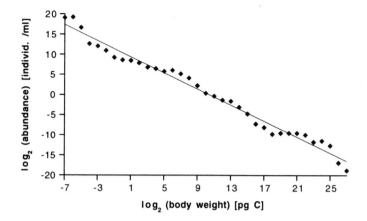

Figure 6.15   Normalized spectrum obtained for the annual average of the plankton community's biomass in Lake Constance. The approximate abundance is obtained by dividing the biomass in a weight class by the weight range of this class. The linear regression yielded a slope of -1 and a coefficient of determination $r^2 = 0.98$. A detailed discussion is given by Gaedke (1992b)

spectrum. In these systems the pelagic habitat is large enough to be looked upon as an autonomous ecosystem. The continuous biomass distributions in pelagic systems seems to reflect the continuous nature of the physical environment. In contrast, the biomass size distributions of marine benthic environments show pronounced modes corresponding with the sizes of benthic bacteria, meiofauna and macrofauna. Schwinghamer (1981) explained such patterns by arguing that organisms perceive the sedimentary environment on three scales: they may either colonize the particle surfaces, inhabit the interstices between particles or regard the sediment as effectively nonparticulate. The transition from one scale to another is discontinuous rather than gradual. Warwick (1984) and Warwick and Joint (1987) found that the species size distribution followed a similar modal pattern to the biomass spectrum. He invoked evolutionary optimization of size-related life history and spatial resource partitioning to explain the pattern. Interestingly, Rodríguez *et al.* (1990) detected major gaps in the spectrum of small mountain lakes. In such systems where the pelagic environment is small we have to take into account a coupling between pelagial and benthos. The presence of such a coupling is further supported by the observation that pelagic larvae of the macrobenthos just fit in size in the trough in the biomass spectrum (Warwick, 1984; Warwick and Joint, 1987). It would be a fascinating endeavour to investigate in more detail whether further regularities can be detected in such distributions and whether they are related to the nature of the physical environment, in particular to the power spectrum of its variance in space and time.

Slowly our picture of ecosystem organization starts to take shape. The feedback pattern seems to be highly dynamic and to extend across a continuum of temporal and spatial scales. We may discern a close relationship between the spatio-temporal organization of the biological community and the physical environment on both ecological and evolutionary time scales. In the absence of physical constraints there seems to be a tendency to evenly fill the available spatio-temporal niche space. Such observations lead us to think about the adequacy of describing ecosystems in terms of defined trophic levels. Should we rather talk about a continuum in trophic function as well?

6.5   A CONTINUUM IN TROPHIC FUNCTION?

Oksanen (1991) argued in favour of the trophic level concept as a major step towards a holistic description of ecosystems. A high level of abstraction was achieved with the introduction of trophic levels that were assumed to have an identity of their own regardless of fluctuations in their component populations. Such a well-defined structure with clear pathways of interaction suggests itself for developing hypotheses on community regulation. In Chapter One we discussed the concept of cascading trophic interactions and its application in developing management schemes for the biomanipulation of lakes. This concept is based on feedback effects cascading along a linear food chain as depicted in Figure 1.2 for a pelagic system. The discussions about the dominance of top-down (grazing and predation) versus bottom-up (resources) control has proceeded along similar lines of reasoning.

In contrast, Cousins (1985, 1987) suggested food webs to represent a trophic continuum in which body size plays a crucial role. He argued that in any real system it was not possible to assign a well-defined trophic level to a species. Correspondingly, he suggested abandoning the trophic level concept.

We discussed earlier that the idea of trophic levels operating on well-defined temporal and spatial scales does not coincide with reality where we encounter rather a continuum in the dynamic characteristics. Do we find as well evidence for a continuum in trophic function? I am going now to illustrate for the example of a pelagic system how we could imagine a progressive increase in the complexity of trophic interactions approaching gradually a continuum in trophic positions. Accounting in all network schemes for feedback cycles would result in an excessive complexity of the pictorial representations. Therefore, I limit myself to depicting the direct trophic transfers, omitting the feedback cycles. Nevertheless, we should keep their presence always in the back of our minds.

To begin with, let us start again with an ensemble of independent linear food chains as represented in Figure 6.16a. Each circle denotes a weight class. The chains being shifted along the body weight axis generate a continuum of time scales. However, we can still discern well-defined trophic levels that are distinguished by the different shadings. Figure 6.16b shows that the food chains become linked when we take into account that a consumer's prey window extends over more than one body weight class. In the situation depicted the identity of trophic levels is still retained. The example given in Figure 6.16c illustrates the breakdown of defined trophic levels, when we continue further along these lines of reasoning. The shaded area denotes the window of a consumer in a higher weight class selecting organisms according to their size irrespective of their being autotrophs or heterotrophs. The trophic position of such a consumer ranges somewhere between 2 and 3. We may thus conceive of a continuum of trophic positions to be generated by shifts in the prey window and variations in the relative contribution of different prey to a consumer's overall diet.

An example for such a behaviour is provided by two taxonomic groups of zooplankton species, daphnids and cyclopoid copepods. Figure 6.17a shows the prey window of daphnids along the body weight axis derived for the data from Lake Constance (U. Gaedke, pers. comm.). The dark shaded circles refer to phytoplankton and spotted circles refer to herbivorous zooplankton. Daphnids are not very selective and feed on a wide range of weight classes. At the level of the species as a whole cyclopoid copepods have a

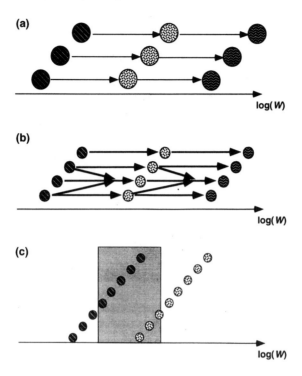

Figure 6.16   Scheme illustrating the progressive increase in the complexity of trophic interactions in a pelagic food web. (a) Independent parallel food chains with defined trophic levels denoted by the different shadings. (b) The chains become linked due to overlapping prey windows. (c) Breakdown of defined trophic levels when the prey window of a consumer comprises both primary producers and herbivores

similarly large prey window. However, Figure 6.17b shows that the wide range of the prey window for copepods derives from the large ontogenetic shift in body weight during maturation. Being raptorial feeders the range of individuals at a given size is much smaller than the one of daphnids preying as filter feeders effectively on a large size range. For daphnids the ontogenetic shift in size and hence in the range of the prey window is of minor importance.

These differences are summarized in Figure 6.18 which sketches approximately the corresponding life history strategies in a diagram of trophic position and body weight. The trophic positions were derived from carbon flow data for Lake Constance (Gaedke *et al.*, 1995). The assignment of weight class corresponds to the classification in the biomass–weight spectrum in Figure 6.15. Cyclopoid copepods effectively move one trophic level due to the more than 100-fold increase in body size during maturation from nauplii to the adult stage. This is reflected in the concomitant shift in the prey window. In the language of trophic dynamic modules it implies that the organisms move up the dynamic classes thereby gradually changing their dynamic and their functional niches.

The two types of behaviour have largely different effects on the temporal organization of an ecological network. Daphnids couple scales thereby effectively reducing the variability and the dynamic diversity. To form an idea of how this works, let us assume

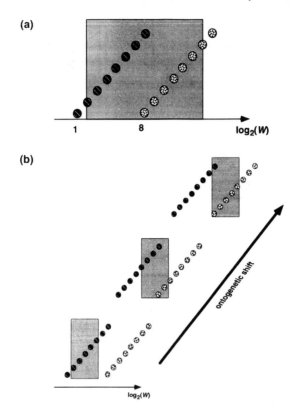

Figure 6.17   (a) The prey window of daphnids along the body weight axis. Dark shaded circles refer to phytoplankton and spotted circles refer to herbivorous zooplankton. (b) At the level of the species as a whole, cyclopoid copepods have a similarly large prey window generated by the ontogenetic shift in size during maturation. The weight classes correspond to the classification in the biomass weight spectrum in Figure 6.15

that the primary producers and herbivores depicted in Figure 6.17a represent an ensemble of differently sized predator–prey pairs. A comparison with Figure 5.17 reminds us that such a configuration was basic to the simulation model discussed in section 5.4. This model serves to illustrate the effect of a consumer that, like daphnids, feeds across a wide range of body weights. I performed simulations with a model version including eight differently sized pairs. Figure 6.19a represents the temporal variations of the total biomass of all phytoplankton and all herbivorous species obtained in the absence of an omnivorous consumer. Figure 6.19b shows the effect of introducing a consumer that feeds indiscriminately on all weight classes. The variations in the trophic position of the omnivorous consumer reflect the dynamics of the predator–prey pairs. We note that the dynamic diversity is efficiently suppressed upon introduction of the omnivorous consumer. We obtain also a stronger coupling between trophic levels. Hence, it is not too astonishing that the success of biomanipulation was observed to depend on the presence of large Daphnia as major predators (Hosper and Meijer, 1993; Reynolds, 1994).

In contrast, cyclopoid copepods may increase rather than decrease the dynamic richness. Such an increase in variability is even more pronounced for fish that from their

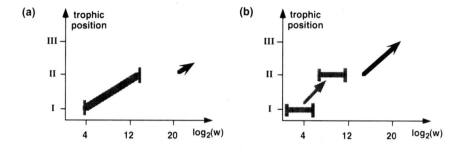

Figure 6.18   Approximate sketch of the life history strategies (a) for daphnids, (b) for cyclopoid copepods. The weight classes correspond to the classification in the biomass weight spectrum in Figure 6.15. The arrows denote the ontogenetic shifts in trophic position and dynamic class. In the case of cyclopoid copepods this leads to a pronounced shift in the prey windows that are denoted by the shaded lines

larval up to the adult stage move even further up the trophic gradient. In addition, they effectively transfer variability from one year to the next. Carpenter and Kitchell (1993) discussed several examples for pronounced interannual effects caused by the waxing and waning of fish cohorts. Such effects add considerably to the unpredictability of ecosystem dynamics especially when we take into account the extreme variability in reproduction. Reproductive success depends on the vagaries of the physical environment. Adverse climatic conditions may prevent larvae from growing and may lead to a mismatch between the period of larval growth and the availability of prey, the abundance of which is influenced by the physical environment as well (Bollens *et al.*, 1992; Cushing, 1990; George *et al.*, 1990). In addition to time lags mediated by the environment, reproductive success depends as well on a favourable timing with respect to the overall community dynamics – whether the temporal niche is favourable for juveniles to thrive. Recall the example of invasion in a group of coexisting species discussed in Chapter Five. The success of invasion was observed to be critically dependent on a favourable timing (cf. Figure 5.10). Similar effects can be expected for the reproductive success of larvae that have to find a suitable temporal niche to survive. Let us again use the model with differently sized phytoplankton–herbivore pairs to illustrate the importance of a favourable timing. I used the same configuration comprising eight pairs shifted along the body weight axis that was used to demonstrate the effect of the feeding behaviour of daphnids (cf. Figure 6.19a). Figure 6.20 shows the results obtained in model simulations for the reproductive success of juveniles that are characterized by a behaviour as depicted in Figure 6.17b. The three representative results, ranging from a complete failure to good success, were obtained by introducing juveniles at different time intervals chosen at random. To focus on the effects of community dynamics I assumed the environment to be constant – the variations are generated by the interaction among the pairs. The seasonality of the environment adds a regular element to the overall pattern of community dynamics by triggering a pulse of activity in spring. Interestingly, Gaedke (1992b) mentioned that perch larvae hatch in spring when large crustaceans that are of similar size are yet absent. Obviously, life history strategies are adapted towards increasing the chances of exploiting ephemeral temporal niches. Such observations call attempts of predicting community dynamics into question.

We can conclude that in pelagic systems there are strong indications that trophic function is not adequately described by the trophic level concept but that it is rather of a

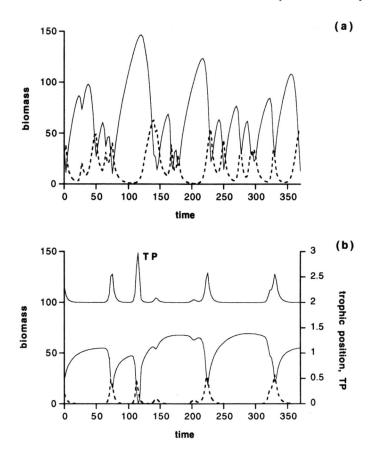

Figure 6.19   Temporal variations of the total biomass of phytoplankton (full line) and herbivores (dashed line) in (a) the absence and (b) the presence of an omnivorous consumer feeding indiscriminately on all weight classes. (b) Shows in addition the variations in the trophic position of the omnivore. The results were obtained with the model introduced in Chapter Five describing the dynamics of ensembles of predator–prey pairs distributed along the body weight axis ($n = 8, q = 8$). The omnivorous consumer was assigned a half saturation constant eight times higher than the herbivores

more continuous nature. A static trophic position is not sufficient to understand the impact of a group of organisms on ecosystem organization. We have as well to take the details of the embedding in a dynamic network context into account. A description in terms of trophic dynamic modules seems to be more appropriate to describe essential properties of ecosystem dynamics and organization. In the light of spatio-temporal organization we can also understand that concepts of trophic cascading and biomanipulation have more successfully been applied to small than to large lakes (Reynolds, 1994). In the latter, we have to consider pronounced effects of spatial heterogeneity in ecosystem dynamics in addition to temporal variations. We should also expect strategies for species in large lakes or marine environments to differ from those in small lakes where, especially for fish populations, spatial organization is largely impossible since the whole system has to be looked upon as the sphere of influence of one social group of the top predators.

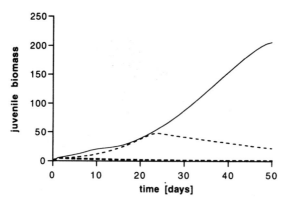

Figure 6.20   Simulation of the reproductive success of juveniles characterized by a behaviour as shown in Figure 6.17b. The juveniles were introduced at different times chosen at random. The results were obtained with the same model used to illustrate the effect of an omnivorous consumer (cf. Figure 6.19)

To summarize, I contrast in Figure 6.21 the three types of biomass distributions that we have been talking about. The relative biomass is depicted as a function of trophic position and dynamic class (= weight class). The figure shows the various types of distribution in their extremes. We may expect more structure and variations in distributions obtained from real ecosystems. In particular, such a distribution may be interpreted as a fingerprint for a food web's internal dynamic structure, thus rendering possible comparative analyses as, for example, among similar ecosystems under different environmental conditions. The questions arise: when do we obtain continuous and when discrete biomass distributions, and what consequences should we draw.

In Figure 6.21a the biomass in each well-defined trophic level is confined to separate narrow weight intervals. Empirical evidence shows that this is hardly ever the case in aquatic systems. In terrestrial systems such situations seem to be encountered in either managed or in depauperate natural systems only. We could conceive of managed meadows, forests or agroecosystems where species and functional diversity are reduced and where management practices result in spatio-temporal synchronization and the selection of single temporal and spatial scales.

In Figure 6.21b discrete trophic levels can still be distinguished but each level extends over a wide range of dynamic classes and hence spatio-temporal scales. The centre of the biomass distributions shifts to higher dynamic classes with increasing trophic level. This corresponds, for example, to the situation depicted in Figure 6.16a where in a pelagic system food chains are shifted along the body weight axis. Our further considerations revealed that the real situation in pelagic systems rather resembles the one depicted in Figure 6.21c. Here the biomass is distributed over a continuum of trophic positions rather than being confined to discrete trophic levels. The distinction between the levels of primary producers and herbivores is still well defined reflecting the major physiological difference between autotrophs and heterotrophs. (Little is yet known about the importance of mixotrophy.) At higher trophic positions we encounter a trophic continuum. This pattern arises from the size-dependent increase in trophic position. It is reinforced by life history strategies, in particular of fish, that move up a body weight and hence a trophic gradient during maturation.

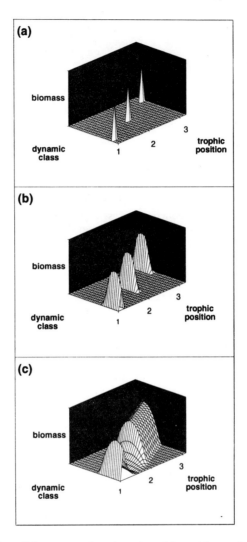

Figure 6.21   Distribution of biomass as a function of trophic position and dynamic class (= weight class). (a) Discrete trophic levels with defined spatio-temporal scales. (b) Discrete trophic levels with a continuum in spatio-temporal scales. (c) A continuum in both trophic position and spatio-temporal scales

Such a continuum in both spatio-temporal scales and function does not imply that at any moment in time and space all of these feedback pathways are active. It is the nature of spatio-temporal organization that we expect a dynamic picture of variations in time and space regarding the dominance pattern of dynamic and functional niches and the quantitative importance of different energy channels. However, we do not have defined pathways of ecosystem control. We have to give up the idea that interaction and feedback can be represented by well-defined arrows, by straight lines linking state variables with a defined function and defined spatio-temporal scales. Feedback in natural systems is rather of a continuous nature. Instead of viewing interactions among populations in a mechanistic fashion like interactions among defined particles we should develop a different perspective viewing them rather as spheres of interactions among clouds.

What can we conclude from this regarding predictability? I conjecture that any predictions at the level of single populations and the details of community dynamics are futile. Such a claim seems to be supported by the apparent failure of dynamic simulation models to predict population dynamics for fisheries management (Kitchell, 1992; Ludwig *et al.*, 1993). Nevertheless, simulation models may be helpful to devise better management strategies that reduce undesirable interference with the natural dynamics of the systems. Interestingly, results from model simulations indicate that current management practices lead rather to an increase in the variability of the fish populations than to the desired stabilization (Post and Rudstam, 1992). In the light of these discussions it seems to be more promising to shift the emphasis from mechanistic details to typical patterns of diversity and organization and to develop schemes for the evaluation of how management practices interfere with any such patterns. To do so we need to broaden our understanding of what the notion of diversity may imply.

6.6   ESSENTIAL COMPONENTS OF ECOSYSTEM DIVERSITY

The importance of biodiversity for ecosystem function is a major issue in evaluating human impacts and in questions related to biological conservation. Recently, it was highlighted in a collection of essays edited by Schulze and Mooney (1993a). The numerous examples chosen from a wide range of different ecosystems also made it clear that few guidelines exist on how to assess diversity. The traditional and yet most common use identifies diversity with the richness in taxonomic species. However, there seems to exist no direct relationship between properties at the community or ecosystem level and the number of the component species. Correspondingly, it is by no means evident what should be protected and conserved. The foregoing considerations seem to suggest we should aim to conserve dynamic instead of static attributes. Rather than conserving single species we have to conserve ecosystems, or better we have to conserve the potential of ecosystems to function, to adapt to changes and to evolve. To pursue these goals we need to assess what the essential components of ecosystem diversity are to maintain species coexistence and ecosystem function and organization. In the end we have to answer questions of the type: What are the *mutual* relationships between the functional properties of an ecosystem and the functional and dynamic diversity of its components, its composition in terms of taxonomic species, and the nature of the physical environment? And finally – how and at which levels does human influence interfere with these relationships and what are the consequences? In order to make any progress towards answering these intricate questions we need a coherent framework. We may formulate the task to comprise an outline of the various components of diversity that have to be expressed in a way rendering them amenable to experimental and modelling investigations. The derivation of quantitative measures may assist in the detection of interdependencies and in making cross-system comparisons. I am going now to outline that the concepts introduced here may provide a base for addressing these issues in a more systematic fashion.

6.6.1 Diversity in dynamic characteristics and function

Based on our previous considerations I conjecture that ecosystem performance, and the adaptive and evolutionary potential of an ecosystem are directly related to its dynamic and functional diversity. The concept of trophic dynamic modules allows a coherent description of ecological networks regarding their dynamic and functional characteristics. We can use such a network description to continue more systematically investigating the ecological importance of dynamic and functional richness along the lines of reasoning outlined in Chapter Five. There we noted that dynamic diversity was important for a network's spatio-temporal organization that was quantified by the flow based measures of organization introduced in Chapter Four. However, it is also important to devise some measures that allow us to assess quantitatively the dynamic and functional diversity at the population level. In particular it would be important to have some measures that are amenable to experimental investigation. A temporally and spatially resolved description in terms of TDMs with respect to both biomass and material exchanges requires a detailed knowledge about a system that may hardly ever be available for any real system. Thus the flow-based measures may find more application to evaluate results from simulation models. The network concept outlined in this chapter can serve as guideline to derive measures applicable to real systems, to design appropriate schemes for partial investigations that allow an integration and comparison of observations made at different spatio-temporal scales and for different functional levels.

Investigations may be performed at different levels of resolving the network structure. Recall that in the stepwise procedure of defining a network, the organisms are first assigned to dynamic (weight) classes according to their characteristic dynamic properties (body weight). Each dynamic class is then resolved into a temporal and spatial succession of dynamic modules on relative temporal and spatial scales. Each dynamic module consists of one or several trophic dynamic modules that are distinguished by their functional niches. The trophic dynamic modules are an observational grid imposed onto an ecosystem in order to receive a coherent resolution along the dimensions of time, space and function. To simply enumerate all the TDMs would thus give us little information. We have to give a TDM a quantitative measure by accounting for its biomass or metabolic activity. Biomass distributions can be replaced by metabolic activities by scaling the biomass in a weight or dynamic class by its allometric factor or the inverse of the turnover time, respectively. I am going now to discuss a variety of diversity measures that can be derived for different levels of resolution along the dimensions outlined. Regarding their mathematical formulation all measures are based on the Shannon entropy that increases with the logarithm of the number of entities that can be distinguished and the evenness of their distribution (cf. Eqn 2.1). Details about the mathematical expressions are summarized in Box 6.4.

The results from model simulations discussed in Chapter Five revealed an improvement of system performance with increasing dynamic diversity. Increasing the number of differently sized predator–prey pairs resulted in an increase in the efficiency of nutrient utilization and in a decrease in the loss of nutrients from the system (cf. Figure 5.22). These results confirm the experimental work by Van Voris *et al.* (1980) who observed in mesocosm experiments that calcium leakage upon cadmium treatment decreased with increasing dynamic richness. They used the number of distinct peaks in the power spectrum of the experimentally determined time series as a measure for dynamic richness.

Using the framework outlined here, the dynamic richness of an ecosystem can now be assessed in a first approximation by the biomass diversity of its component dynamic classes, D_{dyn}, defined in Eqn (6.8) in Box 6.4. To develop a better understanding of its meaning let us apply D_{dyn} to evaluate the biomass distribution in pelagic systems.

We discussed in detail that pelagic systems tend towards a state with an even distribution of the biomass across weight and hence dynamic classes. This results in a high dynamic diversity of the ensemble of the component organisms. Figure 6.22 contrasts two examples of how the biomass may be distributed among weight classes. The rather even distribution depicted in (a) was derived for the annual biomass of the plankton community in Lake Constance by Gaedke (1992b); (b) is a hypothetical distribution with pronounced modes. Such a discontinuous distribution would be obtained for a food chain as depicted in Figure 6.21a. The biomass diversity yields (a) $D_{dyn} = 4.6$ and (b) $D_{dyn} = 2.1$. The maximum for an ideal uniform distribution would be equal to $Log_2 35 = 5.1$, due to the 35 weight classes that were distinguished. The closeness of D_{dyn} to a uniform distribution obtained for Lake Constance confirms our evaluation of the normalized biomass spectrum in Figure 6.15.

In section 6.5 we noted that a consumer feeding like daphnids on a large range of body weights effects a decrease in dynamic diversity. The model results depicted in Figure 6.19 can now be evaluated quantitatively. D_{dyn} dropped from 2.9 (Figure 6.19a) to 1.8 (Figure 6.19b) upon introduction of the omnivorous consumer. In contrast, D_{dyn} remained largely unchanged when a consumer like the cyclopoid copepods was introduced that integrates into the temporal organization of the existing community.

We may now further resolve the network structure by taking into account the richness in function. This can be done by distinguishing for each dynamic class the biomass associated with the different functional niches. The diversity with respect to dynamic attributes and network function can then be quantified by the biomass diversity of all functional niches, D_f, defined in Eqn (6.9) in Box 6.4. D_f has a lower bound equal to D_{dyn} that is obtained for the case that each dynamic class comprises one functional niche only. Such a configuration is given by the simple food chains in Figures 6.2 and 6.21a. D_f is higher than D_{dyn} when for each dynamic class different functional niches can be distinguished, as was the case for the simple network depicted in Figures 6.10 and 6.11.

If we continue along the same lines of reasoning we would in the next step resolve the network as well along the dimensions of space and time to have a full description in terms of TDMs. The corresponding measure of the biomass diversity of all component TDMs, D_{tot}, is defined in Eqn (6.10) in Box 6.4. D_{tot} is high when the total biomass is distributed uniformly across all dynamic classes, when the biomass in a dynamic class is distributed uniformly across all functional niches, and when the biomass assigned to a functional niche is distributed uniformly over the whole temporal and spatial periods of observation. What should we now expect in the light of spatio-temporal organization? Resolution along the axes of the dynamic and functional niches gives us an idea about the dynamic and functional diversity of the network structure. We expect thus a high spatio-temporal organization to be associated with an increase in the corresponding diversity measures D_{dyn} and D_f. If we resolve now the patterns in time and space, a uniform spatio-temporal distribution of the biomass and/or the activities across all dynamic classes and functional niches would correspond to a high degree of redundancy and the virtual absence of spatio-temporal organization. Hence, we would expect a decrease in the difference

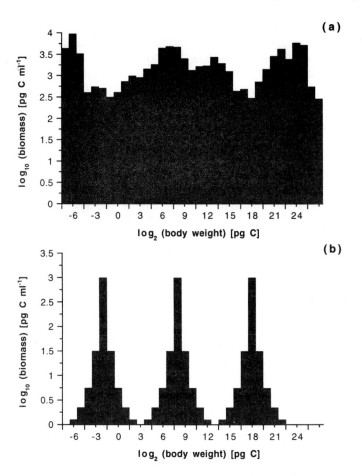

Figure 6.22 Examples for biomass weight class distributions. (a) Continuous distribution obtained for the annual biomass in Lake Constance (Gaedke, 1992b). (b) Hypothetical model distribution that would be obtained for a food chain as depicted in Figure 6.21a. The diversity measures can be calculated to yield (a) $D_{dyn} = 4.6$ and (b) $D_{dyn} = 2.1$. The maximum for $D_{dyn} = \log_2 35 = 5.1$, due to the 35 weight classes distinguished

$D_{tot} - D_f$ with increasing spatio-temporal organization. This difference equals zero when each functional niche is observed over a single interval in time and space only. Such would indicate a highly dynamic network structure.

These diversity measures complement the flow-based measures of organization introduced in the previous chapter. Measures based on the distributions of biomass have the major advantage of being more amenable to empirical investigations than measures requiring the quantification of numerous flows with a high temporal and spatial resolution. However, in order to delineate functional niches we need as well to know the pattern of energy and material transfers in a network. It would further be advisable to have at least some quantitative estimates. However, we do not need to determine the flow magnitudes with the same accuracy that would be required for analysing the spatio-temporal flow patterns. Hence, it might be useful to assess spatio-temporal organization by devising some measures quantifying the spatial and temporal diversity associated with the spatio-temporal segregation of functionally redundant populations. To clarify what I mean by this, I am going to discuss an example below.

For comparative purposes, it may be of interest to assess the distribution of biomass and activities for a functional group, in particular for primary producers. To give an example, I contrast in Figure 6.23 an intensively exploited agroecosystem with a diverse natural ecosystem. The diagrammatic representation emphasizes the major differences in diversity attributes. In intensive agricultural systems spatial and temporal niches are in

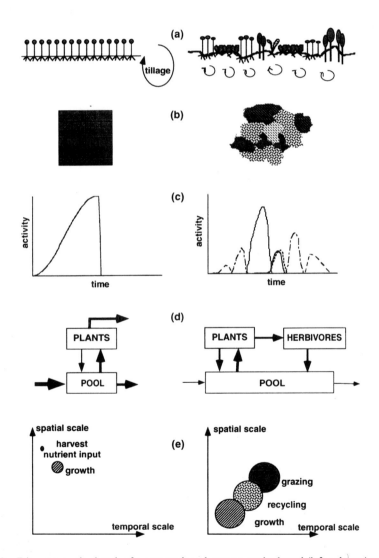

Figure 6.23   Diagrammatic sketch of a comparison between agricultural (left column) and natural (right column) ecosystems. (a) Diversity of species and life forms (b) spatial pattern, (c) temporal pattern, (d) important pathways of nutrient transfers, (e) spatio-temporal scales of activities. Further explanation in the text

general eradicated by homogenizing the environment and the dynamics of the system in space, time and function. Figure 6.23a contrasts the uniformity in species and life forms of artificial monocultures with the diversity of natural communities. In addition I indicated the influence of tillage disrupting the spatial organization below-ground (Hendrix *et al.*, 1986). In natural systems localized activities preserve the spatial integrity. The diversity of plant morphologies allows for a horizontal and vertical spatial organization. Moore and de Ruiter (1991), for example, reported major differences between natural and convential agricultural systems in the temporal and spatial heterogeneity of below-ground food webs. Contrasting with convential agricultural systems, natural food webs displayed pronounced spatio-temporal variations regarding the importance of functionally distinct energy channels. Figure 6.23b compares the horizontal spatial and Figure 6.23c compares the temporal patterns. In agricultural systems, plant activities proceed synchronized over the whole spatial area to be stopped abruptly at the same time by harvesting. Any organization by feedback through nutrient recycling is prevented by the addition of fertilizer and by replacing the impact of herbivory with a spatially and temporally synchronized harvesting by humans. The pathways of nutrient transfers shown in Figure 6.23d indicate the agricultural system to be dominated by an external management scheme in contrast to the natural system which is structured by feedback processes. An important aspect is as well the high leaching of nutrients from agricultural in comparison to natural ecosystems. Figure 6.23e shows the overall difference in the temporal and spatial scales of important functional attributes. Management results in a complete loss of dynamic and functional diversity. Harvesting and fertilization occur on short temporal but large spatial scales. In natural systems we encounter a higher dynamic diversity and a balance between temporal and spatial scales. The details of where growth, grazing and recycling may be located with respect to their spatio-temporal scales remain to be determined for each particular system. I only intend here to emphasize the overall difference in pattern between natural and intensively managed systems. A rigid scheme is imposed by excessive inputs of energy and matter to render agroecosystems more predictable and amenable to external control. Whether such practices are compatible with any idea of sustainability is more than questionable. I will come back to this point in the last chapter.

The differences outlined in Figure 6.23 can also be evaluated in a quantitative fashion. The diversity measure, D_p, defined in Eqn (6.12) in Box 6.4 quantifies the diversity of primary producers with respect to the dynamic and functional characteristics. Remember that terrestrial plants were distinguished into functional types with respect to their life forms and their tissue distribution amounting essentially to a distinction by generation time. For the agricultural system we expect a D_p close to zero. The D_p of the natural system depends on the diversity in life forms and successional stages. In general it should, for example, be lower for a grassland than for a forest due to the larger range in generation times encountered in the latter. Box 6.4 also lists measures to assess the spatial and temporal diversity regarding the spatio-temporal pattern of dynamic and functional characteristics. Again we expect these measures to be close to zero for intensively managed agricultural systems whereas they increase with increasing spatial and temporal resource partitioning in natural systems. The negative consequences of intensive agricultural practices are already known (e.g. Swift and Anderson, 1993). It might be of major interest to investigate in more detail the relationship between functional properties such as nutrient recycling and the resilience to perturbations as a function of the diversity along

---

Box 6.4   Measures assessing different components of ecosystem diversity

In calculating these measures one has to take into account that units such as DMs or TDMs have an extension in time and/or space. The biomass has to be given a weight in proportion to the absolute temporal (τ_i) and/or spatial (σ_i) extension of the unit under consideration relative to the total periods of observation.

(a) Diversity measures for an increasing resolution of an ecosystem's structure

Diversity of an ecosystem with respect to the distribution of biomass over dynamic (weight) classes:

$$D_{\text{dyn}} = -\sum_i \frac{B_i}{\sum_i B_i} \log \frac{B_i}{\sum_i B_i} \geq 0 \qquad (6.8)$$

where

$$B_i = \frac{\tau_i}{T} \frac{\sigma_i}{S} \sum_k \sum_r \sum_a B_{ikra}$$

B_{ikra} is the biomass of the TDM_{ikra} comprising all organisms that are present over the time interval τ_{ik} at the spatial location σ_{ir} and that share the functional niche C_{ia}. D_{dyn} is based on a resolution to the level of dynamic classes without taking into account any spatio-temporal patterns and/or functional distinctions.

Resolving D_{dyn} with respect to the functional niches:

$$D_f = -\sum_i \sum_a \frac{B_{ia}}{\sum_{i,a} B_{ia}} \log \frac{B_{ia}}{\sum_{i,a} B_{ia}} \geq D_{\text{dyn}} \qquad (6.9)$$

where

$$B_{ia} = \frac{\tau_i}{T} \frac{\sigma_i}{S} \sum_k \sum_r B_{ikra}$$

B_{ia} refers to the averaged biomass of all organisms in the ath functional niche of the dynamic class C_i. To give an example, $B_{11} = 3B_{12}$ in the network depicted in Figure 6.11 assuming that the total biomass of a TDM in C_1 is time invariant.

Biomass diversity of an ecosystem based on a complete description in terms of trophic dynamic modules:

$$D_{\text{tot}} = -\sum_i \sum_k \sum_r \sum_a \frac{B_{ikra}}{B} \log \frac{B_{ikra}}{B} \geq D_f \qquad (6.10)$$

where

$$B = \sum_i \frac{\tau_i}{T} \frac{\sigma_i}{S} \sum_k \sum_r \sum_a B_{ikra} = \sum_{i,a} B_{ia} = \sum_i B_i$$

A simplified measure for the functional diversity can be derived from a topological network description without taking into account quantitative attributes:

$$D_{\text{top}} = \frac{1}{n} \sum_{i=1}^{n} \frac{\text{number of TDMs in class } i}{\text{number of DMs in class } i} \geq 1 \qquad (6.11)$$

Normalization by the total number of dynamic classes, n, renders the measure independent of both the partitioning and the range chosen along the dynamic axis. The contribution of each dynamic class is normalized by the total number of DMs equal to the minimum of distinct TDMs.

(Continued overleaf)

Box 6.4   (*continued*)

(b) Diversity measures for the functional group of primary producers

Dynamic and functional diversity for primary producers

$$D_p = -\sum_i \sum_{a_p} \frac{B_{ia_p}}{B_p} \log \frac{B_{ia_p}}{B_p} \leq D_f \tag{6.12}$$

where

$$B_p = \sum_i \frac{\tau_i}{T} \frac{\sigma_i}{S} \sum_k \sum_r \sum_{a_p} B_{ikra_p}$$

The index a_p refers to the functional niche of terrestrial primary producers with respect to composition and life forms as defined in section 6.2.2.

The spatial diversity can be assessed as:

$$D_{p_s} = -\sum_i \sum_{a_p} \frac{B_{ia_p}}{B_p} \log \frac{S_{ia_p}}{S} \tag{6.13}$$

where

$$S_{ia_p} = \sum_r f_{irp_a} \sigma_i \quad f_{irp_a} = 1 \text{ if } B_{irp_a} > 0, \quad f_{irp_a} = 0 \text{ if } B_{irp_a} = 0$$

The temporal diversity can be assessed as:

$$D_{p_t} = -\sum_i \sum_{a_p} \frac{B_{ia_p}}{B_p} \log \frac{T_{ia_p}}{T} \tag{6.14}$$

where

$$T_{ia_p} = \sum_k f_{ikp_a} \tau_i \quad f_{ikp_a} = 1 \text{ if } B_{ikp_a} > 0, f_{ikp_a} = 0 \text{ if } B_{ikp_a} = 0$$

The spatio-temporal diversity can be assessed as:

$$D_{p_{ts}} = -\sum_i \sum_{a_p} \frac{B_{ia_p}}{B_p} \log \frac{S_{ia_p}}{S} \frac{T_{ia_p}}{T} \tag{6.15}$$

Especially in terrestrial systems where the presence of biomass may not be indicative for activity it would be useful to replace the biomass in the diversity measures by some measure of activity, e.g. nutrient uptake or primary production.

---

the dimensions outlined here. Such information should assist in the derivation of guidelines for sustainable environmental management.

The various measures defined, convey different types of information. Such diversity measures can easily be accommodated to fit the need of a particular situation. The choice of an appropriate measure depends on the question of interest and the information available. We could also conceive of using different analytical tools to evaluate given patterns. What is essential is the description of the network in terms of a coherent framework rendering possible comparisons across spatio-temporal scales and among different systems.

We may now ask how the components of diversity just outlined are related to taxonomic diversity. We should expect mutual interdependencies since species differ, for example, in their life forms, in size, in their response to environmental variations. To simply assume that functional and dynamic diversity always increase with the number of taxonomic species would not be very helpful and even worse is not supported by the few experiments available (Schulze and Mooney, 1993a). We have to become more specific in dealing with the problem of what is the mutual relationship between the types of diversity outlined, the organization of the system as a whole and different aspects of taxonomic diversity. I am going now to sketch first what we might expect regarding the distribution of taxonomic species and how we could proceed in detecting patterns.

6.6.2 The relevance for taxonomic species

At the beginning of my expositions I claimed evolution to proceed by the dualistic interplay and the mutual dependence between organisms and their environment. As a consequence I concluded that in order to reveal general patterns one has to consider an intermediate level of functional aggregations. The network concept in terms of trophic dynamic modules now offers a tool for doing so. The trophic dynamic modules were distinguished by the dynamic characteristics of the component organisms and by the network context. The organisms were thus distributed in a niche space comprising dynamic and functional properties. This concept is based on the premise that the ecological network's functional organization across time and space is central to understanding the mutual relationship between species properties and ecosystem function. If such an assumption is warranted and if the niche space chosen constitutes an appropriate description it must bear relevance at the level of taxonomic species. In addition, since taxonomic species reflect a genetic and thus an evolutionary potential, the pattern of species composition may constitute an essential ingredient for the other components of diversity to persist and to evolve. To become more specific we have to reflect on the relationships we might expect. In the following the notion of species refers to taxonomic species as genetically distinct entities.

Based on our prior discussions, I assume the niche space of dynamic and functional attributes to be continuous which implies that the niches are evenly distributed. The probability for new species to emerge should then change in proportion to the probability for functional niches to arise in the context of the spatio-temporal framework of ecosystem organization. What determines these probabilities? Recall that the dynamic modules were just defined as the units in time and space where changes in the input- and output-environments of an ensemble of organisms are likely to occur. The number of niches and hence the number of species should then be directly related to the number of dynamic modules within a dynamic class. However, some caution is required. The dynamic modules are defined as temporally and spatially distinct ensembles. Should we also assume time and space to be largely independent dimensions for speciation? Or is it more likely that space and time interact in that, for example, the patchiness and the spatial desynchronization of fluctuations in a species abundance prevent extinctions and constitute an essential element of spatio-temporal organization. In Chapter Four, I discussed how the flow-based measures of organization can be used to evaluate the relationship between space and time regarding their functional importance for ecosystem organization. These considerations suggest that we rather should expect both dimensions to be mutually dependent in that spatial patterns vary in time and that temporal patterns vary in space. This should also be reflected in the relationship between the number of species and the number of dynamic modules, DMs.

Let us first derive the scaling relationship for the case that space and time are mutually dependent. This means that the distinction along a single dimension of dynamic characteristics is largely decisive for speciation. Grouping organisms into dynamic classes we would thus expect the number of taxonomic species per dynamic class to change with the number of dynamic modules to be distinguished along the dimension of time (cf. Figure 6.8):

$$\frac{n_{i+1}}{n_i} = \frac{\tau_i}{\tau_{i+1}} = \left(\frac{W_i}{W_{i+1}}\right)^{\varepsilon} = \left(\frac{L_i}{L_{i+1}}\right)^{3\varepsilon} \qquad (6.16)$$

If space and time were independent dimensions for speciation we would expect the number of taxonomic species per dynamic class to change with the number of dynamic modules to be distinguished along the dimensions of time and space (cf. Figure 6.9):

$$\frac{n_{i+1}}{n_i} = \frac{\tau_i \sigma_i^2}{\tau_{i+1} \sigma_{i+1}^2} = \left(\frac{\tau_i}{\tau_{i+1}}\right)^{(1+2\gamma)} = \left(\frac{W_i}{W_{i+1}}\right)^{\varepsilon(1+2\gamma)} = \left(\frac{L_i}{L_{i+1}}\right)^{3\varepsilon(1+2\gamma)} \qquad (6.17)$$

To establish a more direct relation to organismal properties, time and area in Eqns (6.16) and (6.17) are replaced by body weight, W, which may be then be replaced by size, L. Eqn (6.17) was obtained by using Eqn (6.7), the relationship derived between spatial extension, σ, and turnover time, τ. The exponent γ may vary between 0.7 and 1.3. The allometric exponents, ε, is about 0.25.

Based on Eqn (6.16) we would thus expect an approximate decline in the number of species per class by a factor $K_\tau^{-0.25}$ for weight classes and by a factor $K_\tau^{-0.75}$ for size classes. K_τ is the ratio between the average weight or size in neighbouring classes. It is equal to the base of the logarithm chosen for the weight or size classes when the range of a class is equal to one unit on a logarithmic scale. Including space as an independent dimension implies that the slope of the decline would at least double.

Data on species numbers reported by Fenchel (1993) for aquatic communities and by Brown and Maurer (1989) and Brown and Nicoletto (1991) for birds and mammals, display shallow slopes that agree well with a scaling along one dimension. The distributions typically have the shape depicted in Figure 6.24. The number of species increases with decreasing body weight until a threshold is reached where the number is at maximum. For smaller weights the number of species declines rapidly. This may be explained by constraints at the level of the organisms. A given body plan may require a minimum size to be efficiently realized regarding both physiological constraints and ecological requirements to survive. For weights exceeding the minimum the number of species shows a gradual decline. Even when the real data show a considerable scatter the qualitative overall pattern is retained. These observations support the assumption that the observed scaling of taxonomic species with body size can be derived by invoking a single dimension of dynamic characteristics.

Enumerating simply the total number of species does not tell us much about the nature of the species composition. What has intrigued ecologists for a long time is the fact that a few species are dominant whereas the majority of species are rare (Ricklefs and Schluter, 1993; Schulze and Mooney, 1993a). The difference in abundance cannot be explained by fluctuations only since some species seem always to be rare. We could nevertheless conceive of such patterns in the context of spatio-temporal organization when we invoke an organization of taxonomic species into functional guilds. Recall the example of the old-field

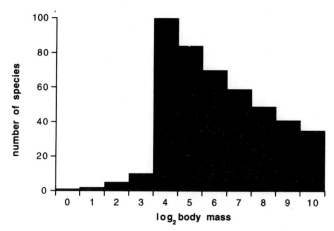

Figure 6.24   Typical shape of the frequency distribution of the body masses of taxonomic species when averaged over a larger spatial scale such as a biome or a whole continent (e.g. Blackburn and Gaston, 1994; Brown and Maurer, 1989; Brown and Nicoletto, 1991)

plant community where two guilds could be distinguished: the early-active, shallow-rooted guild dominated by *Poa* and the late-active, deeper-rooted guild dominated by *Schizachyrium* (cf. Figure 4.7). Such dominance patterns may be explained by changes in competitiveness and by trade-offs along multiple resource gradients (Tilman, 1993). These arguments only give a rationale for the coexistence of many species when we might have expected extinction. It remains open to question whether rare species are of any significance for the ecosystem as a whole. Would the integrity of an ecosystem be impaired if only the dominant species remained? Are rare species of any importance for ecosystem function and organization? Pate and Hopper (1993) advanced by the hypothesis that rare species may be essential as a functional pool to be activated in case of rare perturbations in fluctuating environments. In the light of nonlinear dynamics and spatio-temporal organization it is not difficult to imagine that the impact of a species is not necessarily correlated with its abundance. Species of minor quantitative importance could have important influences on ecosystem dynamics extending from mediating spatio-temporal organization, to system performance, and to the potential for evolution. I suggest that the presence of rare species may be taken as an indication of processes of self-organization playing a dominant role. More specifically, the shape of the distribution may prove to be an important element of characterizing an ecosystem. Instead of asking how many species are required to maintain ecosystem function (Schulze and Mooney, 1993b), we should rather ask: How important is it that ecosystems display self-organization? How can this be assessed based on distribution patterns of different components of ecosystem diversity in relation to the distribution pattern of taxonomic species?

To investigate overall scaling properties in a more systematic way, I suggest resolving species distributions along the main dimensions sketched in Figure 6.25. The dashed arrows indicate that the resolution along single dimensions may be combined as exemplified by the progressive resolution of diversity patterns in the last paragraph. It is not only of interest to determine the number of species but also the shape of the species abundance curves as a function of a certain property, e.g. size class or size class at a spatial location.

Let us for example assume that we wish to characterize the distribution patterns of zooplankton species in a lake. We further assume that the species are organized within functional groups each comprising a certain body size range. Such organization may result in specific abundance patterns within a functional assemblage. How does the distribution change when we start to resolve the whole ensemble along the axis of dynamic characteristics equivalent to body size? In such a case we should expect a discontinuous change in the distribution pattern within a size class, once the degree of resolution along the body size axis exceeds the range of functional groups. We can further ask how the distribution patterns change as a function of the resolution in time at a spatial location (which might be of prime interest in aquatic systems), as a function of the resolution in space at a moment in time (which might be of prime interest in terrestrial systems), and a combination of both. Based on our discussion in Chapter Four regarding the significance of temporal and spatial patterns and their mutual dependence, we might expect large differences between terrestrial and aquatic systems. Are there critical scales in time and space where we observe a transition in the distribution of the number of species across classes and the abundance pattern within a class? Such thresholds would indicate the typical overall temporal and/or spatial scales for the level of organization into functional groups. Brown and Maurer (1989) and Brown and Nicoletto (1991) observed the distribution of species across body size classes within a taxon (mammals and birds) to change when they compared continents, biomes and habitats. The pattern depicted in Figure 6.24 was observed to become progressively shallower with decreasing spatial resolution. Such changes indicate that different spatial scales are effective regarding ecological and evolutionary phenomena. We may distinguish the habitat requirements of a population from the spatial range of a species. It seems that with increasing body weight the habitat increases faster than the range. What is the significance of such patterns and their changes with spatial scale? At present the small data base does not allow any generalizations. We may speculate that the persistence of species depends on a spatial organization of distinct populations preventing thus the spread of local extinctions. The likelihood of local catastrophic breakdowns has recently been emphasized by Mangel and Tier (1994). Their observations confirm the conclusions derived here that local populations may experience large fluctuations caused by the complex endogenous dynamics in an ecosystem. Given the above findings that relative to the habitat

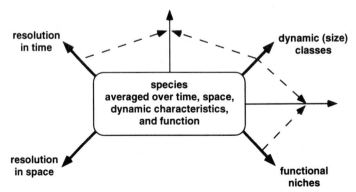

Figure 6.25   Main dimensions for resolving the distribution of taxonomic species in an ecosystem. The resolution along single dimensions may be combined as indicated by the dashed arrows

requirement of a population the geographic range decreases with increasing body weight, we might conclude that large species are more prone to face extinction. To deal with the questions raised in a meaningful way requires a more systematic investigation of general patterns, in particular in the larger context of ecosystem organization.

6.6.3 Patterns in ecosystem organization and diversity

Figure 6.26 sketches different constituent elements of ecosystem diversity. The arrows indicate the interdependence among these elements and their mutual relation with system level properties. An ecosystem may be characterized by the organization of its network, the functioning in terms of processing energy and matter, and the adaptability referring to the potential to preserve function in a variable environment. The meaning of dynamic and functional diversity has been dealt with in detail in section 6.6.1. Biodiversity refers to taxonomic diversity which includes not only the number of species but also the shape of their distribution.

The heterogeneity in the physical environment is another essential component of ecosystem diversity. The importance of environmental variability for species coexistence and community organization is now widely recognized (e.g. Chesson and Case, 1986; Levin, 1992). Habitat diversity related to spatial complexity has received considerable attention, since the loss in habitat diversity is important in driving the extinctions of species. In general, anthropogenic influence results most often in a homogenization of temporal and spatial patterns thus destroying habitat diversity. Douglas and Lake (1994) showed, for example, the surface complexity of stones to be an essential factor for maintaining species richness in streams. Managed streams proved to be quite a poor environment in this respect. The surface complexity of stones is an example for spatial heterogeneity that is time-invariant. Another example is given by fixed spatial gradients (e.g. change of climate with altitude). We may further distinguish temporal heterogeneity that is space invariant (e.g. seasonal changes in isolation), and variations comprising both time and space (e.g. seasonal dynamics of flood-plains). The latter type of change may be referred to as a disturbance where there is a trade-off between temporal and spatial scales in the sense of fast changes over large spatial scales. Contrasting to earlier assumptions viewing any disturbance as detrimental, disturbance regimes have nowadays been recognized as an important characteristic of a given habitat. Management methods

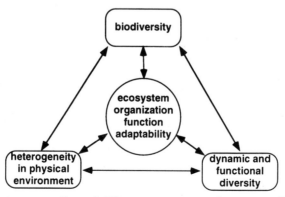

Figure 6.26 Ecosystem properties and different components of ecosystem diversity. The arrows indicate the mutual dependencies

preventing supposedly damaging disturbances such as forest fires have been replaced by prescribed burning (Hansen *et al.*, 1991). In his intermediate disturbance hypothesis, Connell (1978) even stated that there exists an optimum frequency of disturbance maximizing the diversity of a species community.

However, if we look now at an ecosystem as a whole we cannot simply state that the variability is high or low or that there might be an optimum frequency of disturbance. Considering that an ecosystem and its component populations cover a wide range of scales in time and space we rather have to focus on the whole spatio-temporal spectrum of variability in the characteristics of both environment and species. To do so and to be able to compare the results obtained from different investigations requires a comprehensive research protocol. Based on our previous considerations, I list in Box 6.5 a tentative outline of a research protocol for investigations that aim at revealing the relationships outlined in Figure 6.26. These relationships remain largely unknown to date (e.g. Schulze and Mooney, 1993a). Obviously such an extensive research protocol is not feasible for every type of investigation, nor is it the only possible design one might conceive of. However, the way in which it is designed should stress the emphasis on elucidating general patterns, on choosing a coherent resolution for biological and physical variables across time and space, and on documenting relevant information about sampling and patterns. Such practice would facilitate subsequent cross-system comparisons and is essential for the detection of any general pattern.

In many cases it may impossible to realize both the temporal and the spatial resolution required to achieve coherence over the whole range of dynamic classes in an ecosystem – even if it is desirable from a theoretical point of view. To give an example, a sampling interval of 100 days for fish corresponds to a sampling interval of 1 day or less for phytoplankton. The situation becomes even worse when we further take into account that the corresponding spatial resolution for phytoplankton would be in the range of about 10 to 100 m. Terrestrial systems pose problems in the other extreme. A meaningful investigation to monitor changes in a forest might require decades or centuries.

A workable solution lies in performing investigations over short time periods where the spatial dimensions are resolved to the detail required. Such spatial transects have to be complemented by investigations over long time periods with high temporal resolution but over a restricted spatial area only. The introduction of relative temporal and spatial units should be helpful to make comparisons across scales and systems. As Reynolds *et al.* (1993) suggested, a variety of investigations that might be impossible in terrestrial systems may be feasible in pelagic ecosystems where the fast dynamics of the component organisms allows the monitoring of changes over many generation times.

Such investigation can also provide information about the relationship between patterns in time and space and between local and global variability. To determine the variability of, for example, the biomass at the population level we need to resolve the spatio-temporal pattern in the typical scales of the dynamic class the population belongs to. Figure 6.27 shows what we might expect for different levels of resolution if spatial and temporal patterns were mutually dependent. The variability should be high for both resolving time at a single location and for resolving space at a single time. The variability should be low for resolving time averaged over space and for resolving space averaged over time. Such expectations derived from the same lines of reasoning that were basic for discussing the patterns depicted in Figures 4.22 and 4.23.

Box 6.5   Outline of a comprehensive research protocol

This protocol is designed for investigating the relationships depicted in Figure 6.26. It is based on the network concept introduced in this chapter.

(a) Data collection

List the variables to be measured and the resolution desired in time and space, in taxonomic detail, in dynamic characteristics (body weight), based on the following decisions:

Determine the range of spatial and temporal scales of the organisms the dynamics of which should be resolved $\Rightarrow$ dynamic characteristics covered by all component organisms (e.g. body weight range).

Choose the extent of observation in time, T, and space, S: The slowest population with the largest turnover time determines the overall extent of the observation $\Rightarrow T \geq \tau_{max}$ and $S \geq \sigma_{max}$.

Choose the grain of resolution in time, Δt, and space, Δs: The fastest population with the smallest turnover time determines the grain of resolution $\Rightarrow \Delta t < \tau_{min}$, and $\Delta s < \sigma_{min}$.

Choose the resolution along the axis of dynamic characteristics $\Rightarrow$ determine the partitioning into turnover time or body weight classes.

Choose a coherent regime of resolution to monitor biological and physical variables with the same level of resolution across spatial ($\Delta s_i/\sigma_i$) and temporal ($\Delta t_i/\tau_i$) scales. This applies as well to global functional variables such as nutrient levels or primary production.

Monitor transfers across the temporal and spatial boundaries of the total observation period.

(b) Data analysis

Display as a function of the dynamic class:

- biomass, metabolic activity,
- functional niches,
- number and abundance distribution curves of taxonomic species.

Compare variabilities of population and system function variables:

- as a function of the time interval of resolution at a local patch,
- as a function of the spatial interval of resolution at a single time,
- as a function of the simultaneous increase of the interval of resolution along time and space.

Display as a function of the time scale and as a function of the spatial scale:

- the variance in the physical environment (e.g. precipitation, temperature);
- the external import of energy and nutrients (e.g. primary function, allochthonous matter).

Display intensity of disturbances as a function of temporal frequency and spatial extension.

(c) Questions to be addressed

What are the correlations between these distribution patterns – in particular between:

- the spatio-temporal diversity of the ecological networks and the physical environment;
- the species richness and the spatio-temporal and functional diversity?

Can we confirm the hypotheses stated in this chapter regarding:

- the behaviour of the various diversity measures (cf. section 6.6.1).
- the distribution of taxonomic species (cf. section 6.6.2).
- the trade-off between global and local variability (cf. Figures 6.27 and 6.28)?

Is it possible to define a set of patterns indicative for natural ecosystems that can be used to define a reference state the deviation from which could be a measure for an ecosystem's state of "health"?

Do these patterns change with ecosystems being exposed to stress?

Can the shape of these patterns be used for monitoring purposes?

Where and how does human influence interfere with these patterns?

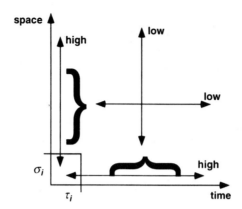

Figure 6.27   Variability of the population biomass (metabolic activity) of a dynamic class at different levels of resolution when the patterns in space and time are mutually dependent. The variability is high for resolving time at a single location or for resolving space at a single time. The variability is low for a resolving time averaged over space or for a resolving space averaged over time

In systems with a high spatio-temporal organization we might further expect a trade-off between global and local variability as sketched in Figure 6.28. Such an expectation derived from the results of the model simulations in Chapter Five where we observed a decrease in the variability at the ecosystem level and an increase in the variability at the population level with increasing dynamic and functional diversity. In particular, we would expect variables at the ecosystem level (e.g. total metabolic activity) to have a lower variability that is largely independent of the time interval of resolution. If such a trade-off can be found in real systems it might provide a powerful tool for characterizing a system's state. In combination with distribution patterns of dynamic and taxonomic diversity we might be able to derive an integrated assessment for monitoring an ecosystem's state of "health". The final goal of extensive measurements of the type outlined in Box 6.5 cannot be to pile up always more and more data. We need to find out what the relevant data are that provide most information with a reasonable sampling effort.

Human impacts and management practices should also be evaluated within a similar framework. I conjecture that human influence is most detrimental in its interfering with temporal and spatial patterns of natural systems. Effects become predominantly apparent at the level of the functional and spatio-temporal organization of ecosystems. The loss in habitat diversity has already been mentioned. Similar effects became evident in the comparison between agricultural and natural systems (cf. Figure 6.23). Instead of fostering dynamic diversity and spatial heterogeneity, human management leads rather to a homogenization and to a selection of dominant temporal and spatial scales.

In Figure 6.29 I tried to sketch a related aspect, the anthropogenic interference with the pattern of variability of natural systems. The dotted line in Figure 6.29a indicates that in natural systems the spatial scale of a phenomenon increases gradually with increasing time scale. The qualitative nature of this relationship applies to both terrestrial and aquatic systems (e.g. Shugart and Urban, 1988; Steele, 1991). The ellipse represents the scale of catastrophies occurring on a large spatial scale but during a very short interval of time. The full line indicates an increasing imbalance between spatial and temporal scales

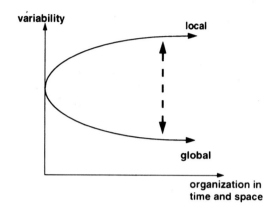

Figure 6.28 Hypothesized trade-off between local variability and global variability of an ecosystem as a consequence of the progressive temporal and spatial organization promoted by the increase in functional and dynamic diversity (Reproduced by permission from Pahl-Wostl, 1993b)

that is caused by anthropogenically induced changes in, for example, biogeochemical cycles or species transfers. The decrease in the time scale of global effects is alarming, especially because there seems to exist no precedent in geological time that would allow humankind to learn from historical comparison. Also, neither the biological world nor the human beings themselves have been exposed to similar effects during the process of evolution.

We may also perceive important changes in the disturbance regimes. The dotted line in Figure 6.29b indicates that in natural systems there is a gradual transition in that the frequencies of disturbances such as fires or floods decrease with increasing intensity and spatial extent. I conjecture that human influence leads to a reversal of this regime caused by the change in the dynamic structure of natural systems, in combination with an efficient suppression of small-scale disturbance. The general direction of change is indicated by the arrows. The full line portrays an extreme case that may be encountered in intensively managed systems. Examples for such shifts are given by flooding events (Knox, 1993) or forest fires (Hansen *et al.*, 1991; Swetnam, 1993) that occur rarely but more and more often with unprecedented intensity.

The elucidation and the investigation of typical patterns should be a fruitful approach to detect regularities in both ecosystem dynamics and the nature of anthropogenically induced changes. In doing so, we may have to face the argument that correlation is not causation. I would be reluctant to dismiss such an objection right away, even when I am not favourable to seeking mechanistic explanations for every pattern observed. Rather I am convinced – and this conviction is basic to the whole approach pursued in this book – that regularities and general patterns exist and can be detected irrespective of the various mechanisms contributing to them. To support such an assumption, we might seek for general explanations and models that render patterns intelligible without emphasizing mechanistic detail. Recent years have witnessed a progressive increase in theoretical ecologists developing new methods to model and investigate spatio-temporal patterns in ecosystems (e.g. Levin, 1992; Turner and Gardner, 1990). The framework introduced here might assist to develop models of a more general nature.

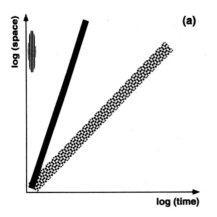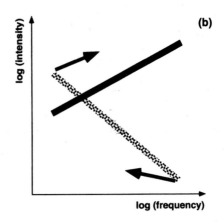

Figure 6.29   (a) Schematic drawing of the relationship between the spatial and temporal scales of a phenomenon. Dotted line – natural biological and physical processes, full line – anthropogenically induced changes; dashed ellipse – catastrophes. (b) The frequency of a disturbance as a function of its intensity. Dotted line – natural systems, full line – anthropogenically induced changes. The arrows denote the direction of change induced by human influence

6.6.4   Generic models of ecosystems

The network concept outlined can serve as a base for the development of generic ecosystem models in which the "species" are represented by ensembles of organisms belonging to the same dynamic class and the same functional niche (Pahl-Wostl and Büssenschütt, in preparation). Figure 6.30 summarizes some essential properties characterizing such model species and their embedding in a network context.

The membership of a dynamic class determines the dynamic properties of a species and hence also the characteristic steps in time, τ_i, and space, σ_i. The internal state may be explicitly referred to by introducing intrinsic variables such as social structure, parasites (cf. section 5.3). To account for preferences with respect to environmental variables, we may specify temporal windows in a seasonal environment and/or spatial windows along spatial gradients or in a patchy environment. Such a distinction may be important to account for the functional diversity of primary producers. As mentioned before primary producers can be looked upon as the major interface of the biological community with the abiotic environment.

The function in the network context is specified by the nature of the material exchanges with groups in other dynamic classes and with the abiotic environment. Links to other dynamic classes may derive from predation (linking in most cases remote classes). They may as well derive from effects of growth or migration. Growth links adjacent dynamic classes over time whereas migration links the same dynamic class over space. Abiotic resource pools serve as input source, mainly for autotrophic organisms, and they receive losses from all organisms. The resource flows provide the main agent of coupling and feedback in the network. Further interactions may be introduced by accounting more explicitly for the impact of the organisms on the abiotic environment. This may, for example, result in a mutual shaping of the species' temporal and spatial windows that were up to now assumed to be determined by the physical environment only. The nature of

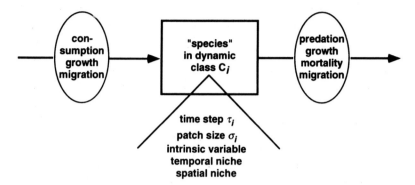

Figure 6.30 Properties characterizing a model species and its embedding in the network context

environmental variability and the diversity of the environment should be expressed in the relative scales of the populations as outlined in the previous paragraphs.

The questions to be addressed by such type of model derive from the same lines of reasoning expounded for the empirical investigations:

What is the relationship between a system's internal dynamic diversity and

* its functional properties,
* its potential to adapt and evolve,
* its potential to absorb perturbations?

What is the influence of the spatio-temporal patterns in the physical environment on a system's spatio-temporal organization and dynamic diversity?
Can one identify general patterns characteristic for a system's self-organization?
What consequences may be expected for anthropogenic changes in spatial and temporal scales and the disturbance regimes as outlined above?

Following the same general guidelines in model development and further in the evaluation of experimental data should allow comparisons among results from different models, data from different systems and results from different spatial and temporal scales and different levels of complexity.

6.7 WHERE ARE WE NOW AND WHERE SHOULD WE HEAD FOR?

In this chapter I introduced a framework on how to delineate the structure of an ecosystem in terms of a network of energy and matter flows. This approach is based on viewing ecosystems as self-organizing systems characterized by a functional organization across a wide range of spatial and temporal scales. Such a perspective gave rise to the assumption that the essential characteristics of the component organisms can be reduced to their dynamic properties and their embedding in the ecological network. Instead of putting the emphasis on specific cause–effect relationships the focus has been directed to typical patterns and distributions.

It has often been argued that correlation is not causation and that we have to aim for a mechanistic understanding to increase explanatory and predictive capacities (Lehman, 1986; Levin, 1992). It is true that sometimes single factors and mechanisms are central to understanding a phenomenon and solving a problem. Phosphorus in lakes is a good example for such a case where a single factor was identified as the main cause for eutrophication (e.g. Vollenweider, 1975). However, I argue that if we wish to really understand the nature of ecosystems and the source for the problems facing us today, we have to choose a more comprehensive approach by looking for regularities at the level of patterns and organization. Let me start to substantiate this point with an example from ecotoxicology and then move on to more general considerations.

Species composition seems to be more sensitive to stress than functional properties of ecosystems. Schindler (1987) observed in experimental acidifications of lakes that functional properties such as primary production or respiration were rather insensitive to monitor the effects of a continued exposure to stress. Early signs of warning could be detected at the level of species composition and morphologies, whereas functional properties showed a lagged but abrupt response. We may interpret these observations as a gradual decline in structural properties to be finally associated with an abrupt decline in function. Correspondingly, we may derive a relationship between diversity and function as sketched in Figure 6.31a. Rather than a linear relationship, indicated by the dashed line, we expect a discontinuous threshold effect. Diversity is not meant to be limited to taxonomic species but embraces the different components of diversity referred to before.

Investigating more systematically the shapes of such relationships would be vital for the evaluation of anthropogenic impacts on ecosystems to provide a scientific base for political decisions and measures of conservation. Pitelka (1993) stated his opinion on what type of ecological information would be relevant for policy decision and for improving what he perceived as major deficiencies in the current credibility of ecological science. In particular, he urged ecologists to focus on quantitively relating effects on specific ecosystem functions to changes in taxonomic diversity. For instance, element cycling might be related to the diversity of plants in a similar fashion to that outlined in Figure 6.31a. The diversity of plants might be related to some measure of conservation such as percentage of habitat area preserved as sketched in Figure 6.31b. Pitelka continued by emphasizing that such information was relevant to weigh costs and benefits of different policy options – provided ecologists were able to eliminate large uncertainties and to define any thresholds within narrow bounds. Correspondingly he defined as criteria for useful information timeliness, specificity and high accuracy.

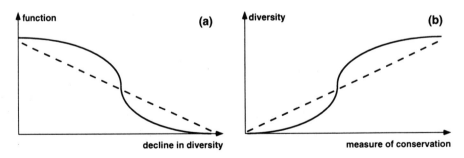

Figure 6.31   Two possible shapes of the relationship between (a) diversity and function and (b) diversity and a measure of conservation such as percentage of habitat area preserved. Diversity has to be seen in a broad context embracing the different elements referred to before

There is no doubt that we need more reliable data. And the data must be of a type to be useful for making decisions in a regional context. There is also no doubt that a lot of progress can be achieved by combining the disparate activities of ecologists within a coherent framework – especially when facing decreasing research budgets. I would also consider it as a major advantage when we could give more evidence for the discontinuous rather than gradual nature of a transition as shown in Figure 6.31. However, I cannot agree to politicians asking for detailed, quantitative predictions without uncertainties in the belief that decisions and plans for the future can thus be optimized. To come back to the example above there is no point in starting to bargain as follows: if ecologists can guarantee that a measure prevents the loss of x species then it is worthwhile to spend an amount of y £ (or $). It is profitable to spend just sufficient money to cross the critical threshold. Beyond this the profit (species saved per money spent) starts to decrease rapidly. Such considerations rely on a rather mechanistic and rigid picture of the world.

The dynamic perspective on the nature of ecosystems developed here leads us to emphasize uncertainty as an essential and positive ingredient instead of viewing it as an obstacle impeding quantitative evaluations and predictions. What has often been perceived as an inherent weakness of ecology as a "soft" science should, to the contrary, turn out to be a major strength. A collaborative effort of empirical and theoretical ecologists, resource managers, and social scientists is required to develop adequate concepts and instruments for decision making along these lines of reasoning.

The concept of spatio-temporal organization may provide a theoretical perspective to assist in such an endeavour. I deliberately use the expression theoretical perspective instead of theory. Most theories comprise as an essential core one or more hypotheses on cause–effect relationships. The validity of an explanation is preferably tested by experimental proof. Such testing is feasible for mechanistic explanations where clear and unequivocal statements can be derived. In the introductory chapters I took issue with such restricted a valuation of theoretical frameworks. As mentioned before I rather favour the idea of theories as maps guiding our perception of the world, assisting in our discerning patterns and context. I suggest we regard a theoretical framework as a specific pair of spectacles that allow us to produce a more coherent picture of the world by organizing the plethora of phenomena perceived and by discerning contexts. Let us now consider whether the spectacles constructed here assist us in finding new approaches on how to deal with our obviously disturbed relationship with natural systems and the problems ensuing therefrom.

Chapter Seven

Attempt at a synthesis and some speculations

From where did we start? How did we proceed? What has been achieved? What are the conclusions to be drawn? What further questions should be addressed? In a nutshell, these are the main issues I am going to deal with in this last chapter.

7.1 A SYNOPSIS

The starting point of our considerations was the observation that ecological knowledge has not been valued to the extent we might have expected in the light of the pressing environmental problems facing humankind today. The origins of the currently rather critical situation of ecology have been traced back to several interrelated facts:

- A fundamental problem lies in our perception of how ecosystems function. The prevailing attitude emphasizing mechanisms and defined cause–effect relationships seems to be inadequate to fully comprehend the nature of ecological systems and to develop meaningful strategies to deal with them.
- Traditional ecological concepts have relied on stability and equilibrium being essential properties of ecological systems. Meanwhile researchers have recognized that variability and change are integral elements of ecological systems. However, preconceived paradigms are not easily overcome, especially when they seem to offer guidance in an overwhelmingly complex world, a guidance that is as yet absent in new ideas emphasizing variability, chaos and chance events.
- Fragmentation prevents ecology finding its own identity. Progress has been hindered, in particular, by the fact that the more "reductionist" research on organisms and populations has proceeded rather independently of the more "holistic" ecosystem research emphasizing process and function.

There is no doubt that even when ecology may be fragmented in methods, approaches and subjects, it could become strongly united by the common goal to improve our understanding of the natural world in order to foster progress towards a sustainable coexistence of human and ecological systems (see also Lubchenco *et al.*, 1991). An analysis of the present situation suggests that progress must be made not only in collaborative efforts to accumulate further knowledge, but at a more fundamental level in the attitude towards deriving and valuing knowledge. Based on these considerations, I stated in the beginning the following objectives (cf. section 3.5):

1. Derive an operational definition of parts and whole in ecosystems.
2. Develop means to describe the spatio-temporal organization of ecosystems that arises from the mutual interaction between parts and whole.
3. Derive conclusions with respect to the predictability and evolution of ecosystems.
4. Outline the role of theory and mathematical models in ecology (and evolution).

The first two points were the central topics of the preceeding chapters. The last two points still deserve a more comprehensive treatment since related comments were just mentioned in passing. However, before doing so, I first summarize the main elements of the argument developed up to now.

To make the distinction with a purely mechanistic description, I claimed that ecosystems must be viewed from what I referred to as a relational perspective. Organisms have no identity in isolation but gain their identity only in the context of their environment. Organisms continuously shape and regenerate their environment that itself provides again opportunities and constraints. Ecosystem organization and evolution arise from this mutual dynamic relationship. Based on such considerations, I suggested we view ecosystems as networks with a triadic structure comprising the level of the organisms, the level of the interacting network compartments, and the level of the system as a whole (cf. Figure 4.2). Depending on our perspective we may, however, come to quite different descriptions. In ecosystem ecology, network compartments are defined from the perspective of the system as functional components such as primary producers or herbivores. In community ecology, the system is defined as a collection of individual units from the perspective of the organisms or taxonomic species. This distinction was just at the origin of the splitting between population/community and ecosystem ecology. I made an attempt to bridge these only seemingly separate schemes by relying on the triadic structure where the compartments are defined from constraints arising from both the level of the organisms and the level of the system as a whole. Since the underlying network concept relies heavily on an understanding of the dynamic nature of ecosystems, I devoted my considerations first to deriving an operational definition of spatio-temporal organization, to understand its meaning and relevance for ecological systems.

The concept of spatio-temporal organization introduced embraces a relational perspective and accounts for the dynamic nature of ecosystems. The rather vague notion of organization was defined in operational terms by introducing measures to quantify spatio-temporal organization in ecological networks. Ecosystems have been depicted as networks of energy and matter flows, assuming thereby that material flows reflect satisfactorily pathways of interaction and information transfer that are required for self-organization to occur.

Investigating the behaviour of multi-species models assisted us in deriving a more tangible picture of what spatio-temporal organization means, and how chaotic dynamics and order come to be entwined. Seemingly negative effects of resource competition turned out to embrace elements of both negative and positive feedback when viewed as a dynamic pattern. Feedback spirals link network compartments across time and space. Studies of single species in isolation proved to be of little value. The fate and success of single species were observed to be critically dependent on the spatio-temporal dynamics of the network as a whole. As ensembles, species had quite a different effect on ecosystem behaviour than would have been expected from studying single species in isolation. Results from model simulations indicate that even, or better particularly, chaotic systems may exhibit functional temporal organization. At first we might feel chaotic dynamics and any type of organization to be incompatible. However, we have to note that chaotic dynamics and spatio-temporal organization pertain to different levels in the triadic structure of an ecosystem. Despite the unpredictable fluctuations of single populations, chaotic dynamics was observed to result in regularities at the level of the system as a whole – provided spatio-temporal organization was possible.

Viewing ecosystems as self-organizing systems embracing a wide range of spatial and temporal scales, I introduced the concept of trophic dynamic modules as an observational grid for structuring an ecological network along the dimensions of space, time and function. The definition of trophic dynamic modules was derived from the level of the organisms and the system as a whole. The dynamic niche was defined based on the organism's perception of spatial and temporal variability. The functional niche was defined by the network context. Spatio-temporal patterns in an ecological network must be expressed in the intrinsic scales of the ensembles of organisms involved, instead of imposing the absolute scales of an external observer. Therefore I introduced relative units in time and space to achieve a coherent description across the whole range of spatio-temporal scales encountered in ecosystems. Such a description is basic for investigating the interrelation among taxonomic, organizational and functional properties. We noted that dynamic characteristics and functional properties extend over a continuum rather than being well defined and discrete. This does not imply that organization is an amorphous continuum. It must rather be viewed as centres of dominance shifting over time and space in a continuum of interactions.

Most of the static concepts such as the niche or species diversity find their counterpart in a dynamic description and must hence be given a new interpretation. The niche was defined to embrace the dynamic and functional characteristics of organisms in combination with the configurations realized in a given network context. Diversity thus comprises different aspects of dynamic, functional and taxonomic characteristics. To elucidate their mutual dependence and the interrelation with system function, I introduced a framework focusing on typical patterns and distributions rather than on mechanisms. Let us now try to sketch the overall picture of ecosystem dynamics thus arising.

7.2   THE UNRULY NATURE OF ECOSYSTEM DYNAMICS

The perspective developed here that views ecological interactions as a continuum across space, time and function runs counter to the traditional ideas of clear interaction pathways and defined cause–effect relationships. Figure 7.1 shows an attempt to sketch the difference between these perceptions of system structure. The system in (a) displays clear pathways of interaction and characteristics also of control. The system in (b) is characterized by delocalized interactions along fuzzy boundaries between functional groups. The interactions may rather be described as spheres of interaction among diffuse clouds. Figure 7.1 is meant to express that an ecological network adapts its organization to escape constraints imposed onto it. Local processes at the species level permanently interact with the macroscopic properties of the system. The broad arrows and the identical shapes of the envelopes indicate this process of permanent mutual interaction. The system continuously explores the envelope of the constraints generated by a network structure in a given environment. By embracing both elements of continuity and change, an

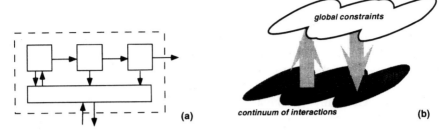

Figure 7.1 Different perceptions of natural systems and their interactions with the environment. (a) A system with clear pathways of interaction and rigid constraints imposed by the environment. (b) A system with delocalized interactions that may rather be described as spheres of interaction among clouds. The system is assumed to be adaptive, trying to escape external constraints in a process of mutual shaping of its internal structure and the environment

organization extending over a multitude of spatio-temporal scales confers on the system the potential for adaptation to changing environmental conditions and for evolution generating qualitatively new types of interaction. Adaptation and evolution must be viewed in a network context not at the level an isolated species on its own.

Until recently ecological theory has tended to portray ecosystems as something akin to large Newtonian systems with unique equilibrium solutions where perturbations are quickly negated by the opposing forces they elicit. Such systems lack the potential for adaptation and evolution which are, however, essential properties of living systems. Arguments for dismissing the possibility of ecosystems exhibiting irregular dynamics have largely been directed against randomness, chaos, stochastic fluctuations and extinctions, that have all been perceived as the inevitable alternative to a stable equilibrium point. Due to their rendering a system's equilibrium point unstable, effects of positive feedback have been discarded as important factors in ecosystem regulation (cf. DeAngelis *et al.*, 1986; Pimm, 1982; Stone and Weisburd, 1992). Amplification of local effects by positive feedback leads to major changes in the whole system in the presence of dominant pathways of interaction as depicted in the system in Figure 7.1a. In a system with delocalized interactions such events are the exception rather than the rule. We may encounter large local population fluctuations whereas global variables referring to system function and averages over spatially segregated populations of the same species may remain rather invariant. In ecology, order and pattern have far too often been associated with the idea of a static equilibrium generated by a rigid control. I take the view that what we perceive as "the balance of nature" is related to the functional and spatio-temporal organization of ecosystems. It is the trade-off between global and local variability arising from the dualistic interplay between the levels of the species and system as a whole that results in functional integrity combined with a high potential for change. It should be emphasized that even when it may only rarely take place, such systems embrace the potential for major changes as well. Owing to alterations in its interaction network, a system may move towards a critical stage where changes radiate over a wide range of the system (e.g. Haken, 1983; Kauffman, 1993). The relative importance of elements of chance, of changes in a system's environment, of evolutionary processes within an ecosystem, remain to be elucidated. In any event, change must be viewed as an essential element of ecosystem dynamics.

We may now ask whether we can integrate such an understanding of ecosystem dynamics in a perspective on ecosystems developing towards increasing spatio-temporal organization. To do so, it may be useful to consider another important aspect influencing a system's development. A system's potential for self-organization depends on what may be referred to as its level of autonomy. The notion of feedback spirals and nutrient recycling become obsolete in ecosystems that are driven by external inputs, as is the case of agroecosystems. Generally speaking, a system's autonomy expresses the relative independence of a system from its environment. In dynamic systems theory a system is called autonomous if the defining equations do not include any external forcing terms. The dynamics of autonomous systems are determined exclusively by the state variables and the internal processes connecting them. In this strict sense all ecosystems are nonautonomous because they are open systems depending on the exchange of energy and matter with the physical environment, the spatio-temporal variations of which are more or less independent of the local ecosystem. However, there may be varying degrees of being autonomous beyond the simple distinction between yes and no. Adopting a more generous terminology, I express the degree of a system's autonomy by the degree of its independence from external influence. External refers to processes that cannot be influenced by the system over the temporal and spatial scales of interest. Concerning the organization of ecological networks, it suggests itself to define the degree of an ecosystem's autonomy to be equivalent to the degree of its independence from external inputs of energy and resources. A quantitative measure is then given by the ratio of the average amount of total energy and/or matter entering the system to the amount being stored within the system. The autonomy relative to energy inputs can be described by the ratio of primary production over total biomass, and by the ratio of the total physical energy entering the system over total biomass. A system with a large ratio of primary production to biomass responds very quickly to changes in its environment since it has little buffering capacity. An excessive increase in the input of physical energy may lead to disruptive mechanical forces – e.g. hurricanes, floods. An excessive input of resources may lead to a disruption of feedback pathways essential for organization. We should be careful to note that all the measures cited above are scale dependent, being meaningful for the temporal and spatial scales chosen.

Autonomy and spatio-temporal organization have to be viewed as being mutually dependent. Results from model simulations showed increasing dynamic and functional diversity to mediate a system's organization that fosters the efficiency of nutrient utilization, increasing thus a system's autonomy. Such a directional change is related to an increase in what has generally been referred to as a system's maturity. Attributes of mature systems were summarized by Odum (1969) in his seminal article, "The strategy of ecosystem development". He noted that mature ecosystems are characterized by a low ratio of production to standing stock biomass (P/B ratio), by closed mineral cycles, and by slow rates of nutrient exchange with their environment. However, these early attempts to devise general strategies of ecosystem development emphasized static attributes and the existence of an optimal state to be finally reached. The characteristics of organization in mature systems were closely tied to an idea of the presence of stability and the absence of fluctuations.

Based on our previous considerations we would not expect anything like a smooth path of ecosystem development. The basic elements of what I suggest here as a general scenario of ecosystem dynamics are sketched in Figure 7.2. The circle must be viewed as a

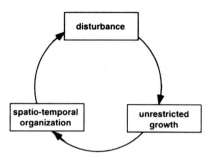

Figure 7.2 Repetitive pattern of ecosystem dynamics. The circle is a projection along the dimension of time. Essentially it must be viewed as a spiral progressing in time

projection along the dimension of time. System change is assumed to proceed along a spiral of phases of unrestricted, quantitative growth, phases of increasing spatio-temporal organization or qualitative growth where a system increasingly comes up against limiting constraints, and disturbance equivalent to the disruption of context by a major perturbation and/or a major innovation. Such a scenario may apply to different levels of an ecosystem's functional organization, albeit on different temporal and spatial scales. It could, for example, relate to a gap in a forest generated by tree fall, or to a large clearing deriving from fire, or to the establishment of a pelagic community in spring or after a major storm event. How should we now conceive of such a process happening?

In the following, I make a distinction between the notions of development and evolution even when such a distinction may sometimes be subtle indeed. Development refers to change within sets of given constraints, whereas evolution is identified with exploring qualitatively new patterns of interaction leading to a change in constraints. Let us assume that a set of material boundary conditions, related to energy supply and resource availability, define the frame for the evolution of an ecosystem. This frame constrains the spatio-temporal dynamic niche space by shaping the envelope of the total activity in the system. In the early stages of ecosystem development quantitative growth prevails. In the latter stages qualitative development, manifested in increasing dynamic and functional diversity mediating spatio-temporal organization, leads towards a minimization of the difference between the system states realized and the envelope of material constraints. In this latter stage, constraints arise as well from the densely occupied space of spatio-temporal functional niches. On the one hand, this niche space is restricted by limitations in the physical environment. On the other hand, the system may be locked in certain patterns of interaction that reflect the path of historical events. Constraints that derive from both material limitations and a filling up of the spatio-temporal niche space leave less and less opportunities for further change. Such constraints may be relaxed by more or less catastrophic events (e.g. storms, fires, epidemic disease) or gradual changes in the physical environment (e.g. climate) or in biological properties (e.g. evolution of physiological or social attributes). This means that most of the time some part of the system is in a transient state. A gradual change in constraints does not imply that the system structure changes gradually as well. Due to the plasticity of a network's organization and due to a certain persistence of a given state, we may rather expect change to occur all of a sudden and unexpectedly once a sufficient potential has accumulated.

Such a perspective on system dynamics embraces uncertainty as an essential element. We discussed this earlier for the variability at a local scale. Uncertainty is inherent as well at the level of the system as a whole with regard to the long-term trajectory of its evolution. In scientific research there has been an emphasis on reducing uncertainties and in striving for predictability. Even in modern quantum physics where uncertainty in the molecular realm is nowadays an accepted phenomenon, such uncertainties are reasonably well defined and they can be overcome in practical applications. Judged from a traditional point of view, uncertainty and the lack of predictive capabilities equal ignorance. Such thinking still pervades most scientific practice. It determines how knowledge is valued, what type of knowledge is required for decision making. It has shaped both scientific and political institutions. Such a view is inadequate to deal with the complexity of the environmental problems facing us today.

To substantiate this claim there is really no need to invoke presumably vexed arguments on the true nature of ecosystem dynamics – whether internal interactions continuously drive a system far away from an equilibrium point or not. I will illustrate for the example of climate change that we need only to carefully evaluate empirical evidence and accepted knowledge on the behaviour of nonlinear systems, to raise serious doubts regarding current strategies to deal with such global problems. However, I will also show that changing perspective, along the lines of reasoning developed here, suggests new approaches on how to cope with these situations.

7.3   THE ROLE OF UNCERTAINTIES IN DEALING WITH CLIMATE CHANGE

Evaluating the impacts of global climate change has gained particular importance for ecological research since the environmental conference in Rio 1992. The Rio climate convention demands the production of greenhouse gases be stabilized at a level which represents no danger to the Earth's climate system. Attainment of this target is not to be put off until considerable climate-related catastrophes have already occurred as a result of human influence, but is to occur fast enough so that the natural ability of ecosystems to adapt to anthropogenic climate change is not overtaxed. Continuity of food production and environmentally sound economic development should be guaranteed on this basis. Thus an upper limit for the production of greenhouse gases is postulated without being laid down quantitatively. Instead the nature and rate of ecosystem response are explicitly referred to as relative measures.

The question arises how scientific research can assist in meeting the claims made. According to the traditional role of science we may expect scientific results to provide a quantitative, predictive framework on which to base political decisions. Correspondingly, science is called upon to model and quantify the dynamic causal chain: society→greenhouse gases→climate→ecosystems→society, and the relevant feedback mechanisms among the elements of this chain. The task for ecologists seems evident – making every possible research effort to elucidate the mutual relationships between ecosystems and climate. However, we also should take a step back and become aware of what we decided to embark on. Can we really expect that once many scientists have contributed their pieces to the puzzle, we can finally build the overall picture? Such expectations may prove to be rather futile when we have a critical look at the uncertainties involved with these issues.

Uncertainties are common in every scientist's life. Actually, they are the main driving force behind any research endeavour. It is also a fact of scientific life that uncertainties are in general attributed to insufficient knowledge about a specific system or situation and to residual scatter in data and method. Environmental research has largely to rely on observation and comparison because opportunities to perform meaningful experiments are rare. Hence, it is particularly difficult to evaluate the state of present knowledge and the uncertainties associated with it. Two major sources for uncertainties are given by:

- insufficient knowledge;
- characteristics inherent in the dynamics of nonlinear systems.

In questions related to global climate change we have to cope with their combined effects. I will explore possible consequences in quite some detail to reveal the intricacy of the problems facing us. As a start let us examine the difficulties in deriving information about past climate changes and their impacts on ecological systems.

7.3.1 Uncertainties in generating a knowledge base

Due to the long time scales associated with most ecologically relevant phenomena, ecosystem dynamics are hardly amenable to direct experimental investigation. Indirect information must be derived from the past, from paleoecological investigations of environmental archives. Our current knowledge about past changes in both climate and ecosystems relies heavily on such information. The contradictory results obtained for the reconstruction of glacial sea surface temperatures reveal that investigations of this type may not always be unequivocal.

Anderson and Webb (1994) compared reconstructions of historical sea surface temperatures in the tropics. The magnitude of past changes in the tropics is more than a curiosity because it indicates the sensitivity of this region to climate change and it provides an important test for the accuracy of climate models. Box 7.1 lists changes for sea surface temperatures obtained with different methods as summarized by Anderson and Webb (1994).

What strikes the eye is the discrepancy between the methodological uncertainties given by the investigators and the overall scatter of the results from the different methods. Obviously there must be some conceptual problems in the understanding of the system that are beyond measurement error or scatter in the data. It is these types of fundamental uncertainties that we are generally not aware off and that pose the major problems. The generation of a knowledge base has to rely on indirect evidence. Results from multiple sources allowing for comparisons and cross-checks are not always available. To illustrate the potential and the limitations inherent in the information available from environmental archives let us have a closer look at one – lake sediments.

In Figure 7.3 I sketched the main pathways of how information about the long-term dynamics of climate and ecosystems come to be stored in lake sediments. Climate factors such as temperature and precipitation influence a lake and its catchment area. The sediments of a lake collect particulate matter, thus storing information about the lake and its surroundings (review by Anderson, 1993). The dimension of time is conserved in the sequence of sediment layers, in the most favourable cases in annual layers, much as in tree rings. Information about past climate changes and their influence on natural systems may reach the sediment via three different routes:

Box 7.1   Differences between glacial and present sea surface temperature (SST) (data from
Anderson and Webb, 1994)

Method	Change in SST (°C)	Uncertainty (°C)
Coral Sr/Ca	− 5	± 0.5
Coral $^{18}O/^{16}O$	− 5	± 0.5
CLIMAP	− 2	± 1.5
Foraminifer $^{18}O/^{16}O$	− 2	± 1
Algae alkenones	− 1	± 0.5

The fractional incorporation of oxygen isotopes $^{18}O/^{16}O$, of strontium and calcium, and the
saturation of long-chain alkenones are temperature-dependent processes. Correspondingly,
one can make use of these temperature sensitivities to reconstruct historical temperatures
from fossil remnants where these parameters are preserved. Sea surface temperatures derived
by CLIMAP (the Climate Long-range Investigation and MAPping programme) were mainly
based on fossils of several plankton groups whose distribution in the tropical ocean have not
changed since glacial times. Judging from modern distributions, a cooling would bring an
entirely different species assemblage.

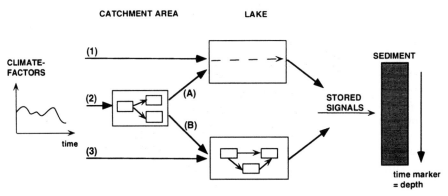

Figure 7.3   Pathways showing how information about climate change and its impact on natural
systems enters lake sediments that may hence serve as an environmental archive. Further
explanation in the text

1. The lake may simply serve as a collector for direct information about local climatic
changes. However, there exists no route where information about climatic changes is
stored in the sediments without being influenced by other processes. The closest is the
information inherent in the ratios of stable isotopes revealing temperature changes and
their influence on isotope fractions in precipitation. But we have for example to assume
a steady state in the water budget of a lake. If we look at the isotopes in fossil remnants,
biological processes come to play an important role. Even when we focus on the
isotopes in abiotic substances such as carbonate we have to take into account: their
origins from biotic and abiotic sources, the discontinuous nature of their deposition

over a seasonal cycle, and the influence of allochthonous inputs via tributaries and/or groundwater.
2. Climate changes may have an influence on the catchment area – e.g. shifts in species composition in the terrestrial vegetation, change in vegetation cover. (A) Again the lake may simply act as a collector by storing for example pollen in the sediment record. (B) Changes in the catchment area may also influence the lacustrine ecosystem itself, e.g. by increased run-off and input of allochthonous material.
3. Finally, climate may affect the lake directly – e.g. changes in water temperature may lead to changes in species composition, changes in the energy input may alter the physical mixing regime and hence influence the ecosystem as a whole.

The sediment record provides the integrated information from these various pathways that may also influence each other. It is not too difficult to imagine that to reconstruct both the climate and its influence on ecosystems from this information is quite a formidable task. There is no doubt that invaluable information can be derived from paleoecological data regarding the nature of the long-term dynamics of ecosystems once the jumble of interdependent pieces of information can be disentangled. However, there is also no doubt that this approach has its limitations. The sources for uncertainties are numerous. Natural archives are no tape recorders where the history of the Earth is written down in an unequivocal fashion.

The difficulties due to the rather limited and fragmented present knowledge about the relationship between the dynamics of climate variables and ecosystem response are compounded by the large gap between the information required for evaluating regional impacts of climate change and the information currently available from climate models. Ecologically relevant climate information is usually given by spatial and temporal distributions of amount and intensity of precipitation, or of averages and extremes of temperature. The spatial resolution required depends on local geographic conditions, being high in complex landscapes such as the Alps. The forecasts currently available from climate models provide a homogenous spatial resolution with a maximum of approximately 100×100 km. In validation runs models performed well at reproducing seasonal surface climates at continental scales (of the order of thousands of kilometres) only (Solow, 1992). It is easy to mention areas of future research to improve this situation by trying to reduce uncertainties. However, we should ask as well to what extent and on which time scale, uncertainties in current knowledge can really be reduced. In particular, there may be serious difficulties arising from the inherent complexity in the behaviour of environmental systems.

7.3.2 The puzzling behaviour of climate

Palmer (1993b) noticed that many of our conceptual paradigms about climate change involve linear dynamical thinking. Linear thinking is implicit in the belief that the anthropogenically imposed effect of greenhouse gases can be separated from the natural endogenous dynamics of climate. This assumption is a necessity for the detection and the prediction of any greenhouse effect. The increasing knowledge about the behaviour of nonlinear systems does not lend much support to such an assumption.

We discussed before that chaotic dynamics pose severe limitations to predicting the details of a system's evolution given a precisely defined initial state. Nevertheless, as chaotic systems tend to be confined to an attractor, at least the qualitative behaviour and some statistical properties are predictable. However, further surprises lurk in the nature of chaotic systems displaying qualitatively different behaviour by jumping between different attractors. Such unpredictable changes in the qualitative characteristics of chaotic systems seem to be by no means uncommon (Palmer, 1993b; Sommerer and Ott, 1993).

The Lorenz model introduced in Chapter Two is a convenient tool to illustrate these points. Despite its simplicity, the Lorenz model contains many qualitative similarities with the large-scale atmosphere (Palmer, 1993a). In the context of the impacts of climate change, studying the behaviour of the Lorenz model is of interest to visualize possible limitations of climate forecasts. Even when it would be pointless to identify the variables with specific climatic features, we could imagine that large-scale patterns of atmospheric circulation and/or ocean currents display similar dynamics.

To visualize what qualitative changes may imply, I chose the parameter r of the Lorenz model to fluctuate around the critical threshold for the onset of chaos (cf. section 2.6). The fluctuations that can be inferred from Figure 7.4a were generated at random by choosing every 100 time units a new value for r with a maximum amplitude of 10%. In addition, low-level random noise with a maximum amplitude of less than 0.1% was superimposed on these macroscopic fluctuations of r. I performed numerous simulation runs that differed only in the low-level noise added to the macroscopic variations of r that are depicted in Figure 7.4a. The results of two selected simulations runs are represented in Figures 7.4b and c. We note that in (b) the system settles into a stable equilibrium point when r drops for the first time under the critical threshold. In (c) the system continues to exhibit chaotic fluctuations. It settles into a stable equilibrium point when r drops again significantly under the critical threshold. In all cases the system remains at the equilibrium point despite the subsequent crossing of r into the range where chaotic behaviour is expected. This lag can be explained by the very slow reactions of the system characteristic of the region close to a critical state – here given by the vicinity of the bifurcation point ($r = 24$). However, the apparent stability of the equilibrium state is deceiving. A small perturbation may lead to the onset of violent oscillations as represented in Figure 7.4d. In this case the system had settled into a labile steady state and it was perturbed at $t = 1500$. Imagine that we had observed the system for the past 500 time steps from 1000 to 1500. We would have been quite surprised indeed by the sudden onset of violent fluctuations.

The sudden jumps in model behaviour derive from the presence of a critical parameter that determines the threshold to chaos. The objection could be raised that the Lorenz model was too simplistic and that the behaviour of real climate was more benign. However, examples for unpredictable behaviour and the jumping between alternate states are ubiquitous in weather phenomena. More recently, researchers have started to become aware of their importance for climate as well. Figure 7.5 show an example for a sudden shift in the state of the Pacific Ocean that was documented by a jump in an indicator derived from the combined records of 40 environmental variables (Kerr, 1992). The shift lasted for about a decade and was accompanied by extreme weather events – storms, unusual cold. The reasons for this shift are largely unknown.

Another well-known example is the ENSO (El Niño – Southern oscillation) cycle. It became evident only recently that the ENSO cycle arises from a coupling of ocean and atmosphere that involves the whole Pacific basin from South America to South East Asia (reviews by Barber and Chávez, 1983, 1986). In normal conditions, the cool surface layer of the coastal ocean off the west coast of South America is relatively thin and the

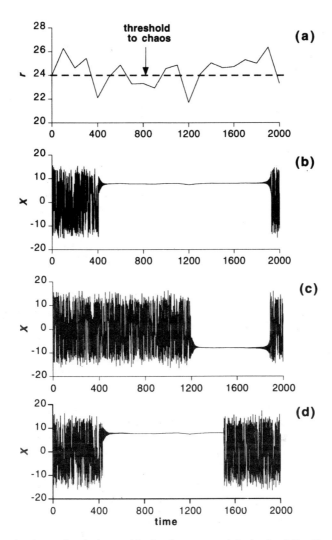

Figure 7.4 Results from simulations with the Lorenz model obtained for fluctuations of the parameter *r* around the critical threshold for the onset of chaos as shown in (a). The time sequences displayed in (a), (b) and (c) were obtained in simulation runs that differed in the low-level noise superimposed on the macroscopic fluctuations of *r*. This noise had an amplitude of less than 0.1% of the absolute value of *r*. (d) shows a simulation where the system was disturbed at $t = 1500$

nutrient-rich waters below are readily upwelled to the surface by winds blowing along the coast. As a result this coastal ocean is one of the most productive regions of the world ocean. At unpredictable intervals this structure breaks down leading to a basin-wide exchange of water and a reversal of the prevailing wind patterns. As a result the nutrient input by upwelling is interrupted with major consequences for the marine ecosystem. This phenomenon is known as El Niño and is well documented due to its impact on the local fisheries (mainly anchovies and sardines) and due to its being generally associated with extreme weather events in the whole Pacific region. Contradicting the historical record and the forecast of climate modellers, the most recent El Niño has persisted over four years

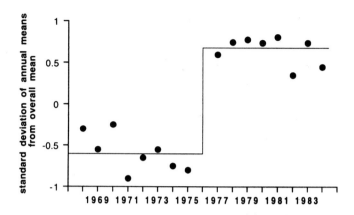

Figure 7.5   Sudden change in climate in the North Pacific region shown in terms of an index composed of 40 environmentally relevant variables (after Kerr, 1992)

now (Kerr, 1993). Instead of reverting to relative cold after an El Niño event the system seems to have locked in the warm "on" mode. Model simulations suggest that the irregularities of the El Niño phenomenon indicate in fact a chaotic process arising from the coupling of an oscillating ocean–atmosphere system with the Earth's annual cycle (Fei-Fei *et al.*, 1994; Tziperman *et al.*, 1994). The speculation suggests that the chaotic nature of ocean phenomena could render climate as unpredictable as the chaotic nature of the atmosphere renders weather unpredictable (see also Covey, 1991). Last, but not least, it remains to mention that even the seeming embodiment of stability and eternity, the sun, is not as stable as we might have believed. Solar activity exhibits considerable regular and irregular fluctuations that influence the Earth's climate on time scales of decades to millennia and may thus further complicate the situation (Beer *et al.*, 1995; Friis-Christensen and Lassen, 1991; Kelly and Wigley, 1992; Schlesinger and Ramankutty, 1992). Rather than decreasing uncertainty, our improved understanding of the climate system has thus revealed the ubiquitous sources for unpredictable behaviour and surprise.

As a consequence of climate's complex behaviour, considerable difficulties arise in providing statistical evidence for anthropogenically induced climate change. It is virtually impossible to make predictions regarding ecologically relevant information on climate variables. With respect to our goal of predicting the effects of climate change on ecosystems, we cannot really expect the situation to improve when we now continue to evaluate the consequences to be expected for ecosystems.

7.3.3   The nature of ecosystem response

What strategy should we follow facing not only insufficiencies in knowledge but also the prospect that further scientific research may even increase uncertainties? Can we still afford to stick to the widely practised policy of waiting, observing and reacting? Instead of considering in detail possible consequences of different climate change scenarios for ecosystems, I demonstrate some problems associated with such a wait-and-see attitude with a simple model. The model describes essential characteristics of the nature of an ecosystem's response to prolonged exposure to stress. It is based on two major simplifying

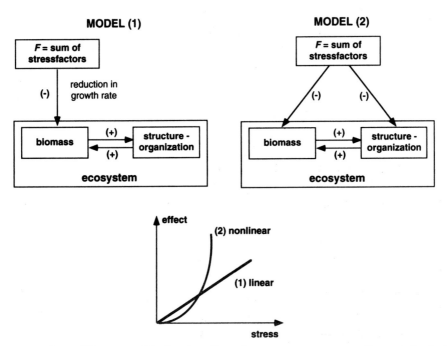

Figure 7.6 Two different models for the effect of stress on ecosystems. See text for further explanations

assumptions. The first is that the effect of stress can be described by a single model accounting for the combined action of stress factors such as exposure to harmful chemicals, deviations of environmental (climatic) conditions from the conditions favourable for growth, habitat fragmentation etc. It is further assumed that an ecosystem has a certain potential for repair and/or adaptation. Such a potential, that may also be referred to as a system's resistance, is meant to reflect a system's internal structure and organization. As discussed in Chapter Six, an ecosystem's organization comprises various aspects such as functional, spatial or taxonomic diversity. In the simple model used here, the effect of organization is summarized in its potential for counteracting the detrimental impacts of stress.

To emphasize the importance of nonlinearities, I designed the two different models that are depicted schematically in Figure 7.6. In model (1), the negative effect of the stress is limited to a reduction in growth rate and increases linearly with the magnitude of the total stress. In model (2), the negative effect of the stress leads to a reduction in growth rate. In addition, stress adversely affects structure and organization and hence a system's potential for repair and adaptation. The effect increases nonlinearly with the magnitude of the total stress. Model (2) thus takes account of feedback effects that we expect to be of major importance in complex environmental systems. The mathematical equations are summarized in Box 7.2.

Figure 7.7a shows the results of model simulations obtained for prolonged exposure to stress that is assumed to increase linearly. Model (1) produces the behaviour expected from linear thinking – the effect increases steadily in proportion to the increase in stress.

Box 7.2   Equations of the simulation models

Dynamics of biomass growth:

$$\frac{dB}{dt} = (g - m)B \quad \text{where} \quad g = 1 - \frac{B}{K} \tag{7.1}$$

The dynamics of the damaging effect, W, of the stress, F, is described by:

$$\frac{dW}{dt} = \varepsilon F - \alpha W \tag{7.2}$$

where αW represents internal mechanisms of compensation.

Two different types of response are considered:

(1) linear, affecting the growth rate only:

$$g_w = \frac{g}{1 + W} \tag{7.3a}$$

(2) nonlinear, affecting the growth rate and the potential for compensation:

$$g_w = \frac{g}{1 + \phi W^2} \quad \text{and} \quad \alpha_w = \frac{\alpha}{1 + \phi W^2} \tag{7.3b}$$

As formulated the models are dimensionless. Biomass densities are expressed in multiples of the capacity K in the logistic growth term. Time is scaled by setting the growth rate equal to 1.

The situation looks quite different in the case of the nonlinear model (2). For low stress levels and short exposure times we may not detect any adverse effects. But then a small additional increase leads to a fast detoriation of the system's condition. Such a lagged response and synergistic effects are typical for systems with negative and positive feedback cycles, as ecosystems typically are. Whereas the system can counteract damaging effects at a low stress level, the combined effects of positive feedback result in a sudden increase of the damage at higher stress levels. Positive feedback means that a variety of interdependent system properties may be adversely affected by the stress, leading thus to a magnification of the overall negative effect.

Figures 7.7b and c show the results from model simulations where, in addition to a continuous increase, stress pulses were given at two different times. Such stress pulses might reflect short time exposure to high doses of harmful chemicals, extreme climate events or parasite attack. We note that in case of model (2), accounting for feedback, the immediate effect of a stress pulse depends largely on the current state of the system. If we had observed a change as depicted in Figure 7.7c, we might have attributed the sudden decline in biomass to the effect of the stress pulse, whereas we would not have done so for a change as depicted in Figure 7.7b. Actually we might have concluded in (b) that the stress pulse had no effect at all. In our model simulations we note for (b) that the system has preserved a memory for the stress pulse since the final breakdown in biomass occurs earlier than in (a). Hence, to derive cause–effect relationships, to make predictions or to set up meaningful experiments is exceedingly difficult in systems where feedback and synergistic effects prevail.

The model was chosen to be simple for illustrative purposes. However, similar results of

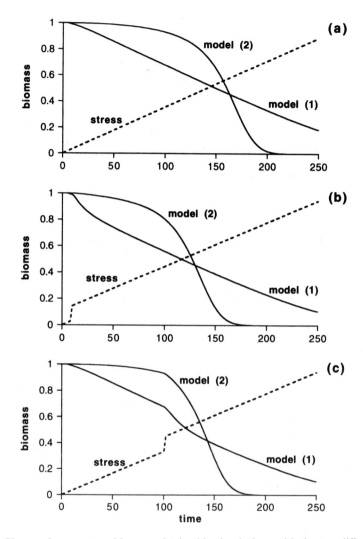

Figure 7.7   Changes in ecosystem biomass obtained in simulations with the two different models for a prolonged exposure to stress. (a) Gradually increasing stress. (b, c) Additional stress pulse are superimposed on the gradual increase

delayed catastrophic breakdowns may be obtained in complex models – e.g. in a forest model for the effect of acid rain (Leo *et al.*, 1993). We can also find support for threshold effects in natural systems. A well-investigated example, interesting for ecological research and management alike, is given by the behaviour of lakes undergoing eutrophication. Figure 7.8 depicts the existence of alternative states in lacustrine ecosystems as a function of the external nutrient load. With increasing nutrient load there is a discontinuity where the system changes from one state to another. At the bottom of Figure 7.8, the two states are roughly characterized for two different types of lakes. Once a critical threshold is reached, a cascade of interrelated effects is triggered off, amplifying the rate of change. Rapid transitions from oxic to anoxic states were monitored in the cores of lake sediments

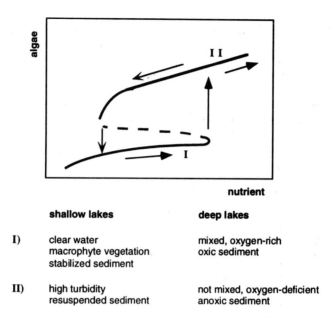

	shallow lakes	deep lakes
I)	clear water macrophyte vegetation stabilized sediment	mixed, oxygen-rich oxic sediment
II)	high turbidity resuspended sediment	not mixed, oxygen-deficient anoxic sediment

Figure 7.8 Alternative states in lacustrine ecosystems as a function of the external nutrient load. The arrows indicate that the recovery for a decreasing nutrient load does not follow the same path obtained for an increasing nutrient load

(e.g. Niessen and Sturm, 1987). The transition may be abrupt – from one year to the next – despite the continuous variation in other lake parameters. At intermediate nutrient levels two alternative stable states exist. The recovery for a decreasing nutrient load does not follow the same path as the one for an increase. This effect is especially pronounced in shallow lakes (Hosper and Meijer, 1993). Algae make the water turbid and cause the disappearance of submerged plants. The turbid-water state is further stabilized by planktivorous fish that suppress the large and efficient filter feeders like *Daphnia*. The clear-water state is stabilized by macrophytes that compete with algae for nutrients and offer spawning ground for predatory fish such as pike. The effects translate thus through the whole ecosystem.

In summary, both model results and experimental data show that ecosystem response is not gradual. To the contrary, once critical thresholds are exceeded phases of deceptively little change may be followed by a rapid breakdown and/or transition to another and possibly undesirable state that is often stabilized by positive feedback preventing recovery. The wait-and-see approach is ill advised in such a situation. Once adverse effects become visible, it is generally too late to take counter-measures.

We may conclude that in nonlinear feedback systems, such as ecosystems:

- Linear extrapolations are not feasible.
- Predictions of threshold effects are exceedingly difficult.
- Relevant factors may not be recognized as such due to their insignificance in an undisturbed situation.
- Clear cause–effect relationships are virtually nonexistent. Effects are contingent on state and context.

7.3.4 What consequences should be drawn?

Returning to the claim of the climate convention to use ecosystem response as a measure, we have to admit that making detailed predictions faces us with considerable difficulties. Ecologists are now called upon to investigate processes that link species and ecosystems with climate and to predict ecological responses under climates that do not even presently exist. The final goal would be to use the different climate change scenarios derived from climate models as bases for correponding scenarios for ecosystem response to sketch the full range of possible impacts. The uncertainties associated with such an endeavour are many. Figure 7.9 demonstrates the fact that the uncertainties involved within a causal chain multiply with each stage, and thus undergo an overall increase. In addition, this figure implies that the uncertainties inherent in the sequence climate – ecosystem – society increase on proceeding from physical systems through biological systems to human systems. Here it should be stressed that this illustration does not yet take into account the effect of feedback between the systems alluded to above. Considering the current attempts of balancing the possible costs of measures counteracting greenhouse gas emissions against costs produced by adverse impacts of climate change, we may become rather doubtful whether some uncertain prospects of damage will really cause any political action.

Current scientific knowledge abounds with evidence for the presence of complex nonlinear dynamics in natural systems and the uncertainties associated with them. Facing such a situation we should be cautious about thinking what conclusions to draw. The scientific community has obviously decided to embark on a gigantic research endeavour. A recent publication of the Internal Geosphere – Biosphere Program entitled "Reducing Uncertainties" lists a huge research agenda for all disciplines of the natural sciences (Williamson, 1992). In an unprecedented effort the Ecological Society of America has identified a broad array of urgent research needs essential for the acquisition of ecological knowledge to utilize and sustain the Earth's resources (Lubchenco *et al.*, 1991). The document puts an overwhelming emphasis on the improvement of predictive capabilities.

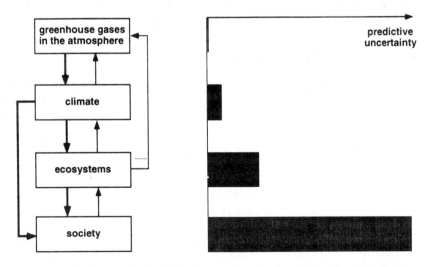

Figure 7.9   Cascade of uncertainties within the causal chain of predicting the effect of anthropogenic greenhouse gas emissions

Uncertainties are only mentioned insomuch as they have to be reduced. Such a traditional scientific attitude can be expected to have some success in dealing with environmental problems where cause–effect relationships are evident and where highly sophisticated techniques can be applied to cure symptoms. Unfortunately, environmental issues hardly ever yield to this demand.

Scientists reducing uncertainties, drawing a clear picture of how the world functions, that is what society expects. That is also how scientists have preferred to perceive themselves and why ecology has not gained a stronger reputation in the scientific community as yet. Most concepts in empirical scientific research are based on experiments where power is exercised to demonstrate and increase knowledge. Hence knowledge that guides a mechanistic and predictive understanding, and that leads to targeted action, is preferred and highly valued. Correspondingly, the degree of knowledge about a system is often valued by one's ability to devise strategies of control and manipulation.

I argue that ecology could make major contributions regarding conceptual approaches in scientific research and consequences for society. Instead of trying to become ever more predictive, to squeeze ecological knowledge into a frame that might not be appropriate, it could be the task of ecologists to develop a new type of knowledge, to assist in establishing means of how to deal with uncertainties that obviously cannot be avoided but only neglected. The urgent need for a change in attitude has become more than evident in reflecting on the ubiquitous presence and the possible consequences of uncertainty related to the evaluation of the impact of climate change on ecosystems. Uncertainties should not always be looked upon as being something inherently negative. At the beginning of this section, I explicitly omitted a further category of uncertainty, namely the uncertainty associated with the fact of evolution and innovative action. Living systems obviously have the potential to create new types of response to unprecedented situations. Should not we explicitly include this property in our considerations?

7.4 ECOLOGICAL RESEARCH, ENVIRONMENTAL MANAGEMENT, AND THE ROLE OF THEORY IN THE LIGHT OF UNCERTAINTY

7.4.1 The role and meaning of a theoretical framework

Even when most scientists might not deal explicitly with models in their daily lives, some theoretical perspective and some view of the world underlies implicitly any research endeavour. The prevalent driving force for setting up research projects lies in the desire to find causal explanations for some phenomena observed. Establishing logical connections among previously unrelated observations provides a feeling of success and of progress in seeing a more coherent picture emerging. To be left with uncertainties is in general experienced as being less pleasant. Uncertainties are already preferrably avoided in experimental design. Our scientific institutions, the exchange of scientific information and the valuation of scientific knowledge are predominantly based on the exchange of solid scientific facts, on thinking in cause–effect relationships in a fragmented world. Fragmentation and isolation are required to analyse the world in mechanistic terms. Nature is preferrably forced into an observational scheme where logical connections can be established, where manipulation can be exercised, and where uncertainties can be circumvented. Even when our highly valued instrumental knowledge leading to action and control has proven to be of little help in dealing with complex environmental problems, there seems to exist no strong incentive to give up established tools especially when alternatives seem to be absent.

Over recent years the application of mathematical modelling in ecology has experienced a real boom, whereas the development of new theoretical concepts has not kept pace. Theoretical concepts should not be equated with mathematical models even when the former find in most cases their formal expression in the latter. Theoretical concepts may be stated verbally, as for example Darwin's principle of natural selection. The subsequent establishing of a mathematical framework and the further development of theory involve processes of mutual exchange. In restrospect it may then be difficult to separate the formal mathematical statements from the underlying theoretical concepts. The latter include as an integral part a certain perspective on the world. Whereas scientific practice allows us to compare and value different approaches that are based on the prevalent framework, it is of little help to test the validity of the underlying world view. I suggested before looking at a theoretical framework as a specific pair of spectacles that determines our selective view of the world. What do we see with the spectacles of the mechanistic framework prevailing in scientific thinking? I argue that the current perception of nature is still dominated by the notion of a machine, albeit a very sophisticated one. Such a machine can be taken apart, reassembled, designed and controlled. I wish to contrast this picture with a different metaphor that may be derived from a relational perspective. Let us conceive of nature as a partner with whom we should like to engage into a dialogue, a partner which is rather powerful and with whom we would be well advised to collaborate. These contrasting perspectives lead to major differences in the questions we pose, the explanations we give, and the way we look at environmental management.

Returning to modelling we may consider the role of mathematical models with special reference to their being part of a mechanistic perception of the world. Like in other fields, the boom in ecological modelling has been fostered by progress in computer technology. The impact of models has further been strengthened by the large efforts devoted to evaluating possible consequences of anthropogenically induced environmental changes. The mathematical models applied in this field are mainly mechanistic models of the process–functional type, which are the type of models most likely to produce a quantitative base for decisions in environmental management. Assisting in the integration of detailed knowledge from various disciplines and the elucidation of complex interdependencies, these models definitely have their virtues. However, what may be looked upon by some people as the major virtues of such models should rather be identified as their major vices: they produce "magic" numbers that may start to lead a life of their own extending far beyond the scope of a model limited by restrictive assumptions and inherent uncertainties. Cause–effect relationships depicted in schemes with boxes and arrows suggest precision and clear routes for influence and control. Numbers and defined cause–effect relationships are easy to communicate what is of particular importance in policy-related and management decisions.

7.4.2   Current strategies of environmental management

Current strategies of environmental management and resource exploitation require predictions on a local scale. To achieve such predictability, one attempts to force systems into desired states by excessive input of energy and/or resources. Control is more easily exercised when a system's diversity is reduced to single spatial, temporal and functional scales. Intensive agroecosystems are the prototype for the implementation of such practices. To a lesser degree they are applied in forestry as well. Aquatic ecosystems are less amenable to external control. We discussed before that in agricultural systems one selects a single species, determines the spatial extent and the boundaries of a field. Activities such as plant growth, harvesting and fertilization are forced into a regular spatial and temporal pattern by an external schedule (cf. Figure 6.23). Such a complete lack of spatio-temporal and functional diversity is a contradiction of what has been identified here as the basic requirements for an ecosystem's organization.

What type of picture emerges from the previous considerations? An increase in ecosystem organization has been characterized to be associated with an increase in functional and dynamic diversity of ecological networks. As a consequence, the diffuse nature of interactions increases and the predictability of events at a local scale decreases. Such changes run counter to any desire to predict, to manage, and to control an ecosystem. However, I argue that any living system will finally escape rigid constraints imposed onto it because *life and externally imposed control towards a defined goal are mutually exclusive properties.* As a metaphor we may draw a comparison with an attempt to form a sculpture out of liquid. You can influence a liquid by, for example, heating or cooling it. However, to keep it in a defined shape you have to enclose it in a rigid box where the original properties of the fluid medium are lost. I am not aware of any successful attempt where a living system, be it from a cell to an ecosystem, could be controlled by an external regime without being mutilated or even destroyed.

The failure of control strategies in current practices of resource management was recently stressed in an illuminating report by Ludwig *et al.* (1993). They substantiated their arguments with examples from fisheries management that has been guided for several years by the concept of maximum sustainable yield. However, scientists have been unable to come up with the data required, natural variability masked the effects of overexploitation, the declines in fish populations did not proceed gradually but as sudden collapses. Such observations sound familiar – are not they typical properties of nonlinear dynamic systems! Ludwig *et al.* (1993) observe rightly that it is more appropriate to think of resources as managing humans than the converse.

In Figure 7.10a I tried to sketch the basic attitude of mind underlying current ideas in environmental management. A managing control that receives information about the state of a natural system initiates corresponding measures to counteract deviations from a desired goal. It has been my intention to stress the resemblance to cybernetic systems such as temperature control by a thermostat. Even when this scheme is rather simplistic, it does emphasize certain aspects that apply to complex management schemes as well: the perception of system dynamics is mechanistic; the system is assumed to be observable – relevant global information can be obtained; based on understanding system function one can interfere by targeted action to counteract deviations from a desired goal; the management control perceives itself largely outside of the system.

7.4.3 New approaches in environmental management

Meanwhile one can observe a change in attitude, particularly in what is nowadays referred to as ecological engineering, that has been defined as the design of ecosystems for the mutual benefit of humans and nature (Mitsch and Jørgensen, 1989). The idea is to keep a

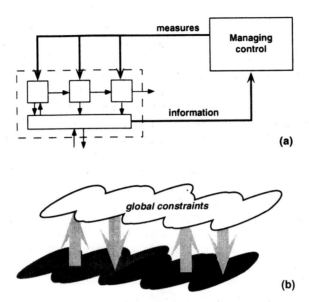

Figure 7.10 Schemes of environmental "management". (a) Current strategies where a human control panel, perceived as external, attempts to direct an ecosystem or even the global system towards a desired goal. (b) Ecosystems (light shaded cloud) and human systems (dark shaded cloud) are closely interwoven in what can be called a human ecological system that organizes itself towards a common goal within common material constraints

system within desired bounds by exercising only subtle influence based on a deeper understanding of the relevant processes. There is no doubt that such an approach is a vast improvement in comparison to the energy- and resource-intensive management schemes of the past and still of the present. Ecological engineering will presumably be basic to dealing with most management problems in the near future.

However, I hold the strong conviction that in the long term we have to develop a completely different understanding of what environmental management implies. We must refrain from thinking that management is identical to an authoritarian gardening of and imposing influence on ecosystems to make them suit our needs. As visualized in Figure 7.10b, the metaphor of a partner suggests a type of mutual management and coexistence between natural and human systems. The light shaded cloud represents natural systems whereas the dark shaded cloud represents human systems. Both have to engage into a tight relationship and a common process of self-organization where the mutual interactions shape the common constraints. We need to develop more tangible ideas on what such combined systems should look like with respect to the spatio-temporal and functional patterns of interaction within and between natural and human systems. Based on our previous considerations we can already say that they must comprise flexible interactions extending over a range of spatial and temporal scales.

The involvement of ecology must go beyond investigating organisms and their natural environment. Ecologists should assist in developing a strategy where socio-economic human societies and natural ecosystems become integrated into what may be called a human–ecological system. To find the right strategy cannot be limited to planning and designing according to predetermined plans, but must involve a better type of communication, a responsive exchange between humans and their environment. What is the type of knowledge that we need for such an endeavour? Definitely a knowledge that extends beyond mechanistic details by focusing on organizational characteristics. What type of research institutions are required? Definitely more flexible and interdisciplinary ones. In particular, theoretical ecologists are challenged to develop new concepts and guidelines. Such a new task requires also a different understanding of the purpose any theoretical framework is designed for. Instead of identifying theories with ultimate modes of causal explanations we should rather perceive them as rule systems. A theoretical framework with rules for the use of concepts is a necessity for improving scientific communication and for enabling scientific progress. It is of paramount importance in a field like ecology where knowledge cannot be "valued" by solid hard facts obtained from unequivocal experiments, but has to rely on establishing a consensus in a scientific discourse. In such a situation it is also obvious that more than one plausible explanation may be given for the same phenomenon depending on the observer's perspective. Hence, I argue in favour of pluralism in scientific explanations. Here ecology could indeed learn from physics. In modern physics it is accepted that the phenomena in the quantum realm escape a unique description. At this level one acknowledged for the first time the necessity of applying complementary, mutually exclusive, ways of describing a phenomenon. Using multiple descriptions that are based on different approaches is essential to do justice to phenomena in the realm of living systems as well. Each description grasps only some facets of reality. Such descriptions may be complementary to a varying degree. Establishing two (or more?) complementary explanations as opposite extremes and organizing other modes of perception in a gradient in between may be an operational way of how to proceed. A clearer picture of a natural system in all its complexity is shaped by viewing it from different perspectives.

Taking such shifts in attitude seriously, we also have to conclude that the societal role of the natural sciences has to change. Traditionally it has been their task to reduce uncertainties, to fight ignorance, and to provide answers to defined problems. In dealing with complex environmental systems, new tasks emerge in developing tools to handle uncertainties, in clearly stating their sources and the possible consequences to be drawn. Scientists should be honest enough to admit that scientific "truth" in general and in relation to complex living systems in particular is to some extent a reflection of an agreement in the scientific community on how reality is currently perceived and on what is considered to be important. As any other human activity, science is a social process. Scientific knowledge does not simply emerge in the brains of ingenious researchers enlightened to reveal the truth in nature. Knowledge must be seen from a relational perspective depending on values, beliefs, perceptions of nature and society (e.g. Feyerabend, 1978; Grene, 1985; Harding, 1991). Developing a self-critical and open attitude is required to foster a dialogue where rational decisions relying on solid facts and predictions from the natural sciences are impossible. Such an attitude is also required to prepare science for engaging into an active social dialogue that can be the only medium to successfully transfer the knowledge acquired. An active dialogue is a necessity for developing sustainable strategies in environmental management. Box 7.3 contrasts, in a comparative overview, the traditional attitude with the one developed here.

We note that the distinctions made follow similar lines of reasoning to the comparisons between a mechanistic and a relational approach (cf. Box 4.1) and between organization and self-organization (cf. Box 4.2). Box 7.3 includes another issue not yet mentioned: the perception and assessment of risks. Risk analysis has traditionally dealt with the failure of

Box 7.3   Role of ecological research in environmental management

	Current attitude	Attitude suggested here
Research	reduces uncertainties	uncovers uncertainties
	makes quantitative predictions	generates innovative, qualitative knowledge
	focuses on mechanism, process	focuses on organization, pattern
	provides expert knowledge	engages as partner in a social dialogue
Management	views nature as machine	views nature as partner
	is rigid, controlling	is flexible, adaptive
	aims at change towards preconceived goals	fosters evolution, innovative action
Risk assessment		
source	= phenomenon or process	system structure, organization
risk	= undesirable events	restriction of evolutionary potential
measure	= probability*damage	decrease in degrees of freedom

technical facilities. Correspondingly, risk assessment has been based on the possible damages and the probabilities of single events. Such practices have also been adopted in environmental risk assessment. It may be possible to estimate the probability that emissions of a certain chemical compound cause damage to human beings or to specific indicator organisms in the environment. Thus one may come up with some numbers as base for an evaluation of what could be an acceptable risk. Typically, complex environmental problems are disentangled to single issues that are dealt with separately and that may be reassembled thereafter. However, I argue that we should pay more attention to hitherto neglected risks that are inherent in our reducing degrees of freedom for future action, in our having legacies from the past. Such risks are associated with the fashion in which a system operates. The questions to be tackled should thus be formulated: given the current fashion of dealing with the environment what are the degrees of freedom left in a certain period of time. This requires that we define more explicitly what should be the degrees of freedom that we wish to preserve for common human ecological systems. It requires as well that we think about what restricts these degrees of freedom. I will still continue the considerations along this line of reasoning in the next section.

Suggestions for a change in management practices were made before. The change towards an adaptive management was earlier propagated by Holling (1978). More recently, Hollick (1993) summarized the large amount of literature about self-organization and management and drew some conclusions for environmental management practices. He suggested that managers should cultivate the capacity of natural systems for self-organization rather than trying to control them. Managers should use technologies that harmonize with the surroundings rather than being imposed on them. To achieve such goals management has to be flexible and decentralized.

We have to be aware that we cannot simply adopt such management practices within the current structures. To do so involves far-reaching changes in perception, in institutions, and in society. A first step towards a change may be given when we start to reflect about the questions we consider as the most urgent. We receive only the answers to the questions we ask. Whereas we can test to some extent the accuracy of the answers, we can test only to a limited extent whether the questions are appropriate. However, we should become suspicious when the answers consistently fail. Let us discuss the questions related to climate change in the light of the previous considerations. The questions posed in relation to the climate convention are framed in the traditional mechanistic framework in line with the prevailing attitude towards environmental management. Focusing on mechanisms, science has largely neglected patterns of organization, and relationship. I claim that we have to approach global environmental problems at a more fundamental level by searching for answers to the questions:

- What has changed in the process of evolution with the appearance of conscious human beings and the advent of human civilization?
- What are the necessary conditions for the global system in general and for ecosystems (including humans) in particular to maintain the potential for evolution and do contemporary human systems bear comparison in this respect?

7.5   EVOLUTION VIEWED AS CHANGE IN ORGANISM–ENVIRONMENT RELATIONSHIPS

Evolution refers here to processes at the level of an ecological network regarding the qualitative nature of the web of interactions. There has always been the argument that ecosystems cannot be evolutionary units because they lack the essential properties of persistence, integrity, and storage of information by inheritance required for natural selection to become effective. This is true if evolution is reduced to a mechanistic perception of natural selection acting on individual units. Such a point of view ignores the fact that ecosystems are the template evolution acts on. Evolution at the level of an ecosystem does not refer to processes of speciation but to qualitative changes in patterns of interaction that may also derive from processes of social learning. Evolutionary information is stored in the configuration of the network, in the functional and dynamic niches that are reflected in the properties of the component species. Any organism gains its identity only by interacting with and by differentiating from its environment. As with a key, which is useless without a lock it fits into, an organism and its network environment are inseparably linked.

Interestingly the notion of evolution is rarely used in discussions about humankind's future in view of global environmental problems. Nowadays the notion "sustainable development" indicates an evolutionary, or at least long-term, perspective. Why is the notion of development preferred to the one of evolution? An answer to this question has to comprise different aspects. First, the distinction may derive from a difference in the time scales under consideration and the understanding of what future change implies. I suggested identifying development with change along preconceived lines of thought within sets of given constraints. Regarding social systems, development refers thus to a rather restricted perception of change largely neglecting the possibility of innovative action. We discussed above that this seems indeed to reflect the prevailing attitude in politics and environmental management. Development can seemingly be predicted, planned and controlled. Evolutionary change with a largely unknown outcome generally does not fit in with such an approach to planning the future. Second, I stressed above that biological evolution has traditionally emphasized the scale of the individual and the species. Sustainable development has its focus on the macroscale of society and economy. Using the notion of evolution in this respect has to face similar reservations to those of considerations at the ecosystem level. Third, the term evolution seems to be reserved for biology, and it is mainly used in historical perspective. The emergence of human civilization is seen as something new and separate that has to be considered in isolation from biological phenomena. Such a distinction becomes blurred when we adopt a relational perspective and view evolution at the level of an interaction network. It may be helpful, even essential to develop an improved understanding of the *combined dynamics of biological and cultural evolution*. Instead of separating the two levels, we should reflect on what any type of "common sustainable evolution" could really imply. I refer here to evolution instead of development to emphasize the long-term perspective, the innovative character of evolutionary change, and the mutual dependence of biological and human systems.

Let us now come back to the first question posed earlier, namely to changes in the evolutionary process brought about by the advent of human civilization. Biological evolution can be viewed as being characterized by an increase in what may be denoted as complexity in the organism–environment relationships. I should mention that fierce debates have raged about invoking progress in relation to evolutionary processes (Nitecki, 1988). Admittedly, progress is a contentious term due to its inevitable connotations of increase in value. Nevertheless, we could as suggested by Gould (1988) support the view of a directionality in evolution. Ayala (1988) emphasized, for example, that a conspicuous feature of the evolutionary process is given by the advances in the ability of organisms to obtain and process information about the state of the environment. I try to sketch now some essential elements of what could be perceived as a directional change, being aware that I cannot do justice to the phenomenon with these brief considerations.

Let us view evolution as proceeding by a mutual relationship between organisms changing their environment and the environment triggering change in organisms. Organisms modify, maintain and create habitats by directly or indirectly modulating the availability of resources to other species and by causing changes in biotic and/or abiotic materials. New species evolve in the context of changes in the interaction with their environment. A more complex environment poses also more sophisticated problems to the organisms which inhabit it. We may discern a change in the quality of this process to have occurred over evolutionary time scales.

The emergence of what is called life was associated with the appearance of function in a network context. A prerequisite was the ability of material entities to gain an identity by isolating themselves from their environment (e.g. cell membranes), by self-reproduction, and by storing information. In an initial stage internal (within an organism) and external (interaction of an organisms with its environment) evolution were mainly based on physical and chemical processes. Subsequently, communication with the environment became progressively specific and selective. Specific receptors and sensory organs allowed enhancement of the information content of environmental (biotic and abiotic) signals beyond mere physico-chemical interactions. Nutrient uptake, for example, can simply proceed by passive osmotic exchanges via the cell membrane. Specific receptors that allow take-up of nutrients against a concentration gradient, cell organelles that store nutrients in times of abundance to release them in times of shortage, active movements of organisms towards regions of nutrient supply, these are all changes that enhance the specificity of interactions and foster the emergence of structured communities. The emergence of a nervous system assisted the coordination of both an increasingly complex body and an increasing complexity of information received from the environment.

Correspondingly, also the quality of an organism's way of responding to and of communicating with its biotic and/or abiotic environment has changed with increasing complexity and longevity. Morphological changes, which may be manifested in the genome, have progressively been replaced by changes at the behavioural level of social contacts and learning. Such a shift in the quality of evolutionary processes has speeded up the rate of evolutionary change. Change refers to the qualitative nature of engaging into interactions with the environment. Organisms started to increasingly shape their environment by purposeful action. Jones *et al.* (1994) recently coined the term "organisms as ecosystem engineers" which emphasizes such processes. The impact of engineering has gained unprecedented dimensions with the advent of *Homo sapiens*. Let us have a closer look at the circumstances of how such a situation came about.

Life thrived first in the protective environment of aquatic surroundings. With the emergence of life on land new dimensions opened for exchange of information and structuring the environment. Contrasting with an aquatic medium, terrestrial environments offer a solid template. Less energy is stored in the physical surroundings. The ratio of the heat capacity per unit volume increases from the atmosphere to the ocean by a factor of 3200! Correspondingly the influence of the biota on its physical environment is small in aquatic environments whereas it is high in terrestrial environments. On land biological structures are not of the ephemeral nature generally encountered in aquatic surroundings where physical forces erode them. Humans have been capable of accessing large energy resources enabling them even to impose structures on their environment. As a consequence, human beings started to actively change their environment to fit their needs. Currently, human beings behave as we would expect from a simple model of pulsed consumption. A consumer loose of constraints depletes the stored resources to finally collapse itself. Since humans have extended their habitat to be identical to the whole planet, such a consumer pulse will presumably be a singular event.

However, humans perceive themselves as rational beings that have surpassed a state where behaviour is directed by biological rules. Human beings have developed their own means for storing products of technical and/or cultural evolution. The natural environment is even perceived as something external to human societies. We may ask whether consciousness is tied to nature and biological evolution or whether humankind has now achieved a "state of autonomy". The latter would allow human beings to cut off their historical roots and to shape nature according to their needs. Nature's only purpose may be perceived as providing the material base in terms of food and resources for human civilization to exist. Obvious ethical arguments related to a responsibility for nature may be raised against such a perspective. However, such a strategy could imply that humans thus impede evolution and the continuation of any kind of life including their own. We may strongly suspect that the importance attributed to ethical considerations may not be independent from what are perceived as possible consequences for humankind. I conjecture that human beings trying to shape their environment according to an internal perception of reality will be caught in a cycle impeding further evolution. In industrial societies the individuals are predominantly surrounded by nonliving technical systems and the world is increasingly shaped by technical constructs arising from rules that emanated from human thought. Hence human beings are increasingly deprived of the interaction with a responsive living environment. That is obvious for technical systems but it applies to a varying degree to social systems and the cultural landscape as well – there is a definite relationship among them. If we recall that evolution is perceived as a mutual interaction and shaping of organisms with their environment, we may be doubtful whether humankind is currently on a promising course.

7.6 A SUSTAINABLE EVOLUTIONARY STRATEGY FOR HUMAN ECOLOGICAL SYSTEMS

7.6.1 A common goal and obstacles in achieving it

An evolutionary strategy for human ecological systems must comprise human society, economy and natural ecosystems that engage into *common and mutual* self-organization. Such an evolutionary strategy should not be viewed as aiming towards the optimization of an elusive future state. It can rather be identified with a strategy to maintain the flexibility and evolutionary potential of human ecological systems. Formulated in more human terms the concept of robust action may be a useful analogy. An action is called robust if it leads to the attainment of short-term goals while preserving long-term flexibility (Eccles *et al.*, 1992). We face the problem to minimize legacies from the past without knowing the future. Hence it is not straightforward to find out what such a strategy implies and how it could be realized.

Even when the way to achieve this goal is yet vague, we may most likely encounter obstacles in the relationships between human society and nature. Let us first deal with material aspects that are prevalent in questions related to economy. Traditional economic theories treat the economy as an isolated system with a circular flow of exchange values between firms and households (e.g. Daly, 1991). Neither matter nor energy enters or exits – it has no exchanges with its environment, and for all practical purposes has no environment at all. Nature may be finite, but it is just a sector of the economy, for which other sectors can substitute. Correspondingly, little care has been given to resource and energy flows and potential feedbacks.

Most industrial societies prove to be highly nonautonomous with respect to energy and resource flows. The notion of autonomy refers to the definition given in section 7.2. Nonautonomous refers thus to the fact that the dynamics of economic and social systems are mainly controlled exogenously by the import of energy and/or resources. In 1990 the annual per capita consumption of primary energy in the OECD countries amounted to 0.2×10^{12} J (Schipper *et al.*, 1992). To develop some feeling for such large a number we have to become aware that highly productive ecosystems have a net annual primary production of about 0.4×10^8 J/m^2. Correspondingly each inhabitant would require an area of 5000 m^2 cultivated land to cover the energy needs by primary production. If we consider that the average population density in the OECD countries amounts to more than 100 inhabitants per km^2, we have to conclude that we live mainly on credit. In addition we shouldn't forget that the consumption of primary energy does not reflect all of the energy consumption – energy is embodied in resources as well. The imbalance of the current situation becomes even more obvious when we compare the primary energy consumption of the individual with the normal individual's food requirement of about $4 \cdot 10^9$ J/a (2500 kcal/d). Energy consumption, and correspondingly the associated consequences, exceed by about two orders of magnitude the human energy consumption ecological systems were designed for by biological evolution.

Such a high energy consumption may not necessarily be perceived as a disadvantage. One may argue that a generous energy supply has been the main driving force for industrial development and that it will be a prerequisite to solve our current environmental problems and to develop new technologies. However, we may suspect that an excessive energy consumption impedes environmentally sound innovative action and the establishment of common socio-economic–ecological systems. Turner (1988) stressed that neither market-based economies nor planned economies seemed to be systems with built-in features that would guarantee sustainability. He referred with sustainability to the conservation of resources for future generations. I claim that sustainability requires as well a qualitative change in the human–nature relationships. A reduction in energy and resource input would foster the establishing of mutual interactions between natural and human systems, by reducing the current imbalance. Rigid states can only be maintained by a high external input of energy and resources. Thus we reduce our degrees of freedom and potential for evolution. Present industrial societies are vulnerable. They are addicted to the consumption of energy and resources rather than being really able to choose the future. The legacy of the past is enormous. The possible future impacts of ill-conceived management practices increase with the amount of energy and resources consumed. We continuously have to react to problems of the past facing us now. The longer we delay a response the fewer degrees of freedom remain.

However, any change of current practice is perceived as a sacrifice for personal life-style and as a threat to economic prosperity (e.g. Daly, 1991: Nordhaus, 1992). As far as traditional economic models are concerned a reduction in the supply of energy and/or resources are always perceived as economic costs detracting from overall welfare. Such narrow a perspective derives from the fact that feedback effects or irreversible damage to the natural resource base are neglected. In addition to these obvious short-comings, traditional economic theory is based on the assumption that an economic system has a tendency to approach an optimum equilibrium point. These considerations sound familiar. More recently this static perspective has been challenged by approaches viewing economic systems as evolving nonequilibrium systems being driven by effects of positive feedback (e.g. Arthur, 1990; Arthur *et al.*, 1987). Instead of viewing change as negative we should see it as a challenge. There is little hope to assume that prospects of future threats can persuade us to reduce our energy consumption as long as this is perceived as a punishment, as a restriction of enjoying life. However, once we realize that a responsible attitude increases our freedom, the quality of life, the problem becomes one of different dimensions. Instead of being addicted to consuming, we take the leadership and decide what to do. We might also start to think about the type of future we want to strive for.

7.6.2   Requirements for progressing towards a beneficial evolution

To achieve the stated goal of common human ecological systems requires changes at different levels – the human mind, the organization of society and our communication with nature, and the material boundary conditions.

Due to human beings' abilities to extend their biological capabilities with technical means, the time scales of the anthropogenically induced environmental changes exceed the time scale of any material biological adaptation and evolution. Hence, human actions being determined by consciousness and technical skills must be balanced by a conscious change in perceptions. Material feedbacks must be replaced by some type of moral order if we wish to maintain the process of evolution. Evolution must proceed in the human being, not in the gene but obviously now in the human mind.

Such evolution is hindered by the current organization of society. We can notice severe imbalances in spatio-temporal scales with respect to an individual's embedding in its environment. The demand of thinking globally and acting locally is wishful thinking. It ignores the psychological and social reality of human beings who are not prepared by their biological heritage to thinking and acting globally. Evolution of consciousness proceeded in a local environment, local with respect to the range of the sensory perceptions. This situation is further aggravated in daily life by the fact that, apart from the private sphere, the individual experiences its possibilities to engage in responsible local action as rather limited. Due to the complexity of the social and economic networks and because human actions are amplified by the input of external energy, large spatial scales separate what an individual perceives as his or her own actions from what are perceived as threatening environmental problems. To realize the links between phenomena at a national or even global scale, and our own localized activities goes beyond human intuition. Therefore, we have to establish causal relationships by reasoning. However, any such relationships are fuzzy due to their being propagated along an intricate and complex web of interactions. The crucial question is how human ecological systems can be structured in a way that this mismatch can be improved and where the benefit to the individual and the benefit to the whole are not in contradiction but are mutually supportive.

Restricting energy input would definitely assist a transformation of human societies. At least in the case of material processes, the amount of energy available determines the spatial scale. We could thus expect an increasing regionalization of material changes. It would admittedly be naive to hope that we can solve global environmental problems on a national or even a regional scale. However, the regional context may be an important one. The dynamics and innovative potential of regions has been noted more recently in the social sciences (Jaeger, 1994). We may perceive regions as innovative units where different strategies could be tested for restructuring social and economic systems. The people in a region could for example decide to engage in a common effort to reduce their energy use and to improve the environmental situation. I cannot conceive of successful global development let alone evolution where a world government chooses what is perceived as an optimal path for all of the Earth's inhabitants. It is also at variance with the perspective on the dynamic nature of ecosystems developed here – a nature that should be preserved in human ecological systems. Evolution needs the continuous testing of different strategies to maintain the process of cooperative self-organization.

7.6.3 The role of scientific institutions in a changing environment

The perception of scientists as experts producing objective facts and knowledge has shaped the communication of scientific information and the role of scientific institutions in society. In many domains, including the one of environmental issues, the dialogue between scientific communities and society proceeds within the framework described in

Figure 7.11. Scientific communities are supported to a large extent by funds provided by policy makers. In traditional settings policy makers also largely decide which questions are important for political and societal decisions. Scientific communities work out answers to these questions and train specialists in various disciplines. Based on the scientific answers they receive and their attitude of mind, the politicians then decide on measures to deal with the problem. Such an approach has already proven to be difficult when applied to comparatively "simple" questions such as the problem of forest decline. Regarding problems associated with global change such a problem-solving strategy does not seem to be promising at all. It is quite obvious that the task of reducing uncertainties lies at the core of the role which scientific communities play in this traditional setting. Scientific research tries to solve questions of truth and falsehood by providing objective knowledge. This has given scientific institutions considerable prestige, but it has also deprived them from becoming involved in questions referring to value judgements. Such may be perceived as an advantage – does not it relieve scientists of being responsible for any unwanted abuse of the products of their research by society or economy. However, regarding environmental problems such a segregation of responsibilities attributes a weak position to the scientific community. For example, the uncertainties associated with climate change make it rather unlikely that scientific research could provide a sufficient rationale for adequate policy regulations in this area, in particular when we consider the conflicts of interests involved in these topics.

I argued before that scientific institutions must change towards a more democratic structure in order to deal with pluralistic approaches and complementary perspectives of the world. One may object that the current method has been successful to maintain scientific quality. Admitting what may be perceived as a kind of relativism, bears the risk that scientific quality and with it scientific reputation drops precipitously. However, I argue that science might just lose its credibility by sticking to traditional methods of argumentation where these are not appropriate. Scientists have to be more clear about where they are certain and where they never will be certain at all. The intensive use of sophisticated modelling tools is a dangerous path since it may blur these distinctions and generate misguided impressions of precision where there is none (see also Funtowicz and Ravetz, 1994; Oreskes *et al.*, 1994). Hence, I conclude that if science is interested in contributing actively to dealing with environmental issues, the role of scientific institutions in society and the societal perception of scientific knowledge have to change.

Political regulation based on expert knowledge is not the only and presumably not the adequate way to deal with environmental problems. Scientific communities could also contribute to fostering innovative action in society and economy and to assisting in building human–ecological systems. This would require that, in the future, science will have to involve itself more actively in an open social dialogue (Pahl-Wostl and Jaeger, 1994). Figure 7.12 represents a possible new framework for new forms of dialogue between scientific communities and society. Special emphasis is given to a regional context that may be perceived as the focal spatial scale for changes to take place. The function of science would not be limited to reducing uncertainties and providing solid knowledge. Scientists would point out important, unavoidable uncertainties and make suggestions how to deal with them. One could say that scientific communities should supply society with assessments of problematic beliefs. The outcome of such a dialogue would aim at encouraging environmentally sound innovative action rather than political regulations. Currently, we lack any institutional setting where such a dialogue could take place. The

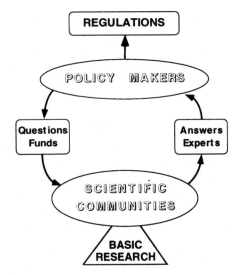

Figure 7.11   The traditional perception of the role of scientific institutions in society

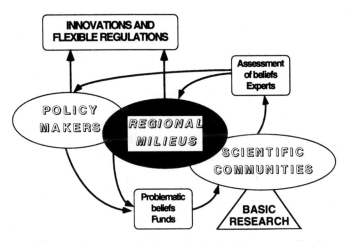

Figure 7.12   A possible new framework for the role of scientific communities in a social dialogue

regional context might be appropriate for innovative research and action to be encouraged within a new interdisciplinary framework fostering an intense communication among different social groups. In the social sciences, regional milieus have a special meaning denoting networks of people and institutions clustered in specific regions and specialized in the generation of innovations in specific fields (e.g. Jaeger, 1990). A well-known example is Silicon Valley in California, but such regions have been identified in many parts of the world. Research about regional milieus has mainly focused on their role in fostering innovations in economy. Silicon Valley is also a good example that such innovations need not necessarily be beneficial for the environment. However, regional contexts could also prove to be the place where innovative approaches to sustainable environmental management and social dialogues are developed. An increasing globalization of politics may find a counterpart in a strengthening of regions to preserve ecological, cultural and economic diversity. To enable such a process, we need to establish structures that encourage a dialogue comprising environmental, social and economic aspects simultaneously.

7.6.4   Strategies for a sustainable regional evolution

These suggestions cannot be but a first tentative outline of an endeavour that requires an intense dialogue among scientists and among social groups. There can be no doubt that we need to formulate goals of a quantitative and of qualitative nature. Quantitative refers to the amount of material and energy exchanged. Such is the general focus of present considerations related to sustainability. In particular we may claim the exchange of energy and resources to be higher within a region than across its boundaries. A region should aim for a high degree of autonomy in this respect. The external input of energy and resources should be less than what is stored within. Regions might collaborate to achieve such goals – on both a national and a global scale. The claim for autonomy is based on the assumption that regions should strive for developing a structure deriving from processes of internal self-organization. Such an assumption is closely related to goals of a qualitative nature referring to a system's organization. Sustainability is not simply a question of reducing material exchanges. A quantitative goal in this respect cannot be imposed onto a social structure in the long-term. Instead we have to find out the types of organization and boundary conditions required for human ecological systems to achieve such goals by themselves. They may for example include a high dynamic and functional diversity to be maintained and/or established for social, economic and ecological structures. We need flexible institutions comprising different social groups, scientists included. Environmental management cannot proceed separately from other planning activities. Nature cannot be restricted to and preserved within isolated "ghettos" as claimed by Eilingsfeld (1989).

Regional changes may have impacts at a national or even global scale. Innovative change could be explored at the regional level and radiate over larger scales. Such a wide-ranging spread of change in attitudes could be caused by a catastrophic breakdown of previous social and economic structures – we may ask whether such a type of change would be entirely desirable. However, such change could also spread when innovative action, initiated locally, finds fertile surroundings.

Figure 7.13 indicates that such a regional strategy must be integrated in a larger context across a range of temporal and spatial scales. The communication with the natural environment must enter at all levels of human societies. Ecological knowledge comprising naturalist, mechanistic and conceptual understanding would be needed at all levels in such an enterprise, albeit with differences in emphasis. Naturalist knowledge of persons being familiar with a particular ecosystem and its species is essential for local sites. Mechanistic knowledge is needed to evaluate acute problems such as the impact of specific measures or potential contaminations by chemicals. However, conceptual knowledge emphasizing pattern and organization is essential to give us general guidance for the type of system design we should strive for. It may not always be possible or even desirable to attempt a strict separation of the different types of knowledge. In the end what we aim for is a different way to interact with nature both qualitatively and quantitatively.

Sceptics may criticize such suggestions to be hopelessly naive whims of a natural

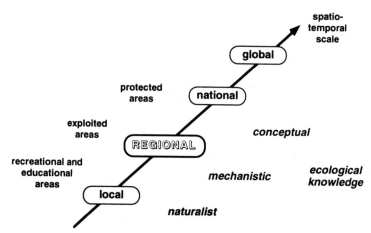

Figure 7.13 Regional activities must be embedded in a larger context of space–time scales. The sequence of naturalist, mechanistic and conceptual ecological knowledge indicates that their relevance for decision making changes with the spatio-temporal scales under consideration

scientist ignorant of social and economic reality. One may further question provocatively whether scientists are really interested in a change taking place. Waterstone (1993) suggested that uncertainty with respect to the greenhouse effect had a very important strategic use for both policy and the scientific communities. For the former, it provides the necessary means for delaying anything other than symbolic action. For the latter, it provides a lucrative source of research funding in times when research budgets are dropping precipitously. Waterhouse stated quite a pessimistic opinion regarding the tackling of global change in general and the greenhouse problem in particular given the current economic and political structures. Nevertheless, he concluded that the challenge posed by the greenhouse warming was to define the issue in such a way that individuals and policy makers saw that it was in their own self-interest to begin to undertake ameliorations and adaptive policies. I am not that pessimistic regarding the number of environmental scientists who really wish their research to have an impact and a change to take place. However, major efforts are required to open a dialogue in order to find out what any such change should imply and how it could be effected. The strategy outlined above is a suggestion how to proceed. It is not the only and it may not be the most realistic one. However, progress can only be achieved by testing different alternatives towards approaching a problem – and at present alternatives are rare.

7.7   A MACROSCOPIC UNCERTAINTY PRINCIPLE?

I have argued that it is impossible to base decisions related to global environmental problems on scientific "facts" proving cause–effect relationships and allowing reliable predictions of future developments. Since we experience only one present and one future, considerations about alternatives become futile. They are intrinsically undefined and unknowable. In analogy to the quantum realm where the presence of uncertainty has grown into an established fact, such a situation could be referred to as a "macroscopic uncertainty principle". The question mark in the paragraph heading indicates that I use the notion of a "macroscopic uncertainty principle" as a metaphor without claiming to achieve or even approach the scientific rigour of the Heisenberg principle. In the hope of not offending the purists among the physicists too much I interpret the microscopic uncertainty principle in physics rather loosely by perceiving as major consequences:

- Energy and matter are equivalent in the microscopic realm resulting in the impossibility to make nature discrete at this level of resolution.
- Quantum phenomena are neither particles nor waves but both. Complementary descriptions are required.
- The nature of an object as perceived by an observer is determined by the context chosen. What is observed depends on the means of interaction.

Regarding the complementary description of matter we may draw an analogy to the two main streams in ecology, population ecology being concerned with single organisms and/or species and system ecology focusing on material flows and function. Actually this splitting may be perceived in the tradition of the difference between mechanistic and thermodynamic descriptions in physics that may in hindsight be seen as a logical consequence arising from the ambiguous nature of material reality where the picture of particle and wave portray only two extremes of a continuum of possible descriptions. In physics one has attempted to reduce the macroscopic thermodynamic descriptions to molecular statistical mechanics. Explaining macroscopic phenomena as collective properties of ensembles of microscopic particles seems to work reasonably well for the description of global variables such as temperature or pressure. Problems arise with the description of patterns, as for example in turbulence. The Lagrangian approach in hydrodynamics describes pattern with ensembles of particle trajectories, whereas the Eulerian approach describes pattern with macroscopic state variables that vary over time and space. However, the two levels of description are only formally equivalent. Applied to even simple practical problems the macroscopic pattern cannot be derived as a collective property of the individual particles – one of the best studied but yet unresolved problems in physics. Perhaps we should look at these descriptions as being complementary rather than being equivalent. In ecological systems the individual "particles" differ largely and statistical properties are hardly ever meaningful. We have to deal with the important phenomenon of function that is not a property of single organisms but arises in a meaningful way in an environmental context only. Entering the realm of human beings we may observe a similar splitting in the social sciences. Economic studies deal mainly with material flows and macroscopic function. They seem to be performed largely independently from investigations of the psychological and social reality of the individual members of a human society. I argue that in all these examples the two levels of descriptions are complementary and each of them is incomplete when viewed in isolation.

Due to the fact that one cannot separate part and whole in the evolutionary process, I conjecture that an uncertainty principle at the macroscopic level of living systems can be postulated especially when the global system as a whole is considered. The trajectory of evolution is largely undefined even when it is constrained by the availability of resources and by the legacy from the past stored in the current interaction network. By adopting a certain perception and mode of interacting with the environment, humanity selects a path within the constraints of material reality. Selection does not refer to a purposeful act when confronted with a number of choices with defined outcome. The notion of selection expresses that it is the essence of evolution to be irreversible and to proceed in time. Since the historical development is unique, we can only experience one pathway of evolution and any alternative remains undetermined.

We may conceive of a logical paradox. Either we accept the phenomenon of evolution to be a true and integral fact of life – then most attempts for decisions based on predictions are obsolete. Or we deny evolution and assume that prediction and control are possible. The latter attitude seems to be dominant at present. When we perform scientific studies we seem to perceive ourselves as external observers of natural, social and/or economic systems, forgetting thereby that we belong to an evolving system, a system to which we are inseparably linked and that we influence and change by all of our actions. Most of our social and economic systems would not exist if we really were able to predict future behaviour. Being part of an evolving system and the ability to control and predict a system's behaviour are mutually exclusive properties. Realizing this and developing it into an incentive for action may prove to be a "major innovative disturbance" in the network of human–environment interactions.

Appendix One

Derivation of measures to quantify a network's spatio-temporal organization

Rutledge *et al.* (1976) were the first to apply measures derived from mutual information to characterizing ecological networks. Doing so, they followed closely the concept of an information channel on which the original definition of mutual information is based. Therefore, I will first summarize the derivation of the mutual information in the context of an information channel and continue by explaining its application to ecological networks. I will focus here on the mathematical steps of the argument. Meaning and applications of the measures were discussed in more detail in Chapter Four.

A1.1 MUTUAL INFORMATION AND INFORMATION CHANNELS

An information channel can be characterized by an information source generating an input alphabet $A = \{a_j\}, j = 1, 2, \ldots, n$; a receiver generating an output alphabet $B = \{b_i\}, i = 1, 2, \ldots, n$; and a set of conditional probabilities $p(b_i/a_j)$ for all i and j. $p(b_i/a_j)$ is the probability for b_i to be generated in the receiver if a_j is generated by the source.

An information channel can thus be depicted schematically as:

$$A \begin{pmatrix} a_1 \\ a_2 \\ . \\ . \\ . \\ a_n \end{pmatrix} \to p(b_i/a_j) \to \begin{pmatrix} b_1 \\ b_2 \\ . \\ . \\ . \\ b_n \end{pmatrix} B$$

The information transmitted depends on the diversity of symbols generated by the source and the reliability of the information transfer. The former can be quantified as:

$$H(A) = -\sum_j p(a_j)\, \log p(a_j) \tag{A1.1}$$

where $p(a_j)$ is the probability for a_j to be sent.

$H(A)$ is often referred to as the information – or Shannon entropy of a set $A = \{a_1, \ldots, a_n\}$. $H(A)$ may be interpreted either as the average information generated by the source or as the average amount of uncertainty which an observer has with respect to an output from the source.

If A is an alphabet with n symbols, then:

$$\log n \geq H(A) \geq 0 \tag{A1.2}$$

To derive inequality (A1.2) we have to make use of the relationship depicted in Figure A1.1 that compares the natural logarithm of x and the line defined by the equation $y = x - 1$.

We may easily verify that for $x > 0$ we obtain the inequality:

$$\ln(x) \leq x - 1 \text{ with equality if and only if } x = 1 \tag{A1.3}$$

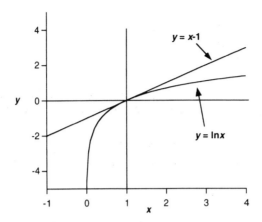

Figure A1.1   Comparison of the functions $y = \ln x$ and $y = x - 1$ shows that $\ln x \le x - 1$

Let $\{p_i\}, \{q_j\}, i, j = 1, 2, \ldots, n$, be any two sets of probability. That is $p_i \ge 0$ and $q_j \ge 0$ for all i and j,
$$\sum_{i=1}^{n} p_i = \sum_{j=1}^{n} q_j = 1.$$
Using inequality (A1.3) we can write:

$$\sum_{i=1}^{n} p_i \ln \frac{q_i}{p_i} \le \sum_{i=1}^{n} p_i \left(\frac{q_i}{p_i} - 1 \right) = \sum_{i=1}^{n} q_i - 1 = 0 \qquad (A1.4)$$

Multiplication with -1 and rearrangement of (A1.4) yields the fundamental inequality:

$$-\sum_{i=1}^{n} p_i \ln q_i \ge -\sum_{i=1}^{n} p_i \ln p_i \qquad (A1.5)$$

Hence

$$\ln(n) \ge -\sum_{i=1}^{n} p_i \ln p_i \text{ for } q_i = \frac{1}{n}$$

Inequality (A1.5) is valid for any type of logarithmic relationship irrespective of the base, since changing the base of a logarithm is identical to a multiplication with a positive constant.

To quantify the reliability of the information transfer, let us assume that the receiver generates a signal b_i. The probability that the symbol a_j was generated by the source before is then:

$p(a_j/b_i)$: conditional probability for a_j to have been sent if b_i is received

The maximum of reliability that is equivalent to a minimum in residual uncertainty is obtained for $p(a_j/b_i) = 1$. The average residual uncertainty can be assessed by averaging over both A and B. First we obtain the average uncertainty $H(A/b_i)$ about a symbol a_j if b_i is received, by averaging over the whole alphabet A:

$$H(A/b_i) = -\sum_j p(a_j/b_i) \log p(a_j/b_i) \qquad (A1.6)$$

Averaging over the whole alphabet B finally results in:

$$H(A/B) = -\sum_i p(b_i) H(A/b_i)$$
$$= -\sum_i p(b_i) \sum_j p(a_j/b_i) \log p(a_j/b_i) \qquad (A1.7)$$

$$= -\sum_j \sum_i p(a_j, b_i) \log p(a_j/b_i)$$

$p(a_j, b_i)$, the joint probability for a_j to be sent and b_i to be received, is defined as:

$$p(a_j, b_i) = p(a_j/b_i)p(b_i) = p(b_i/a_j)p(a_j) = p(b_i, a_j) \tag{A1.8}$$

The conditional entropy $H(A/B)$ is also called the equivocation of A with respect to B. It measures the average uncertainty about a signal a_j provided a signal b_i was received.
$H(A/B)$ is bounded by:

$$H(A) \geq H(A/B) \geq 0 \tag{A1.9}$$

Inequality (A1.9) can be derived by noting that the joint probability is bounded by:

$$p(b_i) \geq p(a_j, b_i) \geq p(a_j)p(b_i)$$

Thus it follows

$$H(A) + H(B) \geq H(A,B) \geq H(B) \geq 0$$

and

$$H(A) \geq H(A,B) - H(B) = H(A/B) \geq 0$$

where $H(A,B)$ is the Shannon entropy of the set of joint probabilities

$$H(A,B) = -\sum_j \sum_i p(a_j, b_i) \log p(a_j, b_i) \tag{A1.10}$$

$H(A/B) = 0$, if for all i there exists $k_i \in \{1, \ldots, n\}$ such that

$$p(a_j/b_i) = \begin{cases} 1 & \text{if } j = k_i \\ 0 & \text{else} \end{cases}$$

After observation of b_i at the receiver no uncertainty about the corresponding signals from the source is left. The information is maximum.

$H(A/B) = H(A)$, if for all j and i: $p(a_j, b_i) = p(a_j)p(b_i)$, hence $p(a_j/b_i) = p(a_j)$ and $p(b_i/a_j) = p(b_i)$. The signal b_i received and the signal a_j sent are statistically independent. In this case no information is transmitted.

A measure for the average amount of information transmitted in a channel is given by the mutual information that is defined as:

$$I(A;B) = H(A) - H(A/B) \tag{A1.11}$$

$H(A)$ units of information are needed on average to specify one symbol a_j generated by the information source. $H(A/B)$ units of information are needed on average to specify one symbol a_j if one observes the symbol b_i produced by a_j in a receiver. On the average, observation of a single b_i provides thus $[H(A) - H(A/B)]$ bits of information (bits refer to logarithms to the base 2). This difference is called the average mutual information $I(A; B)$ of A and B.

Replacement of $H(A)$ and $H(A/B)$ in Eqn (A1.11) by the relationships (A1.1) and (A1.7), respectively, and application of (A1.8) result in:

$$I(A;B) = \sum_j \sum_i p(a_j, b_i) \log \frac{p(a_j, b_i)}{p(a_j)p(b_i)} \tag{A1.12}$$

Using (A1.9) and (A1.11) one can derive that the range of the average mutual information is bounded by:

$$H(A) \geq I(A;B) \geq 0 \tag{A1.13}$$

$$I(A; B) = 0, \text{ if and only if } H(A/B) = H(A)$$

and

$$I(A; B) = H(A), \text{ if and only if } H(A/B) = 0$$

$H(A)$ may also be referred to as the capacity of the channel, because it constitutes the maximum of information that can be transmitted. $H(A/B)$ may also be referred to as the redundancy of the channel. $I(A;B)$ is zero if the redundancy equals the capacity, that means if the signals sent and received are statistically independent. $I(A;B)$ is at maximum equal to the capacity if the redundancy is zero.

The average mutual information is symmetric with respect to the input and output alphabets:

$$I(A;B) = I(B;A) \qquad\qquad (A1.14)$$

since

$$I(B;A) = H(B) - H(B/A) = H(B) + H(A) - H(A/B)$$

hence

$$I(A;B) = H(A,B) - H(A/B) - H(B/A) \qquad\qquad (A1.15)$$

The interchange of A and B leaves the mutual information unaltered. The adjective mutual is therefore reasonable and suggests that $I(A;B)$ is a measure of the information in the joint occurrence of A and B. The mutual information measure may thus be used to quantify the correlation between any two sets of random variables. If the variables are not statistically independent, they will have a positive mutual information.

A1.2   THE CONCEPT OF MUTUAL INFORMATION APPLIED TO ECOLOGICAL NETWORKS

Rutledge *et al.* (1976) used the concept of the average mutual information to derive a measure quantifying the redundancy in ecological networks. To do so they interpreted ecological networks in terms of an information channel as schematically depicted in Figure A1.2a.

The "information source" is given by the sum of all compartmental outputs, the "receiver" is given by the sum of all compartmental inputs. The "information transfer" is identified with the flows between the compartments. Rutledge *et al.* (1976) based their considerations on a network model of a closed system by accounting for internal transfers only. Hirata and Ulanowicz (1984) extended the description to open systems by introducing environmental compartments as shown in Figure A1.2b.

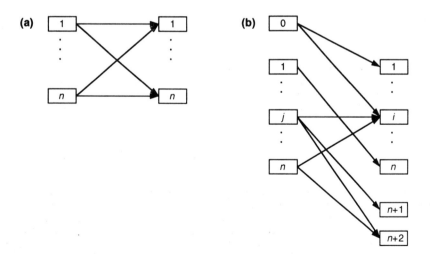

Figure A1.2   Representation of a network as an "information channel". (a) Internal $(1,2\ldots,n)$ compartments only. (b) Including additional compartments to account for exchanges with the environment given by imports (0), exports $(n+1)$ and dissipative losses $(n+2)$

The imports are accounted for by emanating from a compartment with index 0, the exports are directed to a compartment $n+1$. In case of energy flows, the dissipative losses are directed to a compartment $n+2$. Hence, the system includes two or three "asymmetric" external compartments serving as sources or sinks, respectively. The output and input alphabets can thus be identified with the corresponding flow measures:

$$A \left\{ \begin{array}{l} T_{0.} \\ T_{1.} \\ T_{2.} \\ . \\ . \quad \to p(T_{.i}/T_{j.}) \to \quad . \\ . \\ T_{n.} \end{array} \right. \left. \begin{array}{l} T_{.1} \\ T_{.2} \\ . \\ . \\ . \\ T_{.n} \\ T_{.n+1} \\ T_{.n+2} \end{array} \right\} B$$

A point always denotes summation over the corresponding index. $A = \{T_{j.}\}$, $j = 0, 1, ..., n$. and $B = \{T_{.i}\}$, $i = 1, 2, ..., n+2$, where the assymetry derives from the introduction of hypothetical external compartments. The probabilities corresponding to the two alphabets A and B can be expressed in terms of flows as:

$$p(a_j) = p(T_{j.}) = \frac{T_{j.}}{T}$$

$$p(b_i) = p(T_{.i}) = \frac{T_{.i}}{T}$$

$$p(b_i/a_j) = p(T_{.i}/T_{j.}) = \frac{T_{ji}}{T_{j.}}$$

$$p(a_j, b_i) = p(T_{j.}, T_{.i}) = \frac{T_{ji}}{T} \tag{A1.16}$$

$T_{.i}$ total input into compartment i
$T_{j.}$ total output from compartment j
T_{ji} flow from compartment j to compartment i
T total system throughflow

$$T = \sum_{j=0}^{n} \sum_{i=1}^{n+2} T_{ji} \tag{A1.17}$$

From an information theoretical point of view summing over both internal flows and external imports and exports is important because a normalization must be based on the sum of all possible events in the system. At the same time such a description allows the treatment of non-steady-state systems where the inputs and outputs are not balanced.

$p(T_{j.}, T_{.i})$ is now the probability that a unit of matter transferred in the network originates from the compartment X_j and enters the compartment X_i. The average mutual information between compartmental outputs and inputs yields thus in terms of flows:

$$I(A;B) = K \sum_{j=0}^{n} \sum_{i=1}^{n+2} \frac{T_{ji}}{T} \log \frac{T_{ji}T}{T_{j.}T_{.i}} \tag{A1.18}$$

The scaling factor K may impart a physical dimension to the mutual information that itself is independent of the amount of matter transferred in a network.

The average mutual information between outputs and inputs represents a measure for a network's organization, resulting from feedback effects in the system. Outputs and inputs are treated equivalently, no cause–effect relationship is assumed. Thus, the interpretation chosen here differs in two major points from the formulation by Rutledge *et al.* (1976). They assumed in close analogy to the concept of the information channel that an output at a time t causes an input at a time $t + \Delta t$. However, in an ecological network outputs and inputs cannot be looked upon as being isolated, successive events. In most cases they occur simultaneously and cause each other. To emphasize the importance of the equivalence of inputs and outputs the definition of the average mutual information is based here on Eqn (A1.15) instead of (A1.11). In the case of the two-dimensional measure it makes no real difference. It will, however, prove to be of importance when higher dimensional measures of mutual information are introduced. The average mutual information is thus bounded by:

$$H(A,B) \geq I(A;B) \geq 0 \qquad (A1.19)$$

The capacity equals the joint entropy $H(A,B)$ that is expressed in terms of flows:

$$H(A,B) = - \sum_{j=0}^{n} \sum_{i=1}^{n+2} \frac{T_{ji}}{T} \log \frac{T_{ji}}{T} \qquad (A1.20)$$

The redundancy, R, equals the sum of $H(A/B)$ and $H(B/A)$ that is expressed in terms of flows:

$$R = H(A/B) + H(B/A) = - \sum_{j=0}^{n} \sum_{i=1}^{n+2} \frac{T_{ji}}{T} \log \left(\frac{T_{ji}}{T_{.i}} \frac{T_{ji}}{T_{j.}} \right) \qquad (A1.21)$$

$I = H(A,B)$ and $R = 0$. $H(A,B)$ is maximum if all flows are of equal magnitude.

Directing all flows that leave the system to one external compartment $n+1$ (possible in the case of nutrient flows), one can derive that $H(A/B) = H(B/A) = 0$, if and only if, for every $j \in \{0,1,\ldots,n\}$ there exists $\alpha_j \in \{1,2\ldots,n+1\}$ such that:

$$\frac{T_{j\alpha_j}}{T} = \frac{T_{j.}}{T} = \frac{T_{.\alpha_j}}{T} = \frac{1}{n+1} \qquad (A1.22)$$

hence $T_{ji} = 0$ if $i \neq \alpha_j$ and $T_{j.} = T_{.j}$ for all j.

The maximum of the average mutual information between inputs and outputs is achieved if every compartment has only one input and only one output. If the currency of flows is energy, dissipative losses leave the system from every living compartment. In this case, the theoretical maximum cannot be realized due to thermodynamic constraints.

$I = 0$ and $R = H(A,B)$. The redundancy is at maximum and the information is zero if the arguments of all logarithms in (A1.18) are equal to one. Assuming an even flow distribution one obtains:

$$\frac{T_{ji}}{T} = \frac{T_{j.}T_{.i}}{T^2} = \frac{1}{(n+1)^2} \text{ for } 0 \leq j \leq n \text{ and } 1 \leq i \leq n+1 \qquad (A1.23)$$

Every compartment is connected to every other and the flows are of equal magnitude.

The extremes of the possible network configurations with I at minimum and maximum, respectively, were discussed for illustrative purposes. The conditions derived indicate that it is unlikely for any real ecosystem to exhibit such extreme configurations. Real systems may be expected to range somewhere between these extremes.

Comment

It is interesting to note that similar information theoretical descriptions were used for two divergent purposes. Ulanowicz (1980, 1986) argued that the development of an ecosystem was characterized by an increase in organization. Hence, he predicted a reduction in redundancy and an increase in the average mutual information over the course of an ecosystem's development towards a mature state.

Rutledge *et al.* (1976), however, argued that a maximum in choice for the flow pathways in an ecosystem resulted in a higher persistence against perturbations. Hence, they predicted that due to stability reasons an ecosystem should be characterized by a high redundancy. It seems as if stability and development stand in opposition to each other. This apparent contradiction may be resolved by accounting for the diversity in flows along the dimensions of time and space.

A1.3   EXTENSION TO MULTI-DIMENSIONAL NETWORKS
(cf. PAHL-WOSTL, 1992a)

Without loss of generality, let us first focus on deriving a measure for the temporal organization and redundancy in networks resolved along the dimensions of time. Based on the considerations in

Chapter Four, such a measure should have the following properties:

$$I_t \geq I \quad \text{and} \quad I_t - I = R - R_t \tag{A1.24}$$

where I_t and R_t refer to the measures of organization and redundancy, respectively, in a network where the temporal flow pattern is resolved. In addition, the symmetry between inputs and outputs should be retained. The measure should quantify the reduction in functional redundancy of the time-averaged structure (and hence the increase in functional organization) that is obtained by accounting for the temporal flow pattern. The new dimension representing time is thus not independent, but acquires meaning only in relation to how it affects the network structure. More specifically we may ask: What is the reduction in uncertainty about the destination (origin) of a flow emanating from (entering) a certain compartment when we switch from considering the whole observation period to considering a specific time interval.

Dealing with information channels, Abramson (1963) suggested possible extensions of the mutual information measure to further dimensions. Information about the signals produced in the source may be gained by several observations. For every a_j one may, for example, receive two signals b_i and c_k from the alphabets: $B = \{b_i\}$, $i = 1, 2, \ldots, m$, and $C = \{c_k\}$, $k = 1, 2, \ldots, r$, respectively. Without loss of generality, we assume, that the two signals are received in the order b_i, c_k. Hence, $p(a_j)$ changes first to $p(a_j/b_i)$ and then to $p(a_j/b_i,c_k)$.

Correspondingly, the average uncertainty about a signal a_j changes from $H(A)$ defined in Eqn (A1.1) to $H(A/b_i)$ defined in Eqn (A1.7), and finally to $H(A/b_i,c_k)$:

$$H(A/b_i,c_k) = -\sum_j p(a_j/b_i,c_k) \log p(a_j/b_i,c_k) \tag{A1.25}$$

Averaging (A1.25) over all b_i and c_k results in $H(A/B,C)$, the equivocation of A with respect to B and C.

$$
\begin{aligned}
H(A/B,C) &= \sum_i \sum_k p(b_i,c_k) H(A/b_i,c_k) \\
&= -\sum_j \sum_i \sum_k p(a_j,b_i,c_k) \log p(a_j/b_i,c_k)
\end{aligned}
\tag{A1.26}
$$

The average mutual information of A and (B,C) is now defined just as we did when the channel output was a single symbol. That is

$$
\begin{aligned}
I(A;B,C) &= H(A) - H(A/B,C) \\
&= H(A,B) - H(B/A) - H(A/B,C) \\
&= I(A;B) + I(A;C/B)
\end{aligned}
\tag{A1.27}
$$

where

$$I(A;C/B) = H(A/B) - H(A/B, C) \tag{A1.28}$$

Eqn (A1.27) may be generalized to an infinite number of dimensions:

$$
\begin{aligned}
I(A;B, C, \ldots, Y) &= I(A;B) + I(A;C/B) + \ldots + I(A; Y/B, C. \ldots) \\
&= H(A,B) - H(B/A) - H(A/B, C) - \ldots, - H(A/B,C, \ldots, Y)
\end{aligned}
\tag{A1.29}
$$

The term on the left is the average information about the information source A provided by the combined observation of the signals $B, C, \ldots, Y$.

$I(A;B,C)$ defined in (A1.27) can be expressed in terms of probabilities to yield:

$$
\begin{aligned}
I(A;B,C) &= \sum_j \sum_i \sum_k p(a_j,b_i,c_k) \log \frac{p(a_j/b_i,c_k)}{p(a_j)} \\
&= \sum_j \sum_i \sum_k p(a_j,b_i,c_k) \log \frac{p(a_j,b_i,c_k)}{p(a_j)p(b_i,c_k)}
\end{aligned}
\tag{A1.30}
$$

where

$$p(a_j/b_i,c_k) = \frac{p(a_j,b_i,c_k)}{p(b_i,c_k)} \tag{A1.31}$$

One may also consider the reverse situation and assume that an output symbol b_i is determined by a sequence of input symbols a_j and c_k. In analogy to the preceeding derivations the average mutual information between (A,C) and B can be derived as:

$$
\begin{aligned}
I(B;A,C) &= H(B) - H(B/A,C) \\
&= H(A,B) - H(A/B) - H(B/A,C) \\
&= I(A;B) + I(B;C/A)
\end{aligned}
$$
(A1.32)

$I(B;A,C)$ expressed in terms of probabilities yields:

$$
\begin{aligned}
I(B;A,C) &= \sum_j \sum_i \sum_k p(a_j,b_i,c_k) \log \frac{p(b_i/a_j,c_k)}{p(b_i)} \\[2mm]
&= \sum_j \sum_i \sum_k p(a_j,b_i,c_k) \log \frac{p(a_j,b_i,c_k)}{p(b_i)p(a_j,c_k)} \; .
\end{aligned}
$$
(A1.33)

where

$$
p(b_i/a_j,c_k) = \frac{p(a_j,b_i,c_k)}{p(a_j,c_k)}
$$
(A1.34)

The measure thus defined fulfils the requirement stated in (A1.24). However, a comparison of (A1.27) with (A1.32) and of (A1.30) with (A1.33) shows that the symmetry between inputs and outputs is lost since they are not treated as being equivalent:

$$
I(A;B,C) \neq I(B;A,C)
$$
(A1.35)

Abramson (1963) suggested yet another possibility on how to introduce further dimensions into the definition of the average mutual information. He defined the mutual information of A and B and C by:

$$
\begin{aligned}
I(A;B;C) &= H(A,B,C) - H(A/B) - H(B/C) - H(C/A) \\
&= I(A;B) - I(A;B/C)
\end{aligned}
$$
(A1.36)

This definition is symmetric with respect to all three sets. It deviates from the original channel concept since the three sets of events are independent and contribute equally to the total mutual information. This implies a change in reference state from $H(A,B)$ to $H(A,B,C)$. The measure thus defined does not fulfil the requirement stated in (A1.24). If we identify set C with time, $I(A;B;C)$ quantifies the average information provided by the knowledge of an input signal B about the output and the time interval when it was received. Applied to a network, the measure defined in (A1.36) would increase with increasing specificity of the flows with respect to a time interval. It is at maximum if during a single time interval only one single flow in the network is active.

A graphical representation may help to clarify the different measures just discussed. In Figure A1.3 the entropies $H(A)$, $H(B)$ and $H(C)$ are represented as closed circles. The overlaps enclosed in bold face represent the various types of average mutual information among the respective sets. Figure A1.3 depicts (a) $I(A;B)$ of two sets; (b) $I(A;B;C)$ of three sets defined in Eqn (A1.36); (c) the assymetric $I(A,C;B)$ defined in Eqn (A1.32); (d) $((A,B;A,C;B,C)$ yet to be defined in Eqn (A1.37).

A comparison between (a) and (b) reveals that in this case the average mutual information is decreased by the addition of a new dimension. This results from the fact that the three sets are regarded as independent sources of events. The measure depicted in (c) is assymmetric with respect to inputs and outputs. $I(A,B;A,C;B,C)$ shown in Figure A1.3d fulfils all requirements stated before. It is defined as:

$$
\begin{aligned}
I(A,B;A,C;B,C) &= H(A,B) - H(A/B,C) - H(B/A,C) \\
&= I(A;B) + I(A;C/B) + I(B;C/A).
\end{aligned}
$$
(A1.37)

$I(A,B;A,C;B,C)$ is symmetric with respect to inputs and outputs. The reference state $H(A,B)$ is retained and each new dimension may provide information that reduces the remaining redundancy. It represents thus a hierarchical measure. A and B are equivalent, whereas C corresponds to another dimension of resolution with respect to A and B. Even when it is not evident in the pictorial representation of three sets, this ranking is reflected in the additional conditional information

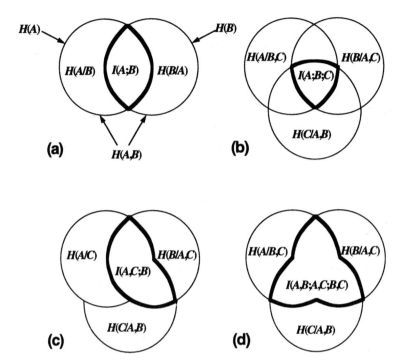

Figure A1.3   Graphical representation of the various measures of mutual information. A closed circle always denotes the Shannon entropy $H(X)$ of the corresponding single set X. Reprinted from *Mathl. Comp. Mod.*, **16**, Pahl-Wostl, C. Information theoretical analysis of functional temporal and spacial organization in flow networks, pp. 35–52, copyright (1992), with kind permission from Elsevier Science Ltd, The Boulevard, Langford Lane, Kidlington OX5 1GB, UK

measures $I(A;C/B)$ and $I(B;C/A)$.

To derive (A1.37) formally $I(A;B)$ is interpreted as the average mutual information of the three sets A, B and (A,B). Following Eqn (A1.36) the average mutual information of three sets of events X, Y and Z is defined as:

$$I(X;Y;Z) = I(X;Y) - I(X;Y/Z) \tag{A1.38}$$

and

$$I(X;Y/Z) = H(X/Z) - H(X/Z,Y) \tag{A1.39}$$

After identifying the correspondence $X \equiv A,B$; $Y \equiv A$; $Z \equiv B$, and by applying Eqns (A1.38), (A1.39) and (A1.11), $I(A,B;A;B)$ can be expressed as:

$$I(A,B;A;B) = H(A,B) - H(A,B/A) - H(A,B/B) + H(A,B/A,B) \tag{A1.40}$$

(A1.40) is trivially equivalent to the original definition of $I(A;B)$ in (A1.15) if one takes into consideration that: $H(A,B/A) = H(B/A)$, $H(A,B/B) = H(A/B)$ and $H(A,B/A,B) = 0$.

Based on the joint entropy $H(A,B)$ as reference state the extension of (A1.40) to another dimension C is defined as:

$$
\begin{aligned}
I(A,B;A,C;B,C) &= H(A,B) - \{(A,B)/(B,C)\} - H\{(A,B)/(A,C)\} \\
&= H(A,B) - H(A/B,C) - H(B/A,C) \\
&= I(A;B) + I(A;C/B) + I(B;C/A)
\end{aligned}
\tag{A1.41}
$$

$$= I(A;B) + H(A/B) + H(B/A) - H(A/B,C) - H(B/A,C)$$

X, Y and Z in Eqns (A1.38) and (A1.39) have to be identified with: $X \equiv A,B$; $Y \equiv A,C$; $Z \equiv B,C$. Thus the measure fulfils the requirements stated in Eqn (A1.24).

$I(A,B; A,C; B,C)$ can now be applied to derive a measure of temporal organization in networks where the flows are resolved in time. The total observation period is divided in r equal time intervals. The dimension of time is then introduced as a new set C of time intervals $\{\Delta t_k\}$, $k = 1, 2, \ldots, r$. The probabilities can be expressed in terms of flows:

$$p(a_j,c_k) = \frac{T_{j.k}}{T}$$

$$p(a_j,b_i,c_k) = \frac{T_{jik}}{T}$$

$$p(b_i/a_j,c_k) = \frac{p(a_j,b_i,c_k)}{p(a_j,c_k)} = \frac{T_{jik}}{T_{j.k}}$$

$$p(a_j/b_i,c_k) = \frac{p(a_j,b_i,c_k)}{p(b_i,c_k)} = \frac{T_{jik}}{T_{.ik}}$$

(A1.42)

where

T_{jik} flow from compartment j to i during time interval k
$T_{j.k}$ total output from compartment j during time interval k
$T_{.ik}$ total input into compartment i during time interval k
$T_{j.i}$ flow from compartment j to i summed over the whole observation period
T total system throughflow:

$$T = \sum_{j=0}^{n} \sum_{i=1}^{n+2} \sum_{k=1}^{r} T_{jik}$$

(A1.43)

$H(A/B,C)$ and $H(B/A,C)$ can thus be expressed in terms of flows:

$$H(A/B,C) = - \sum_{j=0}^{n} \sum_{i=1}^{n+2} \sum_{k=1}^{r} \frac{T_{jik}}{T} \log \frac{T_{jik}}{T_{.ik}}$$

$$H(B/A,C) = - \sum_{j=0}^{n} \sum_{i=1}^{n+2} \sum_{k=1}^{r} \frac{T_{jik}}{T} \log \frac{T_{jik}}{T_{j.k}}$$

(A1.44)

The measure of temporal organization $I(A,B; A,C; B,C) \equiv I_t$ can be expressed in terms of flows as:

$$I_t = \sum_{j=0}^{n} \sum_{i=1}^{n+2} \sum_{k=1}^{r} \frac{T_{jik}}{T} \log \frac{T_{jik}^2 T}{T_{ji.} T_{j.k} T_{.ik}}$$

(A1.45)

As required, the measure defined in (A1.45) quantifies the decrease in redundancy of compartmental outputs and inputs upon resolution of the temporal flow pattern.

The redundancy, $R_t = H(A/B, C) + H(B/A, C)$ is zero if and only if the condition stated in (A1.22) is fulfilled for every $k \in \{1,2, \ldots, r\}$. Hence, during a single time interval each network compartment has one input and one output only: $T_{j.k} = T_{.\alpha_j k}$ for all j, i and k.

The definition of the average mutual information can be extended to further dimensions. For example, a spatial dimension may be introduced as a new set D of all spatial intervals $\{\Delta s_l\}$, $l = 1, 2, \ldots, 9$. The time- and space-dependent average mutual information is then defined as:

$$I(A,B; A,C,D; B,C,D) = H(A,B) - H(A/B,C,D) - H(B/A,C,D)$$
$$= I(A,B; A,C; B,C) + I(A; D/B,C) + I(B; D/A,C)$$

(A1.46)

The time- and space-dependent average mutual information, I_{ts}, defined in (A1.46) can be expressed in terms of flows as:

$$I_{ts} = \sum_{j=0}^{n} \sum_{i=1}^{n+2} \sum_{k=1}^{r} \sum_{l=1}^{9} \frac{T_{jikl}}{T} \log \frac{T_{jikl}^2 T}{T_{ji..} T_{j.kl} T_{.ikl}}$$

(A1.47)

Eqn (A1.46) may be generalized to

$$I(A,B;A,C, ..., Y;B,C, ..., Y) = H(A,B) - H(A/B,C, ..., Y) - H(B/A,C, ..., Y)$$

$$(A1.48)$$

The extension to further dimensions reveals clearly the hierarchical nature of the measures defined. The capacity $H(A,B)$ is always retained. Each new dimension acquires meaning only in relationship to its providing information about the specificity of the flow pattern connecting network compartments expressed by the dependence of the sets A and B.

A1.4 DECOMPOSITION OF THE MEASURES OF ORGANIZATION

The measure defined in Eqn (A1.18) may be decomposed into terms stemming from inputs (I_o), internal exchanges (I_i), exports (I_e) and dissipation (I_d) (cf. Ulanowicz, 1986):

$$I = I_o + I_i + I_e + I_d \qquad (A1.49)$$

where

$$
\begin{aligned}
I_o &= \sum_{i=1}^{n} \frac{T_{0i}}{T} \log \frac{T_{0i}T}{T_{0.}T_{.i}} \\
I_i &= \sum_{j=1}^{n} \sum_{i=1}^{n} \frac{T_{ji}}{T} \log \frac{T_{ji}T}{T_{j.}T_{.i}} \\
I_e &= \sum_{j=1}^{n} \frac{T_{j(n+1)}}{T} \log \frac{T_{j(n+1)}T}{T_{j.}T_{.(n+1)}} \\
I_d &= \sum_{j=1}^{n} \frac{T_{j(n+2)}}{T} \log \frac{T_{j(n+2)}T}{T_{j.}T_{.(n+2)}}
\end{aligned}
\qquad (A1.50)
$$

I_t, I_s and I_{ts} may be decomposed accordingly by replacing each flow T_{ji} with the temporal and/or spatial sequence of the T_{jik}, or T_{jil}, or T_{jikl}, and by summing over the dimensions of time and/or space.

Accordingly one can account for the contribution of different types of flow to the redundancy of a system. The redundancy R defined in Eqn (A1.21) may be decomposed into terms stemming from inputs (R_o), internal exchanges (R_i), exports (R_e) and dissipation (R_d):

$$R = R_o + R_i + R_e + R_d \qquad (A1.51)$$

where

$$
\begin{aligned}
R_o &= -\sum_{i=1}^{n} \frac{T_{0i}}{T} \log \frac{T_{0i}^2}{T_{0.}T_{.i}} \\
R_i &= -\sum_{j=1}^{n} \sum_{i=1}^{n} \frac{T_{ji}}{T} \log \frac{T_{ji}^2}{T_{j.}T_{.i}} \\
R_e &= -\sum_{j=1}^{n} \frac{T_{j(n+1)}}{T} \log \frac{T_{j(n+1)}^2}{T_{j.}T_{.(n+1)}} \\
R_d &= -\sum_{j=1}^{n} \frac{T_{j(n+2)}}{T} \log \frac{T_{j(n+2)}^2}{T_{j.}T_{.(n+2)}}
\end{aligned}
\qquad (A1.52)
$$

R_t and R_{ts} may be decomposed accordingly.

By selecting only subsets of flows one can assess the contribution of different functional elements to the spatio-temporal organization. In particular it may be of interest to focus on the internal exchanges expressing the organization of the food web or on the imports expressing the organization of primary producers. Functional changes due to spatio-temporal patterns may be expressed either in terms of a reduction in the measures of redundancy or in terms of an increase in the measures of organization.

Appendix Two

Bridging levels of organization in ecological networks

A2.1 INTRODUCTION

Historically, ecology has been divided along the lines of autecology, or the study of individual organisms, and synecology, the study of groups of organisms associated together as a unit. We discussed before that such categorization is often an impediment to investigating the relationships between different levels of organization. Another example is given by the burgeoning field of ecotoxicology that is still mostly concerned with the effects of toxics on particular target organisms, when it should be devoted more to describing how such impacts are manifested as changes in the organizational pattern of the entire community.

It is a cliché that, in ecosystems, each component can affect every other. But as anyone who has studied whole systems is aware, any impact upon a particular component is propagated forcefully to a few others, moderately to many more and almost not at all to the remaining majority. Thus, there have been many efforts to quantify the sensitivity of a given population to an impact occurring somewhere else in the system. Perhaps the most oft-employed technique for this purpose is to perform ecosystems simulation modelling within the framework of which one then may compute the sensitivities of any given population to changes in others.

A problem with all the sensitivity analyses mentioned thus far is that they deal with indirect influence on a bilateral basis, that is, how population x responds to changes in component y. One infers from the aggregate of such pairwise interactions how sensitive the overall system structure might be to changes in each component. It would be preferable to use indices of overall ecosystem structure and investigate how these measures are affected by impacts to each part of the system. The measures of ascendency and organization introduced in Chapter Four may be used for that purpose. Their potential for characterizing system performance and structure was outlined earlier. To study the impacts of system components requires that the measures be decomposed into the contributions of the various flows and/or compartments.

A2.2 DEFINITIONS OF SYSTEM SENSITIVITIES RELATED TO SINGLE FLOWS

The sensitivities of the ascendency, Asc, defined in Eqn (4.29) to minute changes in each component flow are calculated as:

$$\frac{\delta Asc}{\delta T_{ji}} = \log \frac{T_{ji} T}{T_{j.} T_{.i}} \tag{A2.1}$$

Relationship (A2.1) was derived recently by Gannon (1992). In what follows the partial derivative in (A2.1) will be called the structural weight, W_{ji}, of a flow as distinct from its straightforward physical magnitude, T_{ji}. Accordingly, W_{ji} attains high values if

$$\text{(a)} \ \frac{T_{ji}}{T_{j.}} \ \text{or} \ \frac{T_{ji}}{T_{.i}} = 1 \qquad \text{and/or} \qquad \text{(b)} \ \frac{T}{T_{j.}} \ \text{or} \ \frac{T}{T_{.i}} >> 1$$

Each of these conditions describes a type of flow specificity: (a) when only a single exchange leaves or enters a particular compartment, and (b) when the flow in question leaves or enters a compartment that makes but a minor contribution to the total system throughput. The two cases pertain to different levels of organization. Condition (a) refers to the specificity which a single flow exhibits relative to the level of the individual compartment. Condition (b) quantifies the specificity of an individual compartment relative to the level of the whole system. This latter condition implies that respirations and other flows that enter high throughflow compartments along with many other contributors will have small weights.

As an example of the rather low weights assigned to respirations, let us consider the network of carbon flows among the 36 major compartments of the mesohaline Chesapeake ecosystem (Baird and Ulanowicz, 1989). The first two columns of Table A2.1 list the aggregated weights for inputs and outputs, respectively. We note that the mean weight of one unit of carbon being respired is 1.1, whereas that for a unit entering the top predator bluefish is 11.5. Because they usually comprise small fractions of the total system activity, exchanges occurring at higher trophic levels tend to be highly specific and therefore are usually weighted more. Despite their high weightings, the contributions of the upper trophic level flows to the system's ascendency usually are not large, because the flows themselves are disproportionately small. This becomes evident from the definition of ascendency which may be rewritten as:

$$Asc = \sum_{j=0}^{n} \sum_{i=1}^{n+2} T_{ji} \log \frac{T_{ji}T}{T_{j.}T_{.i}} = \sum_{j=0}^{n} \sum_{i=1}^{n+2} T_{ji} W_{ji} \qquad (A2.2)$$

Each flow contributes to the system's ascendency by the product of its physical magnitude and its structural weight.

In analogous manner a sensitivity measure may be derived based on the temporal ascendency, Asc_t, defined in Eqn (4.32). One begins by calculating the sensitivities of Asc_t to minute changes in each component flow:

$$\frac{\delta Asc}{\delta T_{jik}} = \log \frac{T_{jik}^2 T}{T_{j.k}T_{.ik}T_{ji.}} \qquad (A2.3)$$

The partial derivative in (A2.3) will be called the time-resolved structural weight, W_{jik}, of a flow during a time interval t_k as distinct from its straighforward physical magnitude, T_{jik}. Accordingly, W_{jik} attains high values if

$$(a) \ \frac{T_{jik}}{T_{j.k}} \ \text{and/or} \ \frac{T_{jik}}{T_{.ik}} = 1 \ \text{and/or} \ (b) \ \frac{T}{T_{ji.}} >> 1$$

Again each condition describes a type of flow specificity (a) when only a single exchange leaves or enters a particular compartment during a defined interval of time, (b) when the exchange in question is part of a flow that makes only a minor contribution to the total system throughput. Asc_t may as well be reformulated to yield:

$$Asc_t = \sum_{j=0}^{n} \sum_{i=1}^{n+2} \sum_{k=1}^{r} T_{jik} \log \frac{T_{jik}^2 T}{T_{ji.}T_{j.k}T_{.ik}} = \sum_{j=0}^{n} \sum_{i=1}^{n+2} \sum_{k=1}^{r} T_{jik} W_{jik} \qquad (A2.4)$$

The single W_{jik} refer to a limited time interval only. They might prove useful to determine sensitive time periods for a disturbance to occur. However, the prime interest here is in deriving a measure for the sensitivity of the whole system's temporal organization to changes in a pathway T_{ji}. Therefore a weighted temporal mean of the W_{jik} is calculated according to:

$$<W_{jik}> = \sum_{k=1}^{r} \frac{T_{jik}}{T_{ji.}} W_{jik} \qquad (A2.5)$$

A weighted mean as defined in (A2.5) seems to be more appropriate than a simple time average. W_{jik} may be high during time intervals when the flow is extremely small. We may thus obtain a distorted picture of the real contribution of the structural term.

Table A2.1   Results for Chesapeake Bay network

i	NAME	$<W_{*i}>$	$<W_{i*}>$	ΔW_i
0	External input		1.9	
1	Phytoplankton	2.2	1.3	0.9
2	Bacteria in susp. POC	3.4	1.0	2.4
3	Bacteria in sedi. POC	2.7	1.5	1.2
4	Benthic diatoms	2.2	1.9	0.3
5	Free bacteria	4.5	2.7	1.8
6	Heterotr. microflag.	4.5	2.1	2.4
7	Microzooplankton	2.4	1.7	0.7
8	Zooplankton	2.0	2.4	−0.3
9	Ctenophores	3.4	1.7	1.7
10	Seanettle (*Chrysaora quinquecirrha*)	5.4	1.3	4.1
11	Other susp. feeders	2.0	1.7	0.3
12	Clam, soft shell (*Mya arenaria*)	2.0	1.8	0.2
13	Oysters, American (*Crassostrea virginica*)	2.0	2.9	−0.8
14	Other polychaetes	2.7	2.0	0.7
15	*Nereis*	2.7	2.3	0.4
16	*Macoma* spp.	2.7	1.7	1.0
17	Meiofauna	2.9	1.5	1.4
18	Crust. deposit feeders	2.7	1.7	1.0
19	Blue crabs (*Callinectes sapidus*)	4.3	2.0	2.3
20	Fish larvae	5.5	2.3	3.1
21	Alewife (*Alosa pseudoharengus*)	5.5	1.9	3.6
22	Bay anchovy (*Anchoa mitchilli*)	3.7	2.1	1.6
23	Atlantic menhaden (*Brevoortia tyrannus*)	3.5	1.9	1.6
24	Shad, American (*Alosa sapidissima*)	5.5	1.9	3.5
25	Atlantic croaker (*Micropogonius undulatus*)	4.7	1.9	2.8
26	Hogchoker (*Trinectes maculatus*)	4.8	1.8	2.9
27	Spot (*Leiostomus xanthurus*)	4.4	2.0	2.4
28	White perch (*Morone americana*)	5.3	1.9	3.4
29	Sea catfish (*Arius felis*)	4.7	1.9	2.9
30	Bluefish (*Pomatomus saltatrix*)	11.5	1.9	9.6
31	Weakfish (*Cynoscion regalis*)	10.9	2.1	8.7
32	Summer flounder (*Paralichthys dentatus*)	10.7	1.9	8.8
33	Striped bass (*Morone saxatilis*)	10.4	1.8	8.5
34	DOC	1.6	4.5	−2.9
35	Suspended POC	1.2	2.3	−1.1
36	Sediment POC	1.8	2.7	−0.9
37	Exports	5.2		
38	Respiration	1.1		

The definitions of the measures are given in section A2.4. Representations of the network structure can be found in Baird and Ulanowicz (1989).

To summarize, a flow may now be characterized by a variety of measures comprising quantitative and structural attributes as well as combinations of both:

T_{ji} a flow's physical magnitude summed over time
W_{ji} sensitivity of *Asc* to a minute change in T_{ji}
$T_{ji}W_{ji}$ a flow contribution to $Asc \equiv (TW)_{ji}$
$<W_{jik}>$ weighted averaged sensitivity of Asc_t to a minute change in T_{jik}
$\sum_k T_{jik}W_{jik}$ a flow's contribution to $Asc_t \equiv \sum(TW)_{jik}$

Results from model simulations serve to illustrate the potential of the measures derived.

A2.3   APPLICATION TO DATA FROM A SIMULATION MODEL

The model represents a simplified description of a pelagic community consisting of three phytoplankton (P_1, P_2, P_3), and three herbivorous (Z_1, Z_2, Z_3) and one carnivorous (C) zooplankton species. The pairs P_1Z_1, P_2Z_2, and P_3Z_3 are functionally redundant regarding the time-averaged network configuration depicted in Figure A2.1. They all share a common nutrient pool and they all serve as prey to a common predator, C. Whereas the energy passes through the system, the nutrients are recycled. The processes are modelled in quite some detail to reflect the seasonal pattern of lakes in a temperate climate. For details of the mathematical equations and parameter values the reader is referred to Pahl-Wostl (1994).

The algal species differ in their growth optima with respect to light intensity and temperature whereas the zooplankton species are equivalent regarding their environmental requirements. Seasonal changes in the environment in combination with unstable nonlinear population dynamics result in temporal nutrient partitioning. Hence, the functional redundancy of the pairs is partly resolved by their being active during different time periods. Figure A2.2 represents the results from model simulations obtained during the second and third seasonal cycles. One observes spring blooms subsequent to nutrient accumulation during winter and clear-water phases due to zooplankton grazing. The succession of the phytoplankton species is determined mainly by their

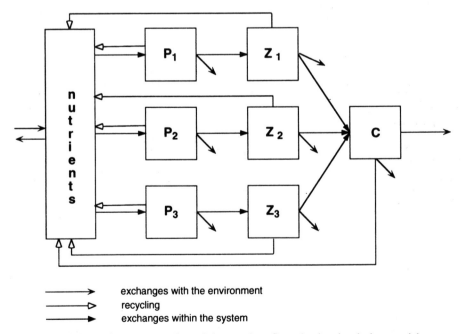

Figure A2.1   Schematic representation of the nutrient flows in the simulation model

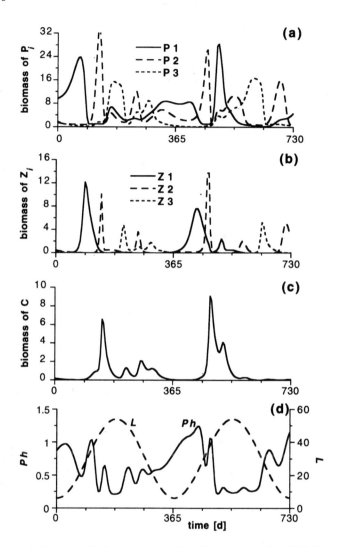

Figure A2.2   Results from model simulations obtained for the second and third period of a seasonal cycle. The time was rescaled to zero. (a) Phytoplankton biomass; (b) zooplankton biomass; (c) carnivore biomass; (d) phosphorus – *Ph* and light intensity – *L* (Reproduced by permission of Elsevier Science from Pahl-Wostl, 1994)

growth optima with respect to light and temperature. However, due to the instability of the internal dynamics, the seasonal pattern varies from one year to the next.

The meaning and the potential of the sensitivity measures derived is illustrated by focusing on the three herbivorous zooplankton species that occupy central positions in the food chains. The flows linking these species to the other members of the food chains are given by the grazing inputs (P_iZ_i) from phytoplankton and by losses due to carnivore predation (Z_iC). Figure A2.3 shows flow characteristics and biomass (B) for Z_1 and Z_3 in percentage of their deviation from the corresponding values obtained for Z_2.

The measures of physical magnitude (energy flow and biomass) suggest Z_2 to be the most important species. Also the combined measures, $(TW)_{ji}$, $\sum_k (TW)_{jik}$, the products of a structural term and a physical magnitude, support such an assumption. The purely structural terms, W_{ji},

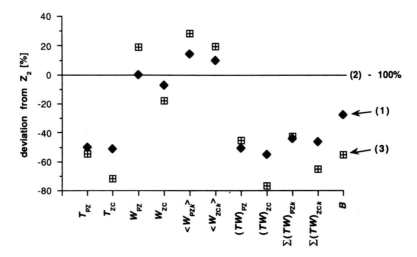

Figure A2.3 Comparison of selected characteristics of the three zooplankton species. The flow characteristics for the grazing inputs (PZ) and for carnivore predation (ZC) and species biomasses obtained for Z_1 ($\equiv 1$) and Z_3 ($\equiv 3$) are depicted as percentages of their deviation from the corresponding values obtained for Z_2 ($\equiv 2$) (Reproduced by permission of Elsevier Science from Pahl-Wostl, 1994)

$<W_{jik}>$, however, show the reverse behaviour. Judged from structural attributes, especially from the weight in the time-resolved network $<W_{jik}>$, Z_3 is the most sensitive and Z_2 the least sensitive species. We may thus derive two contradicting sensitivity scalings:

based on quantitative attributes: $Z_2 > Z_1 > Z_3$
based on qualitative attributes: $Z_3 > Z_1 > Z_2$

To test these predictions, I performed a variety of model simulations with reduced grazing rates of the three zooplankton species. Figure A2.4 shows the changes in some global system properties obtained in different simulation runs where the grazing rate of one zooplankton species at a time was reduced by 50%.

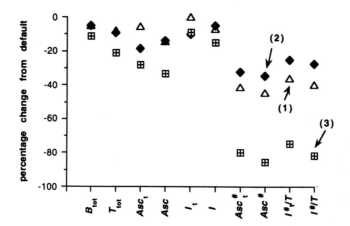

Figure A2.4 Changes in system properties caused by a 50% reduction in the grazing rate of Z_i. i is given in parentheses (Reproduced by permission of Elsevier Science from Pahl-Wostl, 1994)

The superscript # refers to the calculation of the corresponding index over the internal exchanges only. This implies that $j, i = 1, 2 \ldots, n$ instead of $j = 0, 1 \ldots, n$ and $i = 1, 2 \ldots, n + 2$ (cf. section A1.4). The effects observed agree with the predictions based on the structural attributes. Reducing the grazing rate of Z_3 affects the whole system more than reducing the grazing rate of Z_2. The internal measures reflecting the exchanges among the living compartments are affected the most. Similar results were obtained for both less severe and even higher reductions in the grazing rates.

A comparison of the annual variations of the standing stocks obtained for the default parameter set, with those obtained for reduced grazing rates, allows us to follow the changes in system dynamics in more detail. Figures A2.5a and b show the results for a 50% reduction in the grazing rate of Z_2. Figures A2.5c and d show the results for a 50% reduction in the grazing rate of Z_3. The reduction of a grazing rate results in an extension of the temporal fluctuations, which is especially pronounced when the rate of Z_3 is reduced. Carnivore biomass was observed to decline (not shown),

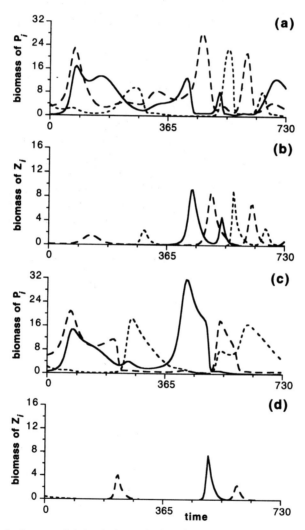

Figure A2.5   Results from model simulations obtained in simulation runs for the second and third period of a seasonal cycle. The time was rescaled to zero. The grazing rate was reduced by 50% (a, b) for Z_2, and (c, d) for Z_3. Figure legends as in Figures A2.2a and b (Reproduced by permission of Elsevier Science from Pahl-Wostl, 1994)

resulting even in the carnivore's extinction in the case of the reduction of Z_3's grazing rate. We may thus conclude that despite its little overall quantitative importance, the presence of Z_3 seems to be essential in critical stages of the system's temporal development. The structural weights proved to be a sensitive indicator for a species' role in total system organization. These simulation results lend support to the hypothesis stated by Carpenter *et al.* (1993). Based on experimental studies of the effects of acidification on lakes, they suggested that pivotal species occupy unique positions in cyclic ecosystem processes.

A2.4   SYSTEM PERFORMANCE AND FLOW STRUCTURE

After having discussed the importance of single flows we may proceed to the level of single compartments. As discussed in more detail in Chapter Four, each compartment may be regarded as being embedded in a structural network environment comprising the flows to and from other compartments (cf. Figure 4.3). The contribution of a particular compartment i to the system's ascendency can thus be described as:

$$Asc_i = \sum_{j=0}^{n} T_{ji}W_{ji} + \sum_{h=1}^{n+2} T_{ih}W_{ih} \text{ hence } Asc = 0.5 \sum_{i=0}^{n+2} Asc_i \tag{A2.6}$$

The various terms in (A2.6) are partitioned into two groupings, the first generated by all flows entering i (its input-environment) and the second by flows leaving i (its output-environment).

As regards the long-term energy balance for a population or a system, there is a basic assymetry between the inputs and the outputs. Whenever inputs exceed outputs, the population in question grows, and vice versa. This amount by which inputs to a compartment exceed its outputs is a measure of its performance and has been termed by Winberg (1960) as its "scope for growth". Genoni and Pahl-Wostl (1991a,b) have suggested that anthropogenic impacts on entire ecosystems could be monitored by tracing changes in its overall energy balance. Using as a straightforward extension of the "scope for growth" concept just the difference between aggregate inputs to and exports from the system, results in a measure that would be insensitive to the structure of the trophic interactions.

One may of introducing the potential for change in trophic structure into the scope for growth is to weight each transfer according to its impact on the whole system (Pahl-Wostl and Ulanowicz, 1993). An appropriate weighting function is assumed to be given by a flow's structural weight, W_{ji}. Accordingly, we may define:

$$fin_i = \sum_{j=0}^{n} \frac{T_{ji}W_{ji}}{Asc_i} \tag{A2.7}$$

$$fout_i = \sum_{h=1}^{n+2} \frac{T_{ih}W_{ih}}{Asc_i} \tag{A2.8}$$

$$\Delta f_i = fin_i - fout_i \tag{A2.9}$$

That is, Δf_i is the difference between the contributions of i's input- and output-environments to the total ascendency, as normalized by i's total contribution. Δf_i should constitute a sensitive measure for characterizing the performance of compartment i within the context of the overall ecosystem network.

In particular, at steady-state the inputs to and outputs from each compartment balance, and the structural differences between the input- and output-environments alone determine the sign and magnitude of Δf_i. These structural differences may be quantified as the numerical difference between the weighted inputs and outputs

$$<W_{\bullet i}> = \frac{\sum\limits_{j=0}^{n} T_{ji}W_{ji}}{\sum\limits_{j=0}^{n} T_{ji}} = \frac{\sum\limits_{j=0}^{n} T_{ji}W_{ji}}{T_{\cdot i}} \tag{A2.10}$$

$$<W_{i^\bullet}> = \frac{\sum\limits_{j=1}^{n+2} T_{ij}W_{ij}}{\sum\limits_{j=1}^{n+2} T_{ij}} = \frac{\sum\limits_{j=1}^{n+2} T_{ij}W_{ij}}{T_{i\bullet}}$$ (A2.11)

$$\Delta W_i = <W_{\bullet i}> - <W_{i\bullet}>$$ (A2.12)

At steady-state $T_{\bullet i} = T_{i\bullet}$, so that under such circumstances (and only then)

$$\Delta f_i = \frac{<W_{\bullet i}> - <W_{i\bullet}>}{<W_{\bullet i}> + <W_{i\bullet}>}$$ (A2.13)

Corresponding measures may be derived from the temporal ascendency, Asc_t:

$$Asc_{ti} = \sum_{k=1}^{r}\left(\sum_{j=0}^{n} T_{jik}W_{jik} + \sum_{h=1}^{n+2} T_{ihk}W_{ihk}\right) \text{ hence } Asc_t = 0.5\sum_{i=0}^{n+2} Asc_{ti}$$ (A2.14)

and

$$fin_{ti} = \sum_{k=1}^{r}\sum_{j=0}^{n}\frac{T_{jik}W_{jik}}{Asc_{ti}}$$ (A2.15)

$$fout_{ti} = \sum_{k=1}^{r}\sum_{h=1}^{n+2}\frac{T_{ihk}W_{ihk}}{Asc_{ti}}$$ (A2.16)

$$\Delta f_{ti} = fin_{ti} - fout_{ti}$$ (A2.17)

$$<W_{\bullet i}> = \frac{\sum\limits_{j=0}^{n}\sum\limits_{k=1}^{r} T_{jik}W_{jik}}{\sum\limits_{j=0}^{n}\sum\limits_{k=1}^{r} T_{jik}} = \frac{\sum\limits_{j=0}^{n}\sum\limits_{k=1}^{r} T_{jik}W_{jik}}{T_{\bullet i\bullet}}$$ (A2.18)

$$<W_{i^\bullet}> = \frac{\sum\limits_{j=1}^{n+2}\sum\limits_{k=1}^{r} T_{ijk}W_{ijk}}{\sum\limits_{j=1}^{n+2}\sum\limits_{k=1}^{r} T_{ijk}} = \frac{\sum\limits_{j=1}^{n+2}\sum\limits_{k=1}^{r} T_{ijk}W_{ijk}}{T_{i\bullet\bullet}}$$ (A2.19)

$$\Delta W_{ti} = <W_{\bullet i\bullet}> - <W_{i\bullet\bullet}>$$ (A2.20)

To summarize, a compartment may be characterized by a variety of measures:

Ain_i	contribution of the sum of a compartment's inputs to Asc
$Aout_i$	contribution of the sum of a compartment's outputs to Asc
Δf_i	$(Ain_i - Aout_i)/(Ain_i + Aout_i)$
Ain_{ti}	contribution of the sum of a compartment's inputs to Asc_t
$Aout_{ti}$	contribution of the sum of a compartment's outputs to Asc_t
Δf_{ti}	$(Ain_{ti} - Aout_{ti})/(Ain_{ti} + Aout_{ti})$
$<W_{\bullet i}>$	mean weight of time-averaged inputs
$<W_{i\bullet}>$	mean weight of time-averaged outputs
ΔW_i	$<W_{\bullet i}> - <W_{i\bullet}>$
$<W_{\bullet i\bullet}>$	mean weight of time-resolved inputs
$<W_{i\bullet\bullet}>$	mean weight of time-resolved outputs
ΔW_{ti}	$<W_{\bullet i\bullet}> - <W_{i\bullet\bullet}>$

A2.5 APPLICATION TO FLOW DATA FROM AQUATIC ECOSYSTEMS

The utilities of the measures defined in (A2.7) through (A2.20) are best demonstrated by way of some examples. One of the simplest input–output configurations is a linear food chain, like the one illustrated in Figure A2.6. Here each compartment receives but a single input from the one preceding it and transfers a fraction ε of this input to the next level. If the initial input to the chain is A, then the difference between the weighted inputs and outputs for compartment 2 become

$$\Delta W_2 = \frac{\varepsilon A}{\varepsilon A} \log \frac{\varepsilon A T}{A \varepsilon A} - \frac{\varepsilon^2 A}{\varepsilon A} \log \frac{\varepsilon^2 A T}{\varepsilon A \varepsilon^2 A} - \frac{(1-\varepsilon)\varepsilon A}{\varepsilon A} \log \frac{(1-\varepsilon)\varepsilon A T}{\varepsilon A T_{.d}}$$

$$= \log \frac{T}{A} \quad - \varepsilon \log \frac{T}{\varepsilon A} \quad - (1-\varepsilon) \log \frac{(1-\varepsilon)T}{T_{.d}}$$

(A2.21)

where $T_{.d}$ represents the sum of all dissipative flows out of the system. ΔW_2 is zero only when $\varepsilon = 1$, a thermodynamic impossibility. Thus, we see that dissipation serves to depreciate the mean weight of all outputs, making $\Delta W > 0$ and $\Delta f > 0$ in most cases where these differences are calculated across living (i.e. respiring) components.

We note in Table A2.1 that ΔW and hence Δf are indeed positive for 34 of the 36 living compartments (cf. Eqn (A2.13)). The two exceptions, however, demonstrate that, although thermodynamics imparts a positive bias to ΔW and Δf, it does not entirely determine the sign of these quantities. As intended, the structural configuration strongly influences their actual values.

The two extreme cases portrayed in Figure A2.7 may serve to illustrate this last point. In case 1 a single input is distributed over many outputs, while in case 2 many inputs are focused into a single (nonrespiratory) output. (Thermodynamically speaking, case 2 is impossible for energy flows and can be only approximated by real components.) For notational convenience α is defined as the fraction which any input comprises of the total output from the compartment in which it originates. (Without loss of generality, this may be assumed equal for all input flows in case 2.) The parameter β is the fraction that an output from the compartment in question comprises of the total input into its recipient. (Again, no generality is lost by assuming β to be equal for all outputs in case 1.) By definition, $0 \le \alpha \le 1$ and $0 \le \beta \le 1$.

Substituting into the definition for ΔW_i, we see, after simplification, that in case 1 $\Delta W_x = \log(n\alpha/\beta)$. Similarly, for case 2 $\Delta W_x = \log(\alpha/n\beta)$. Because $n > 1$, and in most cases $\alpha \cong \beta$, we see that ΔW_x tends to be positive in case 1 and negative in case 2. The reasons for exceptions to this scheme also become apparent. In case 1, ΔW_x would be negative only in the (very rare) circumstances that all the recipients were highly specialized and compartment x were a very minor predator of its prey. Conversely, ΔW_x in case 2 could become positive only in the unlikely event that it were the dominant predator of all its various prey species and a minor element in the diet of its predator.

Returning to the elements of Table A2.1, we note that the two living compartments with negative ΔW_i (and Δf_i) are the mesozooplankton and the oysters, respectively. The remaining negative values belong to abiotic compartments (dissolved and particulate carbon) that exhibit almost no respiration, receive input from many sources and are consumed largely by populations that specialize as detritivores. Hence, they strongly resemble case 2, and all have negative values of ΔW_i. Both living components are filter feeders, having relatively broad diets that include items shared by many other filter feeders (i.e. α is small) and contributing to comparatively few predators, which in turn have

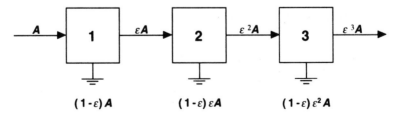

Figure A2.6   A linear food chain representing a simple structural element of a trophic network. An external input of magnitude, A enters compartment 1. Each compartment transfers a fraction ε ($\varepsilon < 1$) to the next higher level and loses a fraction $(1 - \varepsilon)$ in respiration

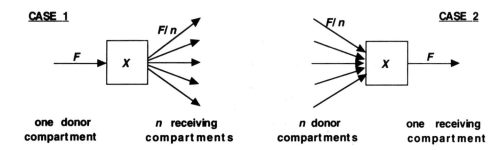

Figure A2.7   Two hypothetical extremes of compartmental flow structure, (a) dispersing a single input among several outputs, and (b) channelling many inputs into but a single outflow (Reproduced by permission of Elsevier Science from Pahl-Wostl and Ulanowicz, 1993)

relatively narrower diets. Hence, the resemblance of these filter feeders to case 2 is strong enough to overcome the positive bias imparted by their respiratory activities. It is also interesting to note that the zooplankton compartment is the main collector to join the two pathways of the "traditional" food chain on the one hand and of the microbial loop on the other hand. In the "traditional" food chain, the energy entering the system via algal primary production is channelled up to fish, whereas in the microbial loop, the energy in the detrital pool is recycled via bacterial production.

A similar pattern can be distinguished for the network of Lake Constance where, contrasting with the Chesapeake Bay, the plankton food web is resolved in more detail whereas the fish are highly aggregated (Straile and Gaedke, submitted). The diagram of the network depicted in Figure A2.8a comprises all feeding relationships that are present over the annual seasonal cycle. The respiratory losses (all living compartments) and the flows to the detrital pool (all living compartments except bacteria) were omitted to avoid further complication of the already complex network. The single compartments with their input- and output-environments are depicted in Figure A2.8b. The numbers at the arrows denote the percentage a flow contributes to the total compartmental input or output, respectively. The values inside the boxes denote a compartment's total annual throughput. The behaviour of the ΔW_i shown in Figure A2.9a as a function of i reflects the changes in the structural embeddings of the single compartments depicted in Figure A2.8b. The character of acting as a collector increases progressively from phytoplankton (1) and bacteria (2) up to the herbivorous crustaceans (6). This is reflected in the decrease of the ΔW_i. The ΔW_{ti} obtained by resolving the total observation period from March to October into 10 time intervals show a similar trend as the ΔW_i with the exception of compartment 3, the heterotrophic nanoflagelates, HNF. The effects of temporal patterns may be traced in more detail by comparing the changes in the structural weight caused by the resolution of the temporal pattern between outputs and the inputs. These may be quantified for a compartment i by the terms, $<W_{i\cdot\cdot}> - <W_{i\cdot}>$ and $<W_{\cdot i\cdot}> - <W_{\cdot i}>$, for the output- and input-environments, respectively. The results obtained for the compartments of the Lake Constance network are represented in Figure A2.9b. Again the HNF show the most pronounced effects. In addition we note that the effect derives from changes in the input-environment. This effect can be attributed to the HNF switching from feeding on phytoplankton in the early phase of the seasonal cycle to feeding exclusively on bacteria at later stages of the seasonal cycle.

It should be mentioned that seasonal changes in the structural indices seem to reflect the switch in the structure of the network's functional configuration. This effect is especially pronounced when the activity of the microbial loop constitutes the major source of energy flow during some phases of the seasonal cycle. However, due to the weakness of the current data base it is not yet possible to draw general conclusions at this stage of our investigations.

Compartments with a positive ΔW_i include the populations at the top of the food chain, e.g., bluefish, summer flounder, weakfish, striped bass and the sea nettle in the network of the Chesapeake Bay. Their inputs are weighted strongly because they tend to be specialists and to be the dominant predators on most of their prey. At the same time, their significant respirations per unit biomass do not weight their outputs proportionately, because aggregate respiration from these

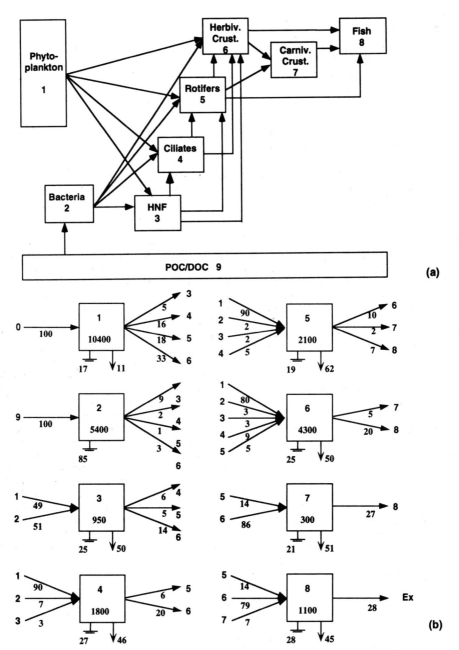

Figure A2.8   Network of Lake Constance in seasonal average, (a) network of all energy flows connecting system compartments, (b) single compartments with their input- and output-environments. The flows at the bottom of a compartment refer to respiration and losses to the DOC pool, respectively

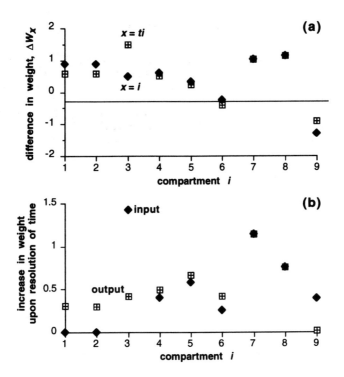

Figure A2.9 Structural indices derived for the network of Lake Constance. (a) ΔW_i and ΔW_{ti}, (b) the difference in the mean weight in the time-resolved and the time-averaged network for outputs: $<W_i^{**}> - <W_i^*>$ and imputs: $<W_i^*> - <W_i^*>$, as a function of i, the compartment's index

species is but a small fraction of total community respiration. We note that it is these compartments that are most likely to exert top-down control on the pyramids of resources that support them.

At the other extreme in ΔW_i are those components that serve to focus and concentrate material and energy before it can flow elsewhere in the system. Such components include the suspended POC, sediment POC, DOC, oysters and mesozooplankton, all with negative ΔW_i. A failure of these system elements to carry out their transfer functions would negatively affect those populations higher up the trophic chain. As Figure A2.7b suggests, case 2 species are capable of acting as a "bottleneck" in the pattern of material and energy flows, so that the system components with negative $\Delta W_{(t)i}$ are those most likely to exert bottom-up control.

These examples should make it clear that the sign of the measure of a component's structural environment does not convey its importance to the functioning of the overall system, but rather indicates how it may contribute. The $\Delta f_{(t)i}$ and the $\Delta W_{(t)i}$ themselves represent analytical tools for the investigation of structural characteristics of a given network configuration indicating:

- the functional role of a compartment (and its seasonal variation);
- the diversity of functional elements within a system;
- the functional importance of seasonal patterns.

Focusing on the presumably anthropogenically induced changes in a network's flow structure over time our hypothesis is that a drop in Δf_i (or ΔW_i) should signify a negative impact of that species on overall system performance. Whenever a large, positive Δf_i decreases, the compartment in question probably is exerting less top-down influence on the trophic web than it previously had been. In corresponding fashion, whenever a component with a large, negative Δf_i becomes even more negative, it is somehow impeding the circulation of material and energy through the remainder of the system.

A2.6 CONCLUSION

Analyses of the type presented in this chapter provide a much-needed quantitative bridge between studies of population-level phenomena and the observations on the behaviour of entire ecosystems. After further testing and application, the measures derived should develop into a fundamental tool useful in synecology.

In analogy to the measures derived for the time-resolved network, corresponding measures can be derived for networks resolved along space or time and space. We thus have a versatile tool to investigate the relative importance between spatial and temporal patterns and their mutual dependence.

References

Abrams, P. and Roth, J. (1994). The effects of enrichment of three-species food chains with nonlinear functional response. *Ecology*, **75**: 1118–1130.

Abramson, N. (1963). *Information Theory and Coding*. McGraw-Hill, New York.

Ahrens, C. and Peters, R. (1991). Patterns and limitations in limnoplankton size spectra. *Can. J. Fish. Aquat. Sci.*, **48**: 1067–1078.

Allen, J.C., Schaffer, W.M. and Rosko, D. (1993). Chaos and extinction in ecological populations. *Nature*, **364**: 229–232.

Allen, T.F. and Hoekstra, T. (1989). Further comment on Carney's article. *Functional Ecology*, **3**: 642–643.

Allen, T.F. and Hoekstra, T.W. (1992). *Toward a Unified Ecology*. Complexity in Ecological Systems Series. Columbia University Press, New York.

Allen, T.F. and Starr, T.B. (1982). *Hierarchy*. University of Chicago Press, Chicago.

Anderson, D.M. and Webb, R.S. (1994). Ice-age tropics revisited. *Nature*, **367**: 23–24.

Anderson, N.J. (1993). Natural versus anthropogenic change in lakes: the role of the sediment record. *Trends Ecol. Evol.*, **8**: 356–361.

Anon. (1989). Chaos and Fraktale. *Spektrum der Wissenschaft*. Special Issue.

Armstrong, R. and McGhee, R. (1980). Competitive exclusion. *Am. Nat.*, **115**: 151–170.

Arthur, B. (1990). Positive feedbacks in the economy. *Scientific American*, February: 80–85.

Arthur, B., Ermoliev, Y. and Kaniovski, Y. (1987). Path-dependent processes and the emergence of macro-structure. *European J. Oper. Res.*, **30**: 294–303.

Auchmuty, J. and Nicolis, G. (1976). Bifurcation analysis of reaction–diffusion equations – III. Chemical oscillations. *Bull. Math. Biol.*, **38**: 325–350.

Ayala, F. (1988). "Can 'progress' be defined as a biological concept?". In M.H. Nitecki, Ed., *Evolutionary Progress*. University of Chicago Press, Chicago, pp. 75–96.

Baird, D. and Ulanowicz, R.E. (1989). The seasonal dynamics of the Chesapeake Bay ecosystem. *Ecol. Monogr.*, **59**: 329–364.

Barber, R. and Chávez, F. (1983). Biological consequences of the El Niño. *Science*, **222**: 1203–1210.

Barber, R.T. and Chávez, F.P. (1986). Ocean variability in relation to living resources during the 1982–83 El Niño. *Nature*, **319**: 279–285.

Beer, J., Baumgartner, S., Dittrich-Hannen, B. *et al.* (1995). "Solar variability traced by cosmogenic isotopes". In J. Pap, C. Fröhlich, H. Hudson and S. Solanki, Eds, *The Sun as Variable Star: Solar and Stellar Irradiance Variations*. Cambridge University Press, Cambridge, in press.

Blackburn, T. and Gaston, K. (1994). The distribution of body sizes of the world's bird species. *Oikos*, **70**: 127–130.

Bollens, S., Frost, B., Schwaninger, H., Davis, C., Way, K. and Landsteiner, M. (1992). Seasonal plankton cycles in a temperate fjord and comments on the match–mismatch hypothesis. *Journal of Plankton Research*, **14**: 1279–1305.

Braitenberg, V. and Schütz, A. (1989). Cortex: hohe Ordnung oder größtmögliches Durcheinander. *Spektrum der Wissenschaft*, May: 74–86.

Brooks, D.R. and Wiley, E. O. (1988). *Evolution as Entropy*. University of Chicago Press, Chicago.

Brown, J. and Maurer, B. (1989). Macroecology: the division of food and space among species on continents. *Science*, **243**: 1145–1150.

Brown, J. and Nicoletto, P. (1991). Spatial scaling of species composition: body masses of North American land mammals. *Am. Nat.*, **138**: 1478–1512.

Calder, W. (1984). *Size, Function and Life History*. Harvard University Press, Cambridge, MA.

Carney, H.J. (1989). On competition and the integration of population, community and ecosystem studies. *Functional Ecology*, **3**: 637–641.

Carney, H.J. (1990). On competition and the integration of population, community and ecosystem studies. II. Replies to Fenchel, Allen and Hoekstra. *Functional Ecology*, **4**: 127–133.

Carpenter, S. and Kitchell, J. (1992). Trophic cascade and biomanipulation: Interface of research and management – A reply to the comment of DeMelo et al. *Limnol. Oceanogr.*, **37**: 208–213.

Carpenter, S.R. and Kitchell, J.F. (1993). *The Trophic Cascade in Lakes*. Cambridge Studies in Ecology, Cambridge University Press, Cambridge.

Carpenter, S., Frost, T., Kitchell, J. and Kratz, T. (1993). "Species dynamics and global environmental change: a perspective from ecosystem experiments". In P. Kareiva, J. Kingsolver and R. Huey, Eds, *Biotic Interactions and Global Change*. Sinauer Assoc. Inc, Sunderland, pp. 267–279.

Cartwright, N. (1983). *How the Laws of Physics Lie*. Clarendon Press, Oxford.

Chesson, P. and Case, T. (1986). "Nonequilibrium theories: chance, variability, history, and coexistence". In J. Diamond and T. Case, Eds, *Community Ecology*. Harper & Row, New York, pp. 229–239.

Clements, F. (1936). Nature and structure of the climax. *J. Ecol.*, **24**: 252–284.

Cody, M. (1986). "Structural niches in plant communities". In J. Diamond and T. Case, Eds, *Community Ecology*. Harper & Row, New York, pp. 381–405.

Cohen, J., Briand, F. and Newman, C. (1990). *Community Food Webs: Data and Theory*. Springer, New York.

Connell, J. (1978). Diversity in tropical rain forest and coral reefs. *Science*, **199**: 1302–1310.

Cousins, S.H. (1985). The trophic continuum in marine ecosystems: structure and equations for a predictive model. *Can. Bull, Fish. Aquat. Sci.*, **213**: 76–93.

Cousins, S. (1987). The decline of the trophic level concept. *Trends Ecol. Evol.*, **2**: 312–316.

Cousins, S.H. (1990). Countable ecosystems deriving from a new food web entity. *Oikos*, **57**: 270–275.

Covey, C. (1991). Chaos in ocean heat transport. *Nature*, **353**: 796–797.

Currie, D.J. (1993). What shape is the relationship between body size and population density? *Oikos*, **66**: 353–358.

Cushing, D.H. (1990). Plankton production and year-class strength in fish populations: an update of the match/mismatch hypothesis. *Adv. Mar. Biol.*, **26**: 249–293.

Cyr, H. and Pace, M. (1993). Allometric theory: Extrapolations from individuals to communities. *Ecology*, **74**: 1234–1245.

Daly, H.E. (1991). *Steady-State-Economy*. Island Press, Washington.

Dayton, P.K. (1980). Citation classic. *Current Contents*, **11**: 18.

DeAngelis, D.L. and Waterhouse, J.C. (1987). Equilibrium and nonequilibrium concepts in ecological models. *Ecological Monographs*, **57**: 1–24.

DeAngelis, D., Post, W. and Travis, C. (1986). *Positive Feedback in Natural Systems*. Springer, New York.

DeMelo, R., France, R. and McQueen, D. (1992). Biomanipulation: Hit or myth? *Limnol. Oceanogr.*, **37**: 192–207.

Diamond, J. and Case, T. (1986). *Community Ecology*. Harper & Row, New York.

Douglas, M. and Lake, P.S. (1994). Species richness of stream stones: an investigation of the mechanisms generating the species–area relationship. *Oikos*, **69**: 387–396.

Eccles, R.G., Nohria, N. and Berkley, J.D. (1992). *Beyond the Hype. Rediscovering the Essence of Management*. Harvard University Press, Cambridge, MA.

Egerton, F.N. (1973). Changing concepts in the balance of nature. *Quart. Rev. Biol.*, **48**: 322–350.

Egerton, F.N. (1976). "Ecological studies and observations before 1900". In B.J. Taylor and T.J. White, Eds, *Issues and Ideas in America*. University of Oklahoma Press, Norman, pp. 311–351.

Eilingsfeld, H. (1989). *Der sanft Wahn: Ökologismus total*. Südwestdeutsche Verlagsanstalt, Mannheim.

Elton, C.S. (1927). *Animal Ecology*. Sidgwick and Jackson, London.

Elton, C.S. (1958). *The Ecology of Invasions by Animals and Plants*. Chapman and Hall, London.

Engelberg, J. and Boyarsky, L. (1979). The noncybernetic nature of ecosystems. *Am. Nat.*, **114**: 317–324.

Fei-Fei, J., Neelin, D. and Ghil, M. (1994). El Niño on the Devil's Staircase: Annual subharmonic steps to chaos. *Science*, **264**: 70–72.

Fenchel, T. (1993). There are more small than large species? *Oikos*, **68**: 375–378.

Feyerabend, P. (1978). *Against Method*. Verso, Thetford.

Feyerabend, P. (1989). *Irrwege der Vernunft*. Suhrkamp, Frankfurt.

Fitter, A.H. (1986). Spatial and temporal patterns of root activity in species-rich alluvial grassland. *Oecologia*, **69**: 594–599.

Freeman, W. (1991). Physiologie und Simulation der Geruchswahrnehmung. *Spektrum der Wissenschaft*, April: 60–69.

Friis-Christensen, E. and Lassen, K. (1991). Length of the solar cycle: an indicator of solar activity closely associated with climate. *Science*, **254**: 698–700.

Funtowicz, S. and Ravetz, J. (1994). The worth of a songbird: ecological economics as a post-normal science. *Ecological Economics*, **10**: 197–207.

Gaedke, U. (1992a). Identifying ecosystem properties: a case study using plankton biomass size distributions. *Ecol. Model.*, **63**: 277–298.

Gaedke, U. (1992b). The size distribution of plankton biomass in a large lake and its seasonal variability. *Limnol. Oceanogr.*, **37**: 1202–1220.

Gaedke, U., Straile, D. and Pahl-Wostl, C. (1995). "Trophic structure and carbon flow dynamics in the pelagic community of a large lake". In G. Polis and K. Winemiller, Eds, *Food Webs: Integration of Pattern and Dynamics*. Chapman and Hall, London, in press.

Gannon, A. (1992). The derivation and application of influence coefficients of the flows in a weighted network. Masters Thesis, University of Maryland, Maryland.

Gardner, M. and Ashby, W. (1970). Connectance of large dynamical (cybernetic) systems: critical values for stability. *Nature*, **228**: 784.

Gaston, K.J., Blackburn, M. and Lawton, J.H. (1993). Comparing animals and automobiles: a vehicle for understanding body size and abundance relationships in assemblages? *Oikos*, **66**: 172–179.

Geller, W. (1986). Diurnal vertical migration of zooplankton in a temperate great lake (L. Constance): A starvation avoidance mechanism? *Arch. Hydrobiol. Suppl.*, **74**: 1–60.

Genoni, G. and Pahl-Wostl, C. (1991a). The measurement of scope for change in ascendency for short-term assessment of community stress. *Can. J. Fish. Aqu. Sci.*, **48**: 968–974.

Genoni, G. and Pahl-Wostl, C. (1991b). "Scope for change in ascendency, a new concept in community ecotoxicology for environmental management". In O. Ravera, Ed., *Terrestrial and Aquatic Ecosystems, Perturbation and Recovery*. Ellis Horwood, Chichester, UK, pp. 69–75.

George, D.G., Hewitt, D.P., Lund, J.W. and Smyly, W.J. (1990). The relative effects of enrichment and climate change on the long-term dynamics of *Daphnia* in Esthwaite Water, Cumbria. *Freshwater Biology*, **23**: 55–70.

Glass, L. and Mackey, M. (1988). *From Clocks to Chaos*. Princeton University Press, Princeton.

Goldberger, A. (1990). *Rythmes et chaos*. Masson, Paris.

Golley, F.B. (1994). *A History of the Ecosystem Concept in Ecology*. Yale University Press, New Haven.

Goodman, D. (1975). The theory of diversity–stability relationships in ecology. *Quarterly Review of Biology*, **50**: 237–266.

Goodwin, B.C. (1990). Structuralism in biology. *Sci. Progress*, **74**: 227–244.

Goodwin, B., Sibatini, A. and Webster, G. (1989). *Dynamic Structures in Biology*. Edinburgh University Press, Edinburgh.

Gould, S.J. (1988). "On replacing the idea of progress with an operational notion of directionality". In M.H. Nitecki, Ed., *Evolutionary Progress*. University of Chicago Press, Chicago, pp. 318–338.

Grene, M. (1985). "Perception, integration and the sciences". In D. Depew and B. Weber, Eds, *Evolution at a Crossroads: The New Biology and the New Philosophy of Science*. MIT Press, Cambridge, MA, pp. 1–20.

Grinnell, J. (1924). Geography and evolution. *Ecology*, **5**: 225–229.

Guckenheimer, J. and Holmes, P. (1983). *Nonlinear Oscillations, Dynamical Systems and Bifurcations of Vector Fields*. Springer, New York.

Haken, H. (1983). *Synergetics, an Introduction*. Springer, New York.

Haken, H. and Stadler, M. (1990). *Synergetics of Cognition*. Springer, Heidelberg.

Hansen, A., Spies, T., Swanson, F. and Ohmann, J. (1991). Conserving biodiversity in managed forests. *Bioscience*, **41**: 382–392.

Harding, S. (1991). *Whose Science? Whose Knowledge?* Open University Press, Milton Keynes.

Hastings, A. (1988). Food web theory and stability. *Ecology*, **69**: 1665–1668.

Hastings, A. (1993). Complex interactions between dispersal and dynamics: lessons from coupled logistic equations. *Ecology*, **74**: 1362–1372.

Hastings, A. and Powell, T. (1991). Chaos in a three-species food chain. *Ecology*, **72**: 896–903.

Hendrix, P.F. *et al.* (1986). Detritus based food webs in conventional and no-tillage agroecosystems. *BioScience*, **36**: 374–380.

Herschkowitz-Kaufman, M. (1975). Bifurcation analysis of nonlinear reaction–diffusion equations – II. Steady state solutions and comparisons with numerical simulations. *Bull. Math. Biol.*, **37**: 589–636.

Hirata, H. and Ulanowicz, R.E. (1984). Information theoretical analysis of ecological networks. *Int. J. Systems Sci.*, **15**: 261–270.

Hobbie, S. (1992). Effect of plant species on nutrient cycling. *Trends Ecol. Evol.*, **7**: 336–339.

Hogg, T. (1990). Control of distributed systems. Progress report, Xerox Research Center, Palo Alto.

Hogg, T., Huberman, B. and McGlade, J. (1989). The stability of ecosystems. *Proc. R. Soc. Lond. B*, **237**: 43–51.

Hollick, M. (1993). Self-organizing systems and environmental management. *Environmental Management*, **17**: 621–628.

Holling, C.S. (1978). *Adaptive Environmental Assessment and Management*. Wiley-Interscience, New York.

Holmes, E.E., Lewis, M.A., Banks, J.E. and Veit, R.R. (1994). Partial differential equations in ecology: spatial interactions and population dynamics. *Ecology*, **75**: 17–29.

Hosper, H. and Meijer, M.-L. (1993). Biomanipulation, will it work for your lake? A simple test for the assessment of chances for clear water, following drastic fish-stock reduction in shallow, eutrophic lakes. *Ecological Engineering*, **2**: 63–72.

Huberman, B.A. (1988). *The Ecology of Computation*. North-Holland, Amsterdam.

Huberman, B.A. (1990). The performance of cooperative processes. *Physica D*, **42**: 38–47.

Huberman, B.A. and Hogg, T. (1988). "The behaviour of computational ecologies". In B.A. Huberman, Ed., *The Ecology of Computation*. North-Holland, Amsterdam, pp. 71–115.

Hurlbert, S. (1978). The measurement of niche overlap and some relatives. *Ecology*, **59**: 67–77.

Huston, M., DeAngelis, D. and Post, W. (1988). New computer models unify ecological theory. *BioScience*, **38**: 682–691.

Hutchinson, G.E. (1957). Concluding remarks. *Cold Spring Harbor Symp. Quant. Biol.*, **22**: 415–427.

Hutchinson, G.E. (1978). *An Introduction to Population Ecology*. Yale University Press, New Haven.

Jaeger, C. (1990). Innovative milieus and environmental awareness. *Sociologia Internationalis*, **28**: 205–216.

Jaeger, C. (1994). *Taming the Dragon: Transforming Economic Institutions in the Face of Global Change*. Gordon and Breach, Philadelphia.

Jones, C.G., Lawton, J.H. and Shachak, M. (1994). Organisms as ecosystem engineers. *Oikos*, **69**: 373–386.

Jørgensen, S.E. (1992). *Integration of Ecosystem Theories: A Pattern*. Ecology & Environment, Kluwer, Dordrecht, The Netherlands.

Kareiva, P. (1990). Population dynamics in spatially complex environments: theory and data. *Phil. Trans. Soc. Lond. B.*, **330**: 175–190.

Kauffman, S.A. (1993). *The Origins of Order*. Oxford University Press, New York.

Kelly, P.M. and Wigley, T. (1992). Solar cycle length, greenhouse forcing and global climate. *Nature*, **360**: 328–330.

Kephart, J., Hogg, T. and Huberman, B.A. (1990). Collective behaviour of predictive agents. *Physica D*, **42**: 48–65.

Kerr, R. (1992). Unmasking a shifty climate system. *Science*, **255**: 1507–1509.

Kerr, R. (1993). El Niño metamorphosis throws forecasters. *Science*, **262**: 656–657.

Kingsland, S.E. (1985). *Modelling Nature*. University of Chicago Press, Chicago.

Kitchell, J. (1992). *Food Web Management: A Case Study of Lake Mendota*. Springer Series on Environmental Management, Springer, New York.

Knox, J.C. (1993). Large increases in flood magnitude in response to modest changes in climate. *Nature*, **361**: 430–432.

Körner, C. (1993). "Scaling from species to vegetation: The usefulness of functional groups". In E.-D. Schulze and H.A. Mooney, Eds, *Biodiversity and Ecosystem Function*. Springer, Heidelberg, pp. 117–142.

Körner, C. and Arnone, J. (1992). Response to elevated carbon dioxide in artificial tropical ecosystems. *Science*, **257**: 1672–1675.

Küppers, M. (1989). Ecological significance of above-ground architectural patterns in woody plants: a question of cost–benefit relationships. *Trends Ecol. Evol.*, **4**: 375–378.

Lawton, J.H. (1989). What is the relationship between population density and body size in animals? *Oikos*, **55**: 429–433.

Lawton, J. (1990). Species richness and population dynamics of animal assemblages. Patterns in body size: abundance space. *Phil. Trans. R. Soc. Lond. B.*, **330**: 283–291.

Lawton, J. and Jones, C.G. (1993). Linking species and ecosystem perspectives. *Trends Ecol. Evol.*, **8**: 311–313.

Lawton, J. and Warren, P. (1988). Static and dynamic explanations for patterns in food webs. *Trends Ecol. Evol.*, **3**: 242–245.

Legendre, L. and Demers, S. (1984). Towards dynamic biological oceanography and limnology. *Can. J. Fish. Aquat. Sci.*, **41**: 2–19.

Lehman, J.T. (1986). The goal of understanding in limnology. *Limnol. Oceanogr.*, **31**: 1160–1166.

Leo, G.D., DelFuria, L. and Gatto, M. (1993). The interaction between soil acidity and forest dynamics: a simple model exhibiting catastrophic behavior. *Theor. Popul. Biol.*, **43**: 31–51.

Levasseur, M., Therriault, J. and Legendre, L. (1984). Hierarchical control of phytoplankton succession by physical factors. *Mar. Ecol. Prog. Ser.*, **19**: 211–222.

Levin, S. (1986). "Random walk models of movement and their implications". In T.G. Hallem and S.A. Levin, Eds, *Mathematical Ecology*. Springer, Berlin, pp. 149–155.

Levin, S. (1989). "Challenges in the development of a theory of community and ecosystem structure and function". In J. Roughgarden, R.M. May and S.A. Levin, Eds, *Perspectives in Ecological Theory*. Princeton University Press, Princeton, pp. 242–255.

Levin, S. (1992). The problem of pattern and scale in ecology. *Ecology*, **73**: 1943–1967.

Levin, S., Powell, T. and Steele, J. (1993). *Patch Dynamics*. Lecture Notes in Biomathematics 96, Springer, New York.

Lewin, R. (1983). Santa Rosalia was a goat. *Science*, **221**: 636–639.

Lidicker, W. (1988). The synergistic effects of reductionist and holistic approaches in animal ecology. *Oikos*, **53**: 278–281.

Loehle, C. (1988). Philosophical tools: potential contributions to ecology. *Oikos*, **51**: 97–104.

Loehle, C. and Pechmann, H. (1988). Evolution: the missing ingredient in systems ecology. *Am. Nat.*, **132**: 884–899.

Lorenz, E. (1963). Deterministic non-periodic flows. *J. Atmos. Sci.*, **20**: 130–141.

Lovelock, J. (1990). *The Ages of Gaia*. Bantam Publishing Group, New York.

Lubchenco, J., Olson, A., Brubaker, L. *et al.* (1991). The sustainable biosphere initiative: an ecological research agenda. *Ecology*, **72**: 371–412.

Ludwig, D., Hilborn, R. and Walters, C. (1993). Uncertainty, resource exploitation, and conservation: Lessons from history. *Science*, **260**: 17–18.

Luenberger, D. (1979). *Introduction to Dynamic Systems*. Wiley, New York.

MacArthur, R. (1955). Fluctuations of animal populations and a measure of community stability. *Ecology*, **36**: 533–536.

Mandelbrot, B. (1985). *The Fractal Geometry of Nature*. Freeman, New York.

Mangel, M. and Tier, C. (1994). Four facts every conservation biologist should know about persistence. *Ecology*, **75**: 607–614.

Martinez, N. (1991). Artifacts or attributes? Effects of resolution on the Little Rock Lake food web. *Ecol. Monogr.*, **61**: 367–392.

Matson, P. and Hunter, D. (1992). The relative contributions of top-down and bottom-up forces in population and community ecology. *Ecology*, **73**: 724–765.

May, R. (1973). *Stability and Complexity in Model Ecosystems*. Princeton University Press, Princeton.

May, R. (1974). On the theory of niche overlap. *Theor. Popul. Biol.*, **5**: 297–332.

May, R. (1976). Simple mathematical models with very complicated dynamics. *Nature*, **261**: 459–467.

May, R. (1981). The role of theory in ecology. *American Zoologist*, **21**: 903–910.

May, R.M. (1986). The search for patterns in the balance of nature: advances and retreats. *Ecology*, **67**: 1115–1126.

Maynard Smith, J. (1975). *The Theory of Evolution*. Harmondsworth, Penguin Book, Baltimore.

Mayr, E. (1970). *Populations, Species and Evolution*. Harvard University Press, Cambridge, MA.

Mayr, E. (1982). *The Growth of Biological Thought*. Harvard University Press, Cambridge, MA.

Mayr, E. (1991). *One Long Argument: Charles Darwin and the Genesis of Modern Evolutionary Thought*. Harvard University Press, Cambridge, MA.

McGowan, J. and Walker, P. (1985). Dominance and diversity maintenance in an oceanic ecosystem. *Ecol. Monogr.*, **55**: 103–118.

McIntosh, R. (1985). *The Background of Ecology – Concept and Theory*. Cambridge University Press, Cambridge.

McIntosh, R. (1987). Pluralism in ecology. *Ann. Rev. Ecol. Syst.*, **18**: 321–341.

McKane, R., Grigal, D. and Russelle, M. (1990). Spatiotemporal differences in ^{15}N uptake and the organization of an old-field plant community. *Ecology*, **71**: 1126–1132.

Mitsch, W. and Jørgensen, S.E. (1989). *Ecological Engineering*. Wiley, New York.

Moloney, C. and Field, J. (1989). General allometric equations of nutrient uptake, ingestion, and respiration in plankton organisms. *Limnol. Oceanogr.*, **34**: 1290–1299.

Moloney, C. and Field, J. (1991). The size-based dynamics of plankton food webs. I. A simulation model for carbon and nitrogen flow. *J. Plankton Res.*, **13**: 1003–1038.

Moloney, C., Field, J. and Lukas, M. (1991). The size-based dynamics of plankton food webs. II. Simulation of three contrasting southern Benguela food webs. *J. Plankton Res.*, **13**: 1039–1092.

Moore, J. and de Ruiter, P. (1991). Temporal and spatial heterogeneity of trophic interactions within below ground food webs. *Agriculture, Ecosystems and Environment*, **34**: 371–397.

Morse, D., Lawton, J., Dodson, M. and Williamson, M. (1985). Fractal dimension of vegetation and the distribution of arthropod body length. *Nature*, **314**: 731–733.

Murray, B. (1986). The structure of theory and the role of competition in community dynamics. *Oikos*, **46**: 145–158.

Nicolis, G. and Prigogine, I. (1977). *Self-Organization in Non-Equilibrium Systems*. Wiley, New York.

Niessen, F. and Sturm, M. (1987). Die Sedimente des Baldeggersees (Schweiz) – Ablagerungsraum und Eutrophierungsentwicklung während der letzten 100 Jahre. *Arch. Hybrodiol.*, **108**: 365–383.

Nitecki, M.H. (1988). *Evolutionary Progress*. University of Chicago Press, Chicago.

Nordhaus, W. (1992). An optimal transition path for controlling greenhouse gases. *Science*, **258**: 1315–1319.

O'Neill, R., DeAngelis, D., Waide, J. and Allen, T.F.H. (1986). *A Hierarchical Concept of Ecosystems*. Princeton University Press, Princeton.

Odum, E.P. (1969). The strategy of ecosystem development. *Science*, **164**: 262–270.

Odum, E.P. and Biever, L.J. (1984). Resource quality, mutualism, and energy partitioning in food chains. *Am. Nat.*, **124**: 360–376.

Odum, E.P. and Odum, H.T. (1959). *Fundamentals of Ecology*. Saunders, Philadelphia.

Odum, H.T. (1971). *Environment, Power, and Society*. Wiley, New York.

Odum, H.T. (1983). *Systems Ecology*. Wiley, New York.

Odum, H.T. (1988). Self-organization, transformity, and information. *Science*, **242**: 1132–1139.

Odum, H.T. and Pinkerton, R.C. (1955). Time's speed regulator: the optimum efficiency for maximum power output in physical and biological systems. *Am. Scientist*, **43**: 331–343.

Oksanen, L. (1991). Trophic levels and trophic dynamics: a consensus emerging. *Trends Ecol. Evol.*, **6**: 58–60.

Okubo, A. (1980). *Diffusion and Ecological Problems: Mathematical Models*. Springer, Berlin.

Oreskes, N., Shrader-Frechette, K. and Belitz, K. (1994). Verification, validation, and confirmation of numerical models in the earth sciences. *Science*, **263**: 641–646.

Ottino, J., Muzzio, F., Tjahjadi, M., Franjione, J., Jana, S. and Kusch, H. (1992). Chaos, symmetry, and self-similarity: exploiting order and disorder in mixing processes. *Science*, **257**: 754–760.

Pahl-Wostl, C. (1990). Temporal organization: a new perspective on the ecological network. *Oikos*, **58**: 293–305.

Pahl-Wostl, C. (1991). Patterns in space and time – a new method for their characterization. *Ecol. Model.*, **58**: 141–158.

Pahl-Wostl, C. (1992a). Information theoretical analysis of functional temporal and spatial organization in flow networks. *Mathl. Comp. Mod.*, **16**: 35–52.

Pahl-Wostl, C. (1992b). The possible effects of aggregation on the quantitative interpretation of flow patterns in ecological networks. *Math. Biosci.*, **112**: 177–183.

Pahl-Wostl, C. (1993a). The hierarchical organization of the aquatic ecosystem: an outline how reductionism and holism may be reconciled. *Ecol. Model.*, **66**: 81–100.

Pahl-Wostl, C. (1993b). Food webs and ecological networks across spatial and temporal scales. *Oikos*, **66**: 415–432.

Pahl-Wostl, C. (1993c). The influence of a hierarchy in time scales on the dynamics of, and the coexistence within, ensembles of predator–prey pairs. *Theor. Popul. Biol.*, **43**: 159–183.

Pahl-Wostl, C. (1994). Sensitivity analysis of ecosystem dynamics based on macroscopic community descriptors: a simulation study. *Ecol. Model.*, **75/76**: 51–62.

Pahl-Wostl, C. and Jaeger, C. (1994). Risk communication: The example of climate change. *EAWAG-News*, **36**: 6–8.

Pahl-Wostl, C. and Ulanowicz, R. (1993). Quantification of species as functional units within an ecological network. *Ecol. Model*, **66**: 65–79.

Paine, R. (1988). Food webs: road maps of interactions or grist for theoretical development? *Ecology*, **69**: 1648–1654.

Palmer, T.N. (1993a). Extended-range prediction and the Lorenz model. *Bull. Amer. Meteorol. Soc.*, **74**: 49–65.

Palmer, T.N. (1993b). A nonlinear dynamical perspective on climate change. *Weather*, **48**: 314–326.

Parker, G.A. and Maynard Smith, J. (1990). Optimality theory in evolutionary biology. *Nature*, **348**: 27–33.

Pate, J.S. and Hopper, S.D. (1993). "Rare and common plants in ecosystems, with special reference to the South-west Australia flora". In E.-D. Schulze and H.A. Mooney, Eds, *Biodiversity and Ecosystem Function*. Springer, Heidelberg, pp. 293–320.

Patten, B.C. (1981). Environs: The superniches of ecosystems. *Amer. Zool.*, **12**: 845–852.

Patten, B.C. and Odum, E.P. (1981). The cybernetic nature of ecosystems. *Am. Nat.*, **118**: 886–895.

Perry, J., Woiwod, P. and Hanski, I. (1993). Using response surface methodology to detect chaos in ecological time series. *Oikos*, **68**: 329–339.

Peters, R. (1983). *The Implications of Body Size*. Cambridge University Press, Cambridge.

Peters, R. (1988). Some general problems for ecology illustrated by food web theory. *Ecology*, **69**: 1673–1676.

Peters, R. (1991). *A Critique for Ecology*. Cambridge University Press, Cambridge.

Pianka, E. (1988). *Evolutionary Ecology*. Harper & Row Publishers, New York.

Pimm, S.L. (1982). *Food Webs*. Chapman and Hall, London.

Pimm, S. (1984). The complexity and stability of ecosystems. *Nature*, **307**: 321–326.

Pimm, S.L. (1991). *The Balance of Nature*? University of Chicago Press, Chicago.

Pimm, S. and Kitching, R. (1988). Food web patterns: trivial flaws or the basis of an active research program. *Ecology*, **69**: 1648–1654.

Pimm, S., Lawton, J. and Cohen, J. (1991). Food web patterns and their consequences. *Nature*, **350**: 669–674.

Pitelka, L.-F. (1993). "Biodiversity and policy decisions". In E.-D. Schulze and H.A. Mooney, Eds, *Biodiversity and Ecosystem Function*. Springer, Heidelberg, pp. 481–493.

Platt, and Denman, K. (1978). The structure of pelagic marine ecosystems. *Rapp. P.-V. Reun., Cons. Int. Explor. Mer.*, **173**: 60–65.

Poincaré, H. (1894). *Les Méthodes Nouvelles de la Mécanique Céleste*. Gauthier-Villars, Paris.

Polis, G. (1991). Complex trophic interactions in deserts: an empirical critique of food-web theory. *Am. Nat.*, **138**: 123–155.

Post, J. and Rudstam, L. (1992). "Fisheries management and the interactive dynamics of walleye and perch populations". In J. Kitchell, Ed., *Food Web Management: A Case Study of Lake Mendota*. Springer, New York, pp. 381–406.

Powell, T.M. (1989). "Physical and biological scale of variability in lakes, estuaries, and the coastal ocean". In J. Roughgarden, R. May and S. Levin, Eds, *Perspectives in Ecological Theory.* Princeton University Press, Princeton, pp. 157–176.

Ramadge, P.J. and Wonham, W.M. (1989). The control of discrete event systems. *Proc. IEEE*, **77**: 81–98.

Redfield, G. (1988). Holism and reductionism in community ecology. *Oikos*, **53**: 276–278.

Reise, K. (1991). "Mosaic cycles in the marine benthos". In H. Remmert, Ed., *The Mosaic Cycle Concept of Ecosystems.* Springer, Berlin, pp. 61–82.

Reiss, M. (1989). *The Allometry of Growth and Reproduction.* Cambridge University Press, Cambridge.

Remmert, H. (1991a). *The Mosaic Cycle Concept of Ecosystems.* Springer, Berlin.

Remmert, H. (1991b). "The mosaic cycle concept of ecosystems – An overview". In H. Remmert, Ed., *The Mosaic Cycle Concept of Ecosystems.* Springer, Berlin, pp. 1–22.

Reynolds, C.S. (1984a). *The Ecology of Freshwater Phytoplankton.* Cambridge University Press, Cambridge.

Reynolds, C.S. (1984b). Phytoplankton periodicity: the interaction of form, function and environmental variability. *Freshwat. Biol.*, **14**: 11–142.

Reynolds, C.S. (1994). The ecological base for the successful biomanipulation of aquatic communities. *Arch. Hydrobiol.*, **130**: 1–33.

Reynolds, C.S., Padisak, J. and Sommer, U. (1993). "Intermediate disturbance in the ecology of phytoplankton and the maintenance of species diversity: a synthesis". In J. Padisak, C.S. Reynolds and U. Sommer, Eds, *Intermediate Disturbance Hypothesis in Phytoplankton Ecology.* Kluwer, Dordrecht, pp. 183–188.

Richardson, J. and Odum, H.T. (1981). "Power and pulsing production model". In W. Mitsch, R. Bossermann and J. Klopatek, Eds, *Energy and Ecological Modelling.* Elsevier, New York, pp. 641–647.

Ricklefs, R.E. (1986). Review of O'Neill et al. *Science*, **236**: 206–208.

Ricklefs, R.E. and Schluter, D. (1993). *Species Diversity in Ecological Communities.* University of Chicago Press, Chicago.

Rodríguez, J.F. and Mullin, M.M. (1986). Relation between biomass and body weight of plankton in a steady state oceanic ecosystem. *Limnol. Oceanogr.*, **31**: 361–370.

Rodríguez, J., Echevarría, F. and Jiménez-Gómez, F. (1990). Physiological ecological scalings of body size in an oligotrophic, high mountain lake (La Caldera, Sierra Nevada, Spain). *J. Plankton Res.*, **12**: 593–599.

Rosenberg, A. (1985). *The Structure of Biological Science.* Cambridge University Press, Cambridge.

Roughgarden, J. (1979). *Theory of Population Genetics and Evolutionary Ecology.* MacMillan Publishing Co, New York.

Rowe, J. (1992). The integration of ecological studies. *Functional Ecology*, **6**: 115–119.

Ruelle, D. (1991). *Chance and Chaos.* Princeton University Press, Princeton.

Ruelle, D. and Takens, F. (1971). On the nature of turbulence. *Comm. Math. Phys.*, **20**: 167–192.

Ruess, R. and Seagle, S. (1994). Landscape pattern in soil microbial processes in the Serengeti National Park, Tanzania. *Ecology*, **75**: 892–904.

Rutledge, R., Basore, B. and Mulholland, R. (1976). Ecological stability: An information theory viewpoint. *J. Theor. Biol.*, **57**: 355–371..

Salomonsen, J. (1992). Examination of the properties of exergy, power and ascendency along a eutrophication gradient. *Ecol. Model.*, **62**: 171–181.

Salthe, S.N. (1985). *Evolving Hierarchical Systems.* Columbia University Press, New York.

Scavia, D. (1980). An ecological model of Lake Ontario. *Ecol. Model.*, **8**: 49–78.

Schaffer, W. (1985). Order and chaos in ecological systems. *Ecology*, **66**: 93–106.

Schindler, D.W. (1987). Detecting ecosystem responses to anthropogenic stress. *Can. J. Fish. Aquat. Sci.*, **44**(suppl.): 6–25.

Schipper, L., Meyers, S., Howarth, R. and Steiner, R. (1992). *Energy Efficiency and Human Activity.* Cambridge University Press, Cambridge.

Schlesinger, M.E. and Ramankutty, N. (1992). Implications for global warming of intercycle solar irradiance variations. *Nature*, **360**: 330–333.

Schmid, B. (1990). Some ecological and evolutionary consequences of modular organization and clonal growth in plants. *Evol. Trends Plants*, **4**: 25–34.

Schoener, T.W. (1989a). "The ecological niche". In J.M. Cherrett, Ed., *Ecological Concepts*. Blackwell Scientific Publications, Oxford, pp. 79–114.

Schoener, T.W. (1989b). Food webs from the small to the large. *Ecology*, **70**: 1559–1589.

Schoenly, K. and Cohen, J. (1991). Temporal variation in food web structure: 16 empirical cases. *Ecol. Monogr.*, **61**: 267–298.

Schulze, E.-D. and Mooney, H.A. (1993a). *Biodiversity and Ecosystem Function*. Springer, Heidelberg.

Schulze, E.-D. and Mooney, H.A. (1993b). "Ecosystem function of biodiversity: A summary". In E.-D. Schulze and H.A. Mooney, Eds, *Biodiversity and Ecosystem Function*. Springer, Heidelberg, pp. 497–510.

Schwinghamer, P. (1981). Characteristic size distributions of integral benthic communities. *Can. J. Fish. Aquat. Sci.* **38**: 1255–1263.

Seasteadt, T.R. and Knapp, A.K. (1993). Consequences of nonequilibrium resource availability across multiple time scales: the transient maxima hypothesis. *Amer. Nat.*, **141**: 621–633.

Shannon, C. (1948). A mathematical theory of communication. *Bell System Tech. J.*, **27**: 379–423; 623–656.

Sheldon, R., Prakash, A. and Sutcliffe, W. (1972). The size distribution of particles in the ocean. *Limnol. Oceanogr.*, **17**: 327–340.

Shephard, G. (1990). *The Synaptic Organization of the Brain*. Oxford University Press, Oxford.

Shugart, H. (1984). *A Theory of Forest Dynamics*. Springer, New York.

Shugart, H. and Urban, D. (1988). "Scale, synthesis, and ecosystem dynamics". In L. Pomeroy and J. Alberts, Eds, *Concepts of Ecosystem Ecology*. Springer, New York, pp. 279–290.

Silvert, W. and Platt, T. (1978). Energy flux in the pelagic ecosystem: a time-dependent equation. *Limnol. Oceanogr.*, **23**: 813–816.

Silvert, W. and Platt, T. (1980). "Dynamic energy-flow model of the particle size distribution in pelagic ecosystems". In W. Kerfoot, Ed., *Evolution and Ecology of Zooplankton Communities*. University Press of New England, Hanover, NH, pp. 754–763.

Simberloff, D. (1980). A succession of paradigms in ecology: essentialism to materialism and probalism. *Synthese*, **43**: 3–39.

Solow, A.R. (1992). "Is there a global warming problem?". In R. Dornbusch and J. Poterba, Eds, *Global Warming: Economic Policy Responses*. MIT Press, Cambridge, MA, pp. 7–28.

Sommer, U. (1984). The paradox of the plankton: fluctuations of phosphorus availability maintain diversity of phytoplankton in flow through cultures. *Limnol. Oceanogr.*, **29**: 633–636.

Sommer, U. (1985). Seasonal succession of phytoplankton in Lake Constance. *BioScience*, **35**: 351–357.

Sommerer, J. and Ott, E. (1993). A physical system with qualitatively uncertain dynamics. *Nature*, **365**: 138–140.

Sparrow, C. (1982). *The Lorenz equations: Bifurcations, Chaos, and Strange Attractors*. Springer, New York.

Sparrow, C. (1986). "The Lorenz Equations". In A. Holden, Ed., *Chaos*. Princeton University Press, Princeton, pp. 111–134.

Sprules, W. and Munawar, M. (1986). Plankton size spectra in relation to ecosystem productivity, size, and perturbation. *Can. J. Fish. Aquat. Sci.*, **43**: 1789–1794.

Steele, J. (1991). Can ecological theory cross the land–sea boundary? *J. Theor. Biol.*, **153**: 425–436.

Steneck, R. and Dethier, M. (1994). A functional group approach to the structure of algal-dominated communities. *Oikos*, **69**: 476–498.

Sterling, A., Peco, B., Casado, M., Galiano, E. and Pineda, F. (1984). Influence of microtopography on floristic variation in the ecological succession in grassland. *Oikos*, **42**: 334–342.

Stone, L. and Weisburd, R. (1992). Positive feedback in aquatic ecosystems. *Trends Evol. Evol.*, **7**: 263–267.

Strong, D.R. (1992). Are trophic cascades all wet? Differentiation and donor-control in speciose ecosystems. *Ecology*, **73**: 747–754.

Sugihara, G. (1980). Minimal community structure: an explanation of species abundance patterns. *Am. Nat.*, **116**: 770–787.

Sugihara, G., Schoenly, K. and Trombla, A. (1989). Scale invariance in food web properties. *Science*, **245**: 48–52.

Swetnam, T.W. (1993). Fire history and climate change in giant sequoia groves. *Science*, **262**: 885–889.

Swift, M. and Anderson, J. (1993). "Biodiversity and ecosystem function in agricultural systems". In E.-D. Schulze and H.A. Mooney, Eds, *Biodiversity and Ecosystem Function*. Springer, Heidelberg, pp. 15–41.

Swihart, R., Norman, S. and Bergstrom, B. (1988). Relating body size to the rate of home range use in mammals. *Ecology*, **69**: 393–399.

Tansley, A.G. (1935). The use and abuse of vegetational concepts and terms. *Ecology*, **27**: 513–530.

Therriault, J.-C. and Platt, T. (1981). Environmental control of phytoplankton patchiness. *Can. J. Fish. Aquat. Sci.*, **38**: 638–641.

Tilman, D. (1993). "Community diversity and succession: The role of competition, dispersal, and habitat modification". In E.-D. Schulze and H.A. Mooney, Eds, *Biodiversity and Ecosystem Function*. Springer, Heidelberg, pp. 327–341.

Tilman, D. (1994). Competition and biodiversity in spatially structured habitats. *Ecology*, **75**: 2–16.

Turchin, P. and Taylor, A. (1992). Complex dynamics in ecological time series. *Ecology*, **73**: 289–305.

Turner, M. and Gardner, R. (1990). *Quantitative Methods in Landscape Ecology*. Springer, New York.

Turner, R.K. (1988). "Sustainability, resource conservation and pollution control: An overview". In R. K. Turner, Ed., *Sustainable Environmental Management*. Belhaven Press, London, pp. 1–28.

Tziperman, E., Stone, L., Cane, M. and Jarosh, H. (1994). El Niño chaos: overlapping of resonances between the seasonal cycle and the Pacific Ocean–atmosphere oscillator. *Science*, **264**: 72–74.

Uhl, C. (1987). Factors controlling succession following slash-and-burn agriculture in Amazonia. *J. Ecol.*, **75**: 377–407.

Ulanowicz, R.E. (1980). An hypothesis on the development of natural communities. *J. Theor. Biol.*, **85**: 223–245.

Ulanowicz, R.E. (1986). *Growth and Development: Ecosystems Phenomenology*. Springer, New York.

Van Voris, P., O'Neill, R., Emanuel, W. and Shugart, H. (1980). Functional complexity and ecosystem stability. *Ecology*, **61**: 1352–1360.

Venrick, E.L. (1990). Phytoplankton in an oligotrophic ocean: species structure and interannual variability. *Ecology*, **7**: 1547–1563.

Venrick, E.L. (1993). Phytoplankton seasonality in the central North Pacific: The endless summer reconsidered. *Limnol. Oceanogr.*, **38**: 1135–1149.

Villa, F. (1992). New computer architectures as tools for ecological thought. *Trends Ecol. Evol.*, **7**: 179–183.

Vollenweider, R.A. (1975). Input–output models with special reference to phosphorous loading concept in limnology. *Schweiz. Z. Hydrol.*, **37**: 53–84.

Vuorisalo, T. and Tuomi, J. (1986). Unitary and modular organisms: criteria for ecological division. *Oikos*, **47**: 382–385.

Warwick, R. (1984). Species size distributions in marine benthic communities. *Oecologia*, **61**: 32–41.

Warwick, R. and Joint, I. (1987). The size distribution of organisms in the Celtic Sea: from bacteria to Metazoa. *Oecologia*, **73**: 185–191.

Waterstone, M. (1993). Adrift on a sea of platitudes: Why we will not resolve the greenhouse issue. *Environmental Management*, **17**: 141–152.

Weiner, J. and Thomas, S. (1992). Competition and allometry in three species of annual plants. *Ecology*, **73**: 648–656.

West, B. (1985). *An Essay on the Importance of Being Nonlinear*. Lecture Notes in Biomathematics, Spinger, Berlin.

White, J. (1979). The plant as a metapopulation. *Ann. Rev. Ecol. Syst.*, **15**: 233–258.

Wicken, J.S. (1980). Thermodynamic theory of evolution. *J. Theor. Biol.*, **87**: 9–23.

Wiegert, R. (1988). Holism and reductionism in ecology: hypotheses, scale and systems models. *Oikos*, **53**: 267–269.

Wiens, J.A., Stenseth, N.C., VanHorne, B. and Ims, R.A. (1993). Ecological mechanisms and landscape ecology. *Oikos*, **66**: 369–380.

Williamson, M. (1989). "Are communities ever stable?". In A.J. Gray, M.J. Crawley and P.J. Edwards, Eds, *Colonization, Succession, and Stability*. Blackwell Scientific Publications, Oxford, pp. 353–370.

Williamson, P. (1992). *Global Change: Reducing Uncertainties*. International Geosphere–Biosphere Program, Stockholm.

Wilson, D. (1988). Holism and reductionism in evolutionary ecology. *Oikos*, **53**: 269–273.

Wilson, D.S. (1992). Complex interactions in metacommunities, with implications for biodiversity and higher levels of selection. *Ecology*, **73**: 1984–2000.
Winberg, C.G. (1960). Rate of metabolism and food requirements for fish. *Fish Res. Board Can. Transl. Ser.*, **194**: 202.
Winemiller, K. (1990). Must connectance decrease with species richness? *Am. Nat.*, **134**: 960–968.
Wuketits, F.W. (1988). *Evolutionstheorien*. Wissenschaftliche Buchgesellschaft, Darmstadt.

Index

DATE DUE
